T0184352

Götz Alefeld
Jiří Rohn
Siegfried Rump
Tetsuro Yamamoto (eds.)

Symbolic Algebraic Methods and
Verification Methods

SpringerWienNewYork

Univ. –Prof. Dr. Götz Alefeld
Institut für Angewandte Mathematik, Universität Karlsruhe, Deutschland

Prof. Dr. Jiří Rohn
Mathematisch-Physikalische Fakultät, Universität Karlovy, Prag, Tschechien

Prof. Dr. Siegfried Rump
Technische Informatik III, Technische Universität Hamburg, Deutschland

Prof. Dr. Tetsuro Yamamoto
Department of Mathematical Science, Ehime University, Matsuyama, Japan

© 2001 Springer-Verlag/Wien

Typesetting: Camera-ready by authors
Printing: Novographic Druck, A-1230 Wien
Binding: Papyrus, A-1100 Wien

SPIN 10777227

With 40 Figures

CIP data applied for

ISBN 3-211-83593-8 Springer-Verlag Wien New York

Contents

List of Contributors

Albrecht Rudolf, Universität Innsbruck, Naturwissenschaftliche Fakultät,
Technikerstr. 25, A-6020 Innsbruck (A)
e-mail: rudolf.albrecht@uibk.ac.at

Alefeld Götz, Universität Karlsruhe, Institut für Angewandte Mathematik,
Kaiserstr. 12, Postfach 69 80, D-76128 Karlsruhe (D)
e-mail: goetz.alefeld@math.uni-karlsruhe.de

Corless Robert M., University of Western Ontario, Dept. of Appplied Mathematics, Western Science Center, ON-N6A 5B7 London (CDN)
e-mail: rob.corless@uwo.ca

Cuyt Annie, University of Antwerpen, Dept. of Math. and Computer Science,
Universiteitsplein 1, B-2610 Antwerpen (B)
e-mail: cuyt@ uia.ua.ac.be

Decker Thomas, Universität Paderborn, FB 17 - Mathematik/Informatik,
Warburger Str. 100, D-33100 Paderborn (D)

Emiris Ioannis Z., INRIA, Sophia Antipolis, Project SAGA, 2004 Route des
Lucioles, B.P. 93, F-06902 Sophia Antipolis (F)
e-mail: emiris@sophia.inria.fr

Frommer Andreas, Universität Wuppertal, FB Mathematik, D-42907
Wuppertal (D)
e-mail: Andreas.Frommer@math.uni-wuppertal.de

Garloff Jürgen, FH Konstanz, FB Informatik, Postfach 10 05 43, D-78405
Konstanz (D)
e-mail: garloff@fh-konstanz.de

Gay David M., Bell Labs, Room 2C-463, 600 Mountain Avenue, NJ 07974
Murray Hill (USA)
e-mail: dmg@research.bell-labs.com

Heckmann Reinhold, Imperial College of Science, Technology & Medicine, Dept.
of Computing, 180 Queen's Gate, SW7 2BZ London (GB)
e-mail: rh@doc.ic.ac.uk

Jansson Christian, TU Hamburg-Harburg, Technische Informatik III,
Schwarzenbergstr. 95, D-21073 Hamburg (D)
e-mail: jansson@tu-harburg.de

Kapur Deepak, University of New Mexico, Dept. of Computer Science, Farris
Eng. Center # 339, NM 87131 Albuquerque (USA)
e-mail: kapur@cs.unm.edu

Krandick Werner, Universität Paderborn, FB 17 - Mathematik/Informatik,
Warburger Str. 100, D-33100 Paderborn (D)
e-mail: krandick@uni-paderborn.de

Kreinovich Vladik, University of Texas at El Paso, Dept. of Computer Science,
El Paso, TX 79968 (USA)
e-mail: vladik@cs.utep.edu

Lang Bruno, RWTH Aachen, Rechenzentrum, Seffenter Weg 23, D-52074 Aachen
(D)
e-mail: lang@rz.rwth-aachen.de

Mayer Günter, Universität Rostock, FB Mathematik, Universitätsplatz 1,
D-18051 Rostock (D)
e-mail: guenter.mayer@mathematik.uni-rostock.de

Mehlhorn Kurt, MPI für Informatik, Im Stadtwald, D-66123 Saarbrücken (D)
e-mail: mehlhorn@mpi-sb.mpg.de

Minamoto Teruya, Department of Computer Science, Saga University, Saga
840-8502 (J)
e-mail: minamoto@ma.is.saga-u.ac.jp

Neher Markus, Universität Karlsruhe, Institut für Angewandte Mathematik,
Kaiserstr. 12, Postfach 69 80, D-76128 Karlsruhe (D)
e-mail: markus.neher@math.uni-karlsruhe.de

Plum Michael, Universität Karlsruhe, Mathematisches Institut, Postfach 69 80,
D-76128 Karlsruhe (D)
e-mail: michael.plum@math.uni-karlsruhe.de

Rump Siegfried M., TU Hamburg-Harburg, Technische Informatik III,
Schwarzenbergstr. 95, D-21073 Hamburg (D)
e-mail: rump@tu-harburg.de

Schäfer Uwe, Universität Karlsruhe, Institut für Angewandte Mathematik,
Kaiserstr. 12, Postfach 69 80, D-76128 Karlsruhe (D)
e-mail: uwe.schaefer@math.uni-karlsruhe.de

Schirra Stefan, MPI für Informatik, Im Stadtwald, D-66123 Saarbrücken (D)
e-mail: stschirr@mpi-sb.mpg.de

Shakhno Stephan, Fakultät für Angewandte Mathematik und Mechanik,
Universität Lwiw, Universitätska Straße 1, U-290602 Lwiw (U)

Smith A. P., FH Konstanz, FB Informatik, Postfach 10 05 43, D-78405
Konstanz (D)

Warnke I., Universität Rostock, FB Mathematik, Universitätsplatz 1, D-18051
Rostock (D)

Yamamoto Tetsuro, Ehime University, Faculty of Science - Dept. of
    Mathematical Science, 790-8577 Matsuyama (J)
    e-mail: yamamoto@dpc.ehime-u.ac.jp

Zemke Jens-Peter M., TU Hamburg-Harburg, Technische Informatik III,
    Schwarzenbergstr. 95, D-21073 Hamburg (D)
    e-mail: zemke@tu-harburg.de

Introduction

The usual "implementation" of real numbers as floating point numbers on existing computers has the well-known disadvantage that most of the real numbers are not exactly representable in floating point. Also the four basic arithmetic operations can usually not be performed exactly.

For numerical algorithms there are frequently error bounds for the computed approximation available. Traditionally a bound for the infinity norm is estimated using theoretical concepts like the condition number of a matrix for example. Therefore the error bounds are not really available in practice since their computation requires more or less the exact solution of the original problem.

During the last years research in different areas has been intensified in order to overcome these problems. As a result applications to different concrete problems were obtained.

The LEDA-library (K. Mehlhorn et al.) offers a collection of data types for combinatorical problems. In a series of applications, where floating point arithmetic fails, reliable results are delivered. Interesting examples can be found in classical geometric problems.

At the Imperial College in London was introduced a simple principle for "exact arithmetic with real numbers" (A. Edalat et al.), which uses certain nonlinear transformations. Among others a library for the effective computation of the elementary functions already has been implemented.

Using symbolic-algebraic methods the solution of a given problem can be computed exactly. These methods are applied successfully in many fields. However, for large problems the computing time may become prohibitive.

Another possibility is offered by so-called verification methods. These methods give correct results using only floating point arithmetic. Error bounds are computed by a sophisticated combination of error estimators. This idea allows to attack even larger problems without loosing too much time in comparison to traditional methods (without verification).

During the last few years it was already started to combine symbolic-algebraic methods and verification methods to so-called hybrid methods.

Scientists in different fields are working today on the outlined subjects. It was the purpose of a Dagstuhl seminar (with the same title as this book) at the Forschungszentrum für Informatik, Schloß Dagstuhl, Germany, to bring together colleagues from Computer Science, Computer Algebra, Numerical Mathematics, Matrix- and NP-theory, Control Theory and similar fields for exchanging the latest results of research and ideas.

This book contains (in alphabetical order) a collection of worked-out talks presented during this seminar. All contributions have been refereed. We are thankful to the authors for submitting their papers and to the referees for assisting us.

We would like to express our warmest thanks to Professor Dr. Reinhard Wilhelm for giving us the opportunity to run this seminar in Dagstuhl and to the whole crew of Schloß Dagstuhl for presenting a very nice atmosphere which let all participants feel like at home.

Finally we are thankful to Springer-Verlag, Vienna, for publishing the papers in its Springer Mathematics series.

G. Alefeld, Karlsruhe   July 2000
J. Rohn, Prague
S. M. Rump, Hamburg
T. Yamamoto, Matsuyama

2

Topological Concepts for Hierarchies of Variables, Types and Controls

Rudolf F. Albrecht

1 Introduction

Let the mathematical specification of a problem be expressed by a function $f: X \to Y$, X, Y being non-empty sets $X \subseteq \prod_{i \in I} X_{[i]}$, $Y \subseteq \prod_{j \in J} Y_{[j]}$, $f: x \mapsto y$. For constructive computation f is usually represented as a composition of a family $F = (f_k)_{k \in K}$ of given primitive functions $f_{[k]} \in G$, $G = \{g_{[p]}: X_{[p]} \to Y_{[p]} \mid p \in P\}$, P a finite set. All $g_{[q]}$ are assumed to be computable. Composition is a concatenation of the f_k represented by a cycle free relation relating result components of some f_k with argument components of other f_l. On the resulting "algorithmic structure" algorithms can be defined. Depending on the choice of the structure and the choice of the algorithm it is expected that the result is a composite function representing f, which has to be approved by verification.

In section 2 we formalize the concept of algorithmic structures and algorithms. For a suitable description of the structures under consideration we introduce variables, their types and their control- (assignment-) functions in section 3. In actual computations physical objects are employed to represent the functions and the output-input relations. These objects have a time behavior in physical time and, in addition, in general they can only approximate the mathematically exact values and operations. Investigation of these approximations requires topological concepts which will be treated in section 4. For example, if $f: x \mapsto y$ is approximated by $\tilde{f}: \tilde{x} \mapsto \tilde{y}$, we either want a deviation measure for $\tilde{y}$ from y for given deviations $\tilde{x}$ from x and $\tilde{f}$ from f, or a verification that $\tilde{y}$ deviates from y less than a postulated measure for suitably chosen $\tilde{f}$ and $\tilde{x}$. As applications we consider "fuzzy sets", "fuzzy" control, and interval arithmetic.

2 Algorithmic structures and algorithms

For any non-empty sets I, S and any function $\iota: I \to S$ we use the following notations: I is an index set, S is an object set, $s_{[i]} =_{\text{def}} \iota(i)$ is element of S, $s_i =_{\text{def}} (i, s_{[i]})$ is element of $I \times S$. $\{s_{[i]} \mid i \in I\}$ is a subset of S. $(s_i)_{i \in I}$ is a family.

Let there be given a family $F = (f_k)_{k \in K}$ with $K \neq \emptyset$ of functions $f_{[k]}: X_{[k]} \to Y_{[k]}$

with $f_{[k]} \in G$, G the set of primitive functions, domain $X_{[k]} \subseteq \prod_{i \in I_{[k]}} X_{[k,i]}$ and range

$Y_{[k]} \subseteq \prod_{j \in J_{[k]}} Y_{[k,j]}$. Further, let be $J_k =_{def} \{(k,j) \mid j \in J_{[k]}\}$, $I_k =_{def} \{(k,i) \mid i \in I_{[k]}\}$ for all

$k \in K$, and let be $C \subseteq \bigcup_{k \in K} J_k \times \bigcup_{k \in K} I_k$ with the following properties:

$\quad$ C is functional from right to left, i.e. from $((k,j), (l,i))$, $((k'j'), (l',i')) \in C$ and $(k,j) \neq (k'j')$ follows $(l,i) \neq (l',i')$,

$\quad$ C is cycle free, i.e. no chain of form $(((k,j),(l,i)), ((l,j'),(m,i')),..., ((q,j^*),(p,i^*)))$ with $k = p$ exists,

$\quad$ to any $c = ((k,j),(l,i)) \in C$ we associate the identity $y_{[k,j]} = x_{[l,i]}$ under the assumption that $x_{[l]} = (x_{[l,i]})_{i \in I_{[l]}} \in dom\, f_{[l]}$ ("domain condition").

We name C a composite, $c \in C$ a primitive "connector", $F \cup C$ an "algorithmic structure". C consisting of ordered tuples induces an irreflexive, transitive structure $\prec$ on F:

If $((k,j),(l,i)) \in C$ and $((l,j'),(m,i')) \in C$, then by definition $f_{[k]} \prec f_{[l]}$ and $f_{[k]} \prec f_{[m]}$. Components which are not related by $\prec$ are named "parallel". $f_{[k]} \prec f_{[l]}$ means $f_{[l]}$ depends on $f_{[k]}$.

$\quad$ F can be (in general many ways) partitioned by the following procedure

(A): $\quad$ n := 0, A := F; <n an integer variable, A a set variable>,

$\qquad$ while $A \neq \varnothing$ do

$\qquad\qquad$ n := n+1;

$\qquad\qquad\qquad$ select a non-empty set $A_n \subseteq A^* =_{def} \{f \mid f \in A \wedge *\forall g \in A\ (g \prec f)\}$;

$\qquad\qquad\qquad$ A := A\A_n;

$\qquad\qquad$ end while;

end (A).

$\quad$ Because $(F, \prec)$ is finite and $\prec$ is acyclic, (A) terminates after $n = N$ steps with card (maximal $\prec$-chain in $(F, \prec)$) $\leq N \leq$ card F. For any A_n, the components of A_n are parallel.

$\quad$ We set $N = \{1,2,..N\}$. The partitioning $(A_n)_{n \in N}$ can be extended to a partitioning $(B_n)_{n \in N}$ of $F \cup C$ by $\wedge n \in N$ ($B_n =_{def} A_n \cup C_n =_{def} \{((k,j),(l,i)) \mid ((k,j),(l,i)) \in C \wedge f_l \in A_n\}$).

$\quad$ For $n \in N$ we introduce $K_n = \{k \mid f_k \in A_n\}$, $P_n = \bigcup_{k \in K_n} I_k$, $Q_n = \bigcup_{k \in K_n} J_k$, and the

functional assignment s_n: $((x_{[p]})_{p \in P_n}, (y_{[q]})_{q \in Q_n})$ which expresses one "algorithmic

step". We define for $n = 1$, $S_1 = s_1$, $\widetilde{P}_1 = P_1$, $\widetilde{Q}_1 = Q_1$, and for $n = 1,2,..N-1$

$\quad \widetilde{P}_{n+1} = \widetilde{P}_n \cup (P_{n+1} \setminus pr_2\, C_{n+1})$ <indices with "free", "external" arguments>,

$\quad \widetilde{Q}_{n+1} = \widetilde{Q}_n \cup Q_{n+1}$ <indices with so far obtained results>,

$\quad S_{n+1} = K(S_n, s_{n+1}, C_{n+1}) =_{def} ((x_{[p]})_{p \in \widetilde{P}_{n+1}}, (y_{[q]})_{q \in \widetilde{Q}_{n+1}})$

$\quad$ <concatenation of S_n and s_{n+1} with respect to C_{n+1} and identity on C_{n+1}>.

4

The set $(N,<)$ defines "logical time" for the "algorithm" $(S_n)_{n \in N}$, which is a sequential process in logical time N with states S_n. S_N represents the final composite function.

$<$ on N induces an ordering $<$ on F: If $f_{[k]} \in A_n$ and $f_{[l]} \in A_m$ and $n < m$ then by definition $f_{[k]} < f_{[l]}$. $f_{[k]}$ is "earlier" than $f_{[l]}$. From $f_{[k]} \prec f_{[l]}$ follows $f_{[k]} < f_{[l]}$.

3 Variables and their control

We consider a non-empty family $(S_p)_{p \in P}$ of non-empty structures (general relations) S_p. For all p let be $S_{[p]} = K_{[p]}(K, V_{[p]})$, $K_{[p]}$ a concatenation of a p-independent component K and a p-dependent component $V_{[p]}$. K may be empty. To facilitate the representation of $S = \{S_{[p]} \mid p \in P\}$ we define objects variable V on variability domain $V = \{V_{[p]} \mid p \in P\}$ with respect to S, written var $V : (V, S)$, and variable $S(\text{var } V) = K(K, \text{var } V)$ on S. K depends on var V. We make the variables "controllable" by associating to $S(\text{var } V)$ a function val with val: $P \times \{\text{var } V\} \times \{S(\text{var } V)\} \to V \times S$, with val: $(p, \text{var } V, S(\text{var } V)) \mapsto (V_{[p]}, S_{[p]})$, and to var V a function val: $P \times \{\text{var } V\} \to V$ with val: $(p, \text{var } V) \mapsto V_{[p]}$. The val-functions are named "control-" or "assignment" functions, P is the set of control-/assignment parameters. For assignment according parameter p we write also val($S(\text{var } V)$): $p \mapsto S_{[p]}$ and val(var V): $p \mapsto V_{[p]}$, or $S(\text{var } V) :=(p) S(V_{[p]})$, var $V :=(p) V_{[p]}$. $S(V_{[p]})$ and $V_{[p]}$ are "instantiations" of $S(\text{var } V)$ and var V, respectively. Introducing a "reset" function val^{-1} : $(V_{[p]}, S_{[p]}) \mapsto (\text{var } V, S(\text{var } V))$, application of val: $(q, \text{var } V, S(\text{var } V)) \mapsto (V_{[q]}, S_{[q]})$ results in a substitution $(V_{[q]}, S_{[q]})$ for $(V_{[p]}, S_{[p]})$.

In the previous section 2 we considered functions f: $X \to Y$. In the formula $y = f(x)$ we meant x is an element of X, y is the corresponding element of Y. However, the usual meaning of the notation is: x is a variable on X, y is the dependent variable on $f(X)$. We prefer to write var $x : X$, var $y := f(\text{var } x)$, var x serves as assignment parameter, var y depends on var x.

A structure can contain variables $(\text{var } w_j)_{j \in J}$ with var $w_j : W_j = \{w_{j[q]} \mid q \in Q_j\}$. The composite variable var $w = (\text{var } w_j)_{j \in J}$ is in general defined on $W \subseteq \prod_{j \in J} W_j$.

For $W \subset \prod_{j \in J} W_j$ assignments to the var w_j are in general dependent on each other, in this case admissible assignment parameters $q = (q_j)_{j \in J}$ are from a set $Q \subset \prod_{j \in J} Q_j$. This has to be taken into account in case of partial assignments to var w, resulting in the partial variable $(u_j)_{j \in J}$ with $u_j = w_{j[q(j)]}$ for $j \in J_a$, $u_j = \text{var } w_j$ for $j \in J \backslash J_a$.

The parameters $(q_j)_{j \in J}$ can themselves be the result of a computation $\varphi_{[r]} : (p_i)_{i \in I} \to (q_j)_{j \in J}$, and $\varphi_{[r]}$ can be an instantiation of a variable var φ with control function val(var φ). Continuing this way we have a hierarchy of control functions.

More than one variable can be defined on the same V and elements of V can be subject to certain relations $S_{[q]}$, $q \in Q$. $(V, (S_{[q]})_{q \in Q})$ is then also named "type" of the variables. For example, $S_{[q]}$ can be a functional relation.

We admit variables in the domain of a variable. For $n = 1,2,...N$ we consider var $v^{(n)} : \{u^{(n-1)}{}_{[q(n-1)]} \mid q(n-1) \in Q(n-1)\}$ and assume, for $n > 1$ exists $q(n-1)$ with $u^{(n-1)}{}_{[q(n-1)]} = $ var $v^{(n-1)} : \{u^{(n-2)}{}_{[q(n-2)]} \mid q(n-2) \in Q(n-2)\}$, $\{u^{(0)}{}_{[q(0)]} \mid q(0) \in Q(0)\}$ are constants. Then var $v^{(n)}$ is of var-level n over $\{u^{(0)}{}_{[q(0)]} \mid q(0) \in Q(0)\}$.

We generalize the structures of section 2 to variables and extend the dependency relation $\prec$:

For any relation $r \subseteq X \times Y$ we define $X \prec r$, $Y \prec r$ (aspect of definition of r). Then for any function $f: X \to Y$, $X \prec f$, $Y \prec f$. For $f: x \mapsto y = f(x)$, $x \prec f \prec y$ (aspect of application of f). Extended to variables we obtain var $x : X$, var $y : Y$, $f = $ val(var y): var $x \mapsto$ var $y = f($var $x)$, or var $y :=($var $x) f($var $x)$, we have var $x \prec f \prec$ var y. For hierarchical variables var $v^{(n+1)} : \{$var $v_l^{(n)} \mid l \in L\}$ we set $\wedge l \in L$ (val(var $v_l^{(n)}$) $\prec$ val(var $v^{(n+1)}$)) ("bottom up"), in the opposite order paths to the constant terminal elements can be found ("top down").

Let there be given $G(P) = \{g_{[p]} : X_{[p]}(Q_{[p]}) \to Y_{[p]} \mid p \in P\}$. We consider:

var $g : G(P)$, val(var g): $P \to G(P)$, or var $g :=(p) g_{[p]} \cdot p \prec$ val(var g) $\prec g_{[p]}$.

var $x_p : X_{[p]}(Q_{[p]})$, val(var x_p) : $Q_{[p]} \to X_{[p]}(Q_{[p]})$, or var $x_{[p]} :=(q) x_{[pq]} \cdot q \prec$ val(var x_p) $\prec x_{[pq]}$.

var $y_p : Y$, val(var y_p) $= g_{[p]} : x_{[pq]} \mapsto g_{[p]}(x_{[pq]})$, or var $y_p :=(x_{[pq]}) y_{[pq]} = g_{[p]}(x_{[pq]})$. $x_{[pq]} \prec g_{[p]} \prec y_{[pq]}$.

If $X_{[p]}$ depends on p, i.e. $\neg \wedge p',p'' \in P$ $(X_{[p']} = X_{[p'']})$, then $g_{[p]} \prec$ val(var x_p).

We consider the concatenation of two functions. Let there be given $P \subseteq \prod_{m \in M} P_m$, a fixed f_k with $f_{[k]} \in G(P)$, $f_k : X_k \to Y_k \subseteq \prod_{j \in J_{[k]}} Y_{kj}$, for $1 \neq k$ var $f_l : F_l = \{f_{l[p]} \mid f_{[lp]} \in G(P) \wedge p \in P\}$, val(var f_l) : $P \to F_l$. We admit that result components y_{kj} are used as part of the control parameters of succeeding val-functions.

We define a connector variable var $C_{k \text{ ctr}}$: set of admissible connectors $C_{k \text{ ctr}} \subset J_k \times M$ from f_k to val(var f_l), and assign var $C_{k \text{ ctr}} := C_{k \text{ ctr}}$ by val(var $C_{k \text{ ctr}}$). We have by definition of a connector $\wedge c = (kj,m) \in C_{k \text{ ctr}}$ $(y_{[kj]} = p_{[m]})$. Supplemented with external ("free") $p_{m'}$ let be $p = (p_m)_{m \in M} \in P$. Now assignment val(var f_l): $p \mapsto f_{l[p]}$ is possible with $f_l =_{\text{def}} f_{l[p]}$ and $f_l: X_l \to Y_l$. Let be $X_l \subseteq \prod_{i \in I_{[l]}} X_{li}$.

We define a connector variable var C_{kl} : set of admissible connectors $C_{kl} \subset J_k \times I_l$ and assign var $C_{kl} := C_{kl}$ by val(var C_{kl}). We have $\wedge c = (kj,li) \in C_{kl}$ $(y_{[kj]} = x_{[li]})$. Supplemented with external ("free") x_{li} let be $x_l = (x_{li})_{i \in I_{[l]}} \in X_l$. For var $y_l : Y_l$ we have val(var y_l) : $x_l \mapsto y_l = f_l(x_l)$. The dependencies are $y_k \prec C_{k \text{ ctr}} \prec$ val(var f_l) $\prec C_{kl} \prec f_l(x_l)$.

Moreover, val(var $C_{k\,ctr}$) and val(var C_{kl}) can be (partly) controlled by some y_{kj}, val(var C_{kl}) depends on the assignment to var f_l. An illustration is given in Fig.1.

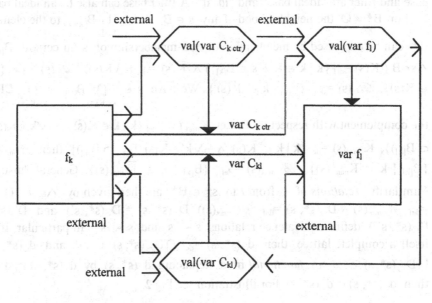

Fig. 1

On $\prec$ - ordered composite structures, algorithms can be defined. In a physical simulation of a functional composition, all control- and production functions and all arguments and results are represented by physical objects with behavior in a physical time $(T,<)$, $<$ a linear or partial ordering. $\prec$ and $<$ are compatible. Depending on the application, after being used at time instant t, an object may be fully or partly re-usable at a later time instant t', $t < t'$. This includes a possible feedback to the same physical object.

4 Topological structures

Structures considered so far are var $p : P$, var $f : F(P)$, var $x :$ var X, var $y :=$ var $f(var\ x)$, the variables defined on types, which are structured sets, possibly containing lower order variables. All assignment functions val are "exact" "point"-functions, assigning to an exact value an exact result. In practice, exact arguments, functions and results can be either not representable or are not available. Approximations to exact objects are in a certain neighborhood to them, which has to be defined. The mathematical tool for this are filter bases, and dual, ideal bases.

We consider a non-empty set S and the complete atomic boolean lattice (pow S, $\subseteq$, $\cap$, $\cup$, $\varnothing$, S) with zero element $\varnothing$ and unit element S. Let $B = \{B_{[k]} \mid k \in K\} \subset$ pow S be a non-empty subset, the indexing bijective, with the following properties:

$$\wedge k \in K\ (B_{[k]} \neq \varnothing) \wedge \wedge k',k'' \in K\ (\vee k''' \in K\ ((B_{[k''']} \subseteq B_{[k']}) \wedge (B_{[k''']} \subseteq B_{[k'']}))).$$

Then $\mathscr{B}$ is a "filter base" on pow S. If in addition $\wedge k \in K \wedge L \subseteq S$ $(B_{[k]} \subseteq L \Rightarrow L \in \mathscr{B})$ then $\mathscr{B}$ is a "filter". We define $B^* = \lim \mathscr{B} =_{\text{def}} \bigcap \mathscr{B}$. The dual notions to filter base and filter are "ideal base" and "ideal". A filter base can also be an ideal base.

For $B^* \neq \varnothing$ the neighborhood of any $s \in B$, $B =_{\text{def}} \bigcup_{k \in K} B_{[k]}$, to the elements of B^* can be expressed by membership or non-membership of s in certain $B_{[k]}$: Let $\wedge s \in B$ $((K(s) =_{\text{def}} \{k \mid k \in K \wedge s \in B_{[k]}\}) \wedge \overline{K}(s) =_{\text{def}} K \setminus K(s))$, $\mathscr{B}_\cap(s) =_{\text{def}} \{B_{[k]} \mid k \in K(s)\}$, $\mathscr{B}_\cup(s) =_{\text{def}} \{B_{[k]} \mid k \in \overline{K}(s)\}$. We have $s \in \bigcap_{k \in K(s)} B_{[k]} \cap \bigcap_{k \in \overline{K}(s)} CB_{[k]}$, C the complement with respect to B. Let $K_{\min}(s) =_{\text{def}} \{k \mid k \in K(s) \wedge \neg \vee k' \in K(s) (B_{[k']} \subset B_{[k]})\}$, $\overline{K}_{\max}(s) =_{\text{def}} \{k \mid k \in \overline{K}(s) \wedge \neg \vee k' \in \overline{K}(s) (B_{[k']} \supset B_{[k]})\}$, then $\mathscr{B}_{\cap\min}(s) =_{\text{def}} \{B_{[k]} \mid k \in K_{\min}(s)\}$, $\mathscr{B}_{\cup\max}(s) =_{\text{def}} \{B_{[k]} \mid k \in \overline{K}_{\max}(s)\}$. General "distance" / "similarity" *relations* of s from / to $s^* \in B^*$ are then given by $\wedge s \in B$ $(D_\cap(s^*, s)$ $=_{\text{def}} \mathscr{B}_{\cap\min}(s) \wedge D_\cup(s^*, s) =_{\text{def}} \mathscr{B}_{\cup\max}(s))$. $D_\cap(s^*, s) = D_\cap(s^*, s')$ and $D_\cup(s^*, s) = D_\cup(s^*, s'')$ define equivalence relations $s \sim_\cap s'$ and $s \sim_\cup s''$. In particular, if $\mathscr{B}$ is itself a complete lattice then $d_\cap(s^*, s) =_{\text{def}} \bigcap D_\cap(s^*, s) \in \mathscr{B}$ and $d_\cup(s^*, s) =_{\text{def}} \bigcup D_\cup(s^*, s) \in \mathscr{B}$ are *functional* in s. Replacing $d_\cup(s^*, s)$ by $d_\cap(s^*, s) \cap d_\cup(s^*, s)$ then $d_\cup(s^*, s) \subset d_\cap(s^*, s)$. For illustration see Fig.2.

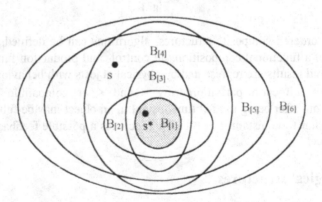

Fig. 2: $\mathscr{B} = \{B_{[1]}, B_{[2]}, B_{[3]}, B_{[4]}, B_{[5]}, B_{[6]}\}$, $\lim \mathscr{B} = B_{[1]}$,
$D_\cap(s^*, s) = \{B_{[4]}, B_{[5]}\}$, $D_\cup(s^*, s) = \{B_{[2]}, B_{[3]}\}$

The notion of filter and ideal bases can be extended to complete lattices. Given a non-empty complete lattice $(V, \leq, \sqcap, \sqcup, \mathbf{o})$, $\mathbf{o}$ the zero element of V, and a $\leq$-homomorphism φ: pow $B \to V$, i.e. $\wedge k, k' \in K$ $((B_{[k]} \subseteq B_{[k']}) \Rightarrow (\varphi(B_{[k]}) \leq \varphi(B_{[k']})))$. With $v_{[k]} = \varphi(B_{[k]})$ it follows from $(B_{[k]} \subseteq B_{[k']}) \wedge (B_{[k]} \subseteq B_{[k'']})$ that $(v_{[k]} \leq v_{[k']}) \wedge (v_{[k]} \leq v_{[k'']})$, hence $\varphi(\mathscr{B})$ is a filter base on V if all $v_{[k]} \neq \mathbf{o}$, and we have $\varphi(\lim \mathscr{B}) \leq \lim \varphi(\mathscr{B})$. In the function $(B_{[k]}, v_{[k]})_{k \in K}$ the elements $B_{[k]}$ of base $\mathscr{B}$ with "support" $B = \bigcup \mathscr{B}$ are valued by $v_{[k]}$.

8

We consider $\overset{-1}{\varphi} : V \to \text{pow } B$ defined by $\wedge v \in V$ ($\overset{-1}{\varphi}$ (v) $=_{\text{def}}$ $\underset{\varphi(U) \le v}{\bigcup} U$). Then $\overset{-1}{\varphi}$ is a homomorphism. If $\mathscr{V} = \{v_{[l]} \mid l \in L\}$ is a filter base on V and $\wedge v \in \mathscr{V}$ ($\overset{-1}{\varphi}$ (v) $\ne \varnothing$), then $\overset{-1}{\varphi}$ ($\mathscr{V}$) is a filter base on pow B.

For $U \subseteq S$ and for $v \in V$ we have $U \subseteq \overset{-1}{\varphi} \varphi(U)$ and, for reversible φ, $v \ge \overset{-1}{\varphi} \varphi$ (v), thus $\mathscr{B} \prec \overset{-1}{\varphi} \varphi(\mathscr{B}))$ and $\varphi \overset{-1}{\varphi}$ ($\mathscr{V}$) $\prec \mathscr{V}$. For two filter bases $\mathscr{A}, \mathscr{B}$, $\mathscr{A} \prec \mathscr{B}$ means $\mathscr{A}$ finer $\mathscr{B}$.

A well known example is: φ the set extension of a function f: $B \to V$, φ (U) = $\bigcup$ $(f(u))_{u \in U}$, in particular, V = pow C, C a non-empty set.

φ being a homomorphism corresponds to the "neighborhood to B^*" interpretation. Choosing φ as antimorphism, $\varphi(\mathscr{B})$ is an ideal base, which corresponds to the "similarity to B^*" interpretation.

In applying these topological concepts, we extend val-functions val(var u), var u : U(V), V the parameter set, val(var u): $V \to U$, to "set"-functions Val(var u), var u : pow U, parameterized by $v \in \text{pow } V$. Given a filter base $\mathscr{V} = \{V_{[k]} \mid k \in K\}$ on pow V, Val(var u)($\mathscr{V}$) is a filter base on pow U. Given a filter base $\mathscr{U} = \{U_{[l]} \mid l \in L\}$ on pow U, we consider Val^{-1}, the reciprocal of Val. If for all $l \in L$ holds Val$^{-1}(U_{[l]}) \ne \varnothing$ then Val^{-1}($\mathscr{U}$) is a filter base. We have topologized ("fuzzy") control.

As examples we consider

(1) F(P) = $\{f_p : X_p \to Y_p \mid p \in P\}$. Let the exact function be f_q, the exact argument be $x_q \in X_q$, f_q: $x_q \mapsto y_q = f_q(x_q)$.

Let $\mathscr{U}_q = \{U_{q[l]} \mid l \in L_q\}$ be a filter base on pow F(P) with $f_q \in \bigcap \mathscr{U}_q$ and F(P) $\in \mathscr{U}_q$. For a fixed $U_{q[l]}$, all $f_p \in U_{q[l]}$ are assumed to be admissible approximations to f_q. We set $X_{q[l]} = \underset{f_p \in U_{q[l]}}{\bigcup} X_p$.

Let $\mathscr{V}_{q[l]} = \{V_{q[lm]} \mid m \in M_{q[l]}\}$ be a filter base on pow $X_{q[l]}$ with $x_q \in \bigcap \mathscr{V}_{q[l]}$ and $X_{q[l]} \in \mathscr{V}_{q[l]}$. For a fixed $V_{q[lm]}$, all $x_p \in V_{q[lm]}$ are assumed to be admissible approximations to x_q. We further assume for all $l \in L_q$ the $\mathscr{V}_{q[l]}$ to be equally fine.

We set $Y_{q[lm]} = \{f_p(x) \mid f_p \in U_{q[l]} \wedge x \in (X_p \cap V_{q[lm]})\}$ and $Y_{q[l]} = \underset{f_p \in U_{q[l]}}{\bigcup} f_p (X_p)$. Then

$\mathscr{W}_{q[l]} = \{Y_{q[lm]} \mid m \in M_{q[l]}\}$ is a filter base on pow $Y_{q[l]}$ with $y_p \in \bigcap \mathscr{W}_{q[l]}$ and $Y_{q[l]} \in \mathscr{W}_{q[l]}$.

For $Y = \underset{p \in P}{\bigcup} f(X_p)$ let there be given a filter base $\mathscr{Z} = \{Z_{[k]} \mid k \in K\}$ on pow Y with $y_q \in \bigcap \mathscr{Z}$ and $Y \in \mathscr{Z}$, to measure distances on Y with respect to y_q.

A $Z \in \mathscr{Z}$ with $Y_{q[lm]} \subseteq Z$ is an estimate with respect to $\mathscr{Z}$. On the other hand, if a $Z' \in \mathscr{Z}$ is given, one can ask for a $U_{q[l']}$ and a $V_{q[l'm']}$ (which need not exist) such that $Y_{q[l'm']} \subseteq Z'$.

(2) "Fuzzy" sets as introduced by L. A. Zadeh in 1965, the underlying topological theory was developed by others much earlier. Let there be given $S \subset R$, the interval $C = [0,1] \subset R$, f: S onto$\to$ C the "membership" function, the monotone filter base $V = \{[\alpha, 1] \mid \alpha \in C\}$ for which $\bigcap V = \{1\}$. Then the "cuts" $f^{-1}([\alpha, 1])$ are considered which are $\neq \varnothing$ and generate a filter base on pow S. Fig. 3 gives a visualization for $S = [a,b]$.

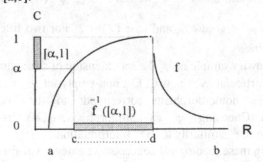

Fig. 3

$S = \{[a,b], [c,d]\}$ is a filter base on pow S, $\bigcap S = [c,d]$, f(S) is a filter base coarser than V. For filter base $E = \{[d-\varepsilon, d+\varepsilon] \mid \varepsilon \to 0\}$ holds lim $E = \{d\}$, f(lim E) = $\{1\}$ $\subset$ lim f(E).

(3) Interval arithmetic on R. Let be IR the set of all intervals on R, $*$ the extension to intervals of the operations $+$ or $-$ or $\times$ or $/$, division by intervals containing 0 excluded. For k=1,2, $I_k \in IR$, $B_k = \{I_k, R\}$ is a filter base with lim B_k = I_k. Then $I_1 * I_2$ is an interval I_3, $B_3 = \{I_3, R\}$ is the resulting filter base. The principle can be extended to less trivial filter bases containing more than 2 intervals, and to any non-empty sets $S_1, S_2, S_3 = S_1 * S_2 \subset R$ with closures I_1, I_2, I_3.

Further details on the topological concepts are given by R. F. Albrecht in 1998, 1999a, 1999b.

References

Albrecht R. F. (1998): On mathematical systems theory. In: R. Albrecht (ed.): Systems: Theory and Practice, Springer, Vienna-New York, 33-86.

Albrecht R. F.(1999a): Topological Approach to Fuzzy Sets and Fuzzy Logic. In: A. Dobnikar, N. C. Steele, D. W. Pearson, R. F. Albrecht: Artificial Neural Nets and Genetic Algorithms, Springer, Vienna-New York, 1-7.

Albrecht R. F. (1999b): Topological Theory of Fuzziness. In: B. Reusch (ed.): Computational Intelligence, Theory and Applications, LNCS vol. 1625, Springer, Heidelberg-New York, 1-11.

This article is part of a research project with the Dipartimento di Automatica e Informatica, Politecnico di Torino, Italy.

Modifications of the Oettli–Prager Theorem with Application to the Eigenvalue Problem

G. Alefeld, V. Kreinovich, G. Mayer

Dedicated to Prof. Dr. Jürgen Herzberger on the occasion of his 60th birthday.

1 Introduction

In this paper we consider the eigenpair set

$$E_{\mathcal{P}} := \{ (x, \lambda) \mid Ax = \lambda x, \ x \neq 0, \ A \in [A], \ A \text{ with property } \mathcal{P} \}, \tag{1}$$

where $[A]$ is a given real $n \times n$ interval matrix (cf. Alefeld and Herzberger (1983), e.g., for interval analysis) and $\mathcal{P}$ is some fixed property such as symmetry, Toeplitz form, etc.. Before we study this set in greater detail we mention other ones which are related to it: When dealing with systems of linear equations

$$\check{A}x = \check{b}, \quad \check{A} \in \mathbb{R}^{n \times n}, \ \check{b} \in \mathbb{R}^{n} \tag{2}$$

($\mathbb{R}^{n \times n}$ set of real $n \times n$ matrices, $\mathbb{R}^{n}$ set of real vectors with n components) there sometimes occurs the problem of varying the input data $\check{A}$, $\check{b}$ within certain tolerances and looking for the set S of the resulting solutions x^*. Examples of this problem are Wilkinson's backward analysis when solving linear systems on a computer (Wilkinson 1963) and an input–output model in economics which is regulated by (2) with input parameters $\check{A}$, $\check{b}$ and output x (Maier 1985). In the first example one solves (2) on a computer (assuming $\check{A}$ to be nonsingular). Due to rounding errors one normally does not obtain the exact solution x^* but another vector $\tilde{x}$. One accepts $\tilde{x}$ as a good approximation of the exact solution x^* if it can be interpreted as a solution of a nearby system $Ax = b$, where 'nearby' means $|A - \check{A}| \leq \Delta$, $|b - \check{b}| \leq \delta$ with given tolerances $O \leq \Delta \in \mathbb{R}^{n \times n}$, $0 \leq \delta \in \mathbb{R}^{n}$. (Here and in the sequel, the absolute value $| \cdot |$ and the inequality sign '$\leq$' are understood entrywise.) In other words, one considers $\tilde{x}$ as a good approximation for x^* if and only if it belongs to the *solution set*

$$S := \{ x \in \mathbb{R}^{n} \mid Ax = b, \ A \in \mathbb{R}^{n \times n}, \ b \in \mathbb{R}^{n}, \ |A - \check{A}| \leq \Delta, \ |b - \check{b}| \leq \delta \} \tag{3}$$

where $\check{A}$, Δ, $\check{b}$, δ are given. Instead of writing $|A - \check{A}| \leq \Delta A$ one often prefers the shorter notation $A \in [A]$ where the bracketed letter denotes the real $n \times n$ interval matrix

$$[A] := [\check{A} - \Delta, \check{A} + \Delta] = [\underline{A}, \overline{A}] = ([a]_{ij}) = ([\underline{a}_{ij}, \overline{a}_{ij}]).$$

Similarly, $|b - \breve{b}| \leq \delta$ is replaced by $b \in [b]$ with the interval vector

$$[b] := [\breve{b} - \delta, \breve{b} + \delta] = [\underline{b}, \overline{b}] = ([b]_i) = ([\underline{b}_i, \overline{b}_i]).$$

The second example – which we called input–output model – can be viewed under two aspects: Firstly, one can vary directly the original input data $\breve{A}$, $\breve{b}$ within given tolerances Δ, δ and consider all solutions outgrowing from the modified systems $Ax = b$ with $A \in [A]$, $b \in [b]$. This means that we look again for the solution set S in (3). Secondly, one can measure the output $x = \tilde{x}$ and ask whether it can be expected to be generated by input data A, b 'nearby' $\breve{A}$ and $\breve{b}$. Again one is led to the problem '$\tilde{x} \in S$?' this time having a sort of inverse problem as in Wilkinson's backward analysis.

A description of S was given in the sixties by Oettli and Prager (1964) in the form

$$x \in S \iff |\breve{b} - \breve{A}x| \leq \Delta \cdot |x| + \delta \tag{4}$$

which is known in the literature as Oettli–Prager Theorem and which was re-formulated since then several times (see Theorem 1 below).

The solution set S covers *all* matrices from $[A]$. Often the matrix $\breve{A}$ in (2) exhibits some particular structure, which should be kept when considering the perturbed systems $Ax = b$; cf. for instance Jansson (1991a,b), Rump (1994) or Alefeld and Mayer (1995). This leads to the modified solution set $S_{\mathcal{P}}$ with some given property $\mathcal{P}$ for A, i.e.,

$$S_{\mathcal{P}} := \{\, x \in \mathbb{R}^n \mid Ax = b, \; A \in [A], \; b \in [b], \; A \text{ with property } \mathcal{P} \,\} \subseteq S. \tag{5}$$

We will present a way how to describe $S_{\mathcal{P}}$ by means of inequalities which involve the bounds $\underline{A}$, $\overline{A}$, $\underline{b}$, $\overline{b}$ of the tolerance intervals. If there are no restrictions on $A \in [A]$ it will turn out that these inequalities reduce to (4). This justifies the title of our paper.

We now come back to the eigenvalue problem

$$Ax = \lambda x, \quad x \neq 0, \tag{6}$$

which we restrict to λ being real. This restriction is not substantial but simplifies matters. When perturbing $A \in \mathbb{R}^{n \times n}$ such that $A \in [A]$ is allowed we are led to the *eigenpair set*

$$E := \{\, (x, \lambda) \in \mathbb{R}^{n+1} \mid Ax = \lambda x, \; x \neq 0, \; A \in [A] \,\}. \tag{7}$$

If we are interested in matrices $A \in [A]$ sharing some property $\mathcal{P}$ we end up with the set $E_{\mathcal{P}}$ which we defined in (1) at the beginning of this section. In order to describe E and $E_{\mathcal{P}}$, respectively, the Oettli–Prager Theorem and its modifications will play a crucial role. As for S and $S_{\mathcal{P}}$ the Fourier–Motzkin elimination process of linear programming forms the basis. We will show that in each orthant of $\mathbb{R}^{n+1}$ and $\mathbb{R}^n$, respectively, the boundary of the eigenpair set E as well as the *symmetric solution set*

$$S_{\text{sym}} := \{\, x \in \mathbb{R}^n \mid Ax = b, \ A = A^T \in [A], \ b \in [b] \,\} \subseteq S \qquad (8)$$

can be described by means of hyperplanes and quadrics. For the *symmetric eigenpair set*

$$E_{\text{sym}} := \{\, (x, \lambda) \in \mathbb{R}^{n+1} \mid Ax = \lambda x, \ x \neq 0, \ A = A^T \in [A] \,\} \qquad (9)$$

one has to enlarge this variety of geometric objects by algebraic surfaces of order 3. If $\mathcal{P}$ means A being skew–symmetric ($A = -A^T$, i.e., $a_{ij} = -a_{ji}$) or persymmetric (A symmetric with respect to the counter–diagonal, i.e., $a_{ij} = a_{n+1-j,n+1-i}$) the boundaries of the corresponding solution sets can be described by the same kind of objects. If one admits more general dependencies in the entries such as A being a Toeplitz matrix (A has constant values along each of its diagonals, i.e., $a_{ij} = c_{j-i}$ with some constants c_k, $k = -(n-1), \ldots, n-1$) or a Hankel matrix (A has constant values along each of its counter–diagonals, i.e., $a_{ij} = c_{i+j-2}$ with some constants c_k, $k = 0, \ldots, 2n-2$) details are more complicated. In these cases – as for general linear dependencies – one can only show that the boundary of $S_{\mathcal{P}}$ and $E_{\mathcal{P}}$, respectively, can be described by means of algebraic equations whose order is unknown up to now (Alefeld et al. 1998).

We have arranged our paper as follows: In Section 2 we shortly describe the Fourier–Motzkin elimination process, in Section 3 we consider the solution set $S_{\mathcal{P}}$ and in Section 4 we study the eigenpair set $E_{\mathcal{P}}$.

2 The Fourier–Motzkin Elimination Process

The Fourier–Motzkin elimination process eliminates parameters in inequalities. We will shortly describe the principle by executing one step when deriving the set of inequalities for $S_{\text{sym}} \cap O_1$. Here, O_1 denotes the closed first orthant of $\mathbb{R}^n$, i.e., $O_1 := \{\, x \in \mathbb{R}^n \mid x \geq 0 \,\}$. For $x \in O_1$ one starts with the trivial equivalences

$$x \in S_{\text{sym}} \Leftrightarrow \ \exists\, A = A^T \in [A], \ b \in [b] : \ Ax = b$$

$$\Leftrightarrow \ \exists\, a_{ij} \in \mathbb{R} \quad (i, j = 1, \ldots, n) :$$

$$\underline{b}_i \leq \sum_{j=1}^{n} a_{ij} x_j \leq \overline{b}_i, \quad \underline{a}_{ij} \leq a_{ij} \leq \overline{a}_{ij}, \quad a_{ij} = a_{ji}$$

$$\Leftrightarrow \ \exists\, a_{ij} \in \mathbb{R} \quad (i, j = 1, \ldots, n) :$$

$$\underline{b}_1 - \sum_{\substack{j=1 \\ j \neq 2}}^{n} a_{1j} x_j \leq a_{12} x_2 \leq \overline{b}_1 - \sum_{\substack{j=1 \\ j \neq 2}}^{n} a_{1j} x_j,$$

$$\underline{b}_2 - \sum_{j=2}^{n} a_{2j} x_j \leq a_{12} x_1 \leq \overline{b}_2 - \sum_{j=2}^{n} a_{2j} x_j,$$

$$\underline{a}_{12} \leq a_{12} = a_{21} \leq \overline{a}_{12},$$

inequalities without a_{12}, a_{21}.

If $x_1 > 0$, $x_2 > 0$ this is equivalent to

$$\exists\, a_{ij} \in \mathbb{R} \quad (i,j = 1,\ldots,n):$$

$$\{\underline{b}_1 - \sum_{\substack{j=1\\j\neq 2}}^{n} a_{1j}x_j\}/x_2 \leq a_{12} \leq \{\overline{b}_1 - \sum_{\substack{j=1\\j\neq 2}}^{n} a_{1j}x_j\}/x_2,$$

$$\{\underline{b}_2 - \sum_{j=2}^{n} a_{2j}x_j\}/x_1 \leq a_{12} \leq \{\overline{b}_2 - \sum_{j=2}^{n} a_{2j}x_j\}/x_1,$$

$$\underline{a}_{12} \leq a_{12} = a_{21} \leq \overline{a}_{12},$$

inequalities without a_{12}, a_{21}.

Here, the first three double inequalities hold if and only if the *maximum* of the first three left–hand sides is less or equal than the *minimum* of the first three right–hand sides. This, however, is true if and only if *each* of the three left–hand sides is less or equal than *each* of the three right–hand sides which results in the following equivalent inequalities:

$$\exists\, a_{ij} \in \mathbb{R} \quad (i,j = 1,\ldots,n,\ (i,j) \neq (1,2),\ (i,j) \neq (2,1)):$$

$$\{\underline{b}_1 - \sum_{\substack{j=1\\j\neq 2}}^{n} a_{1j}x_j\}/x_2 \leq \{\overline{b}_2 - \sum_{j=2}^{n} a_{2j}x_j\}/x_1, \qquad \{\underline{b}_1 - \sum_{\substack{j=1\\j\neq 2}}^{n} a_{1j}x_j\}/x_2 \leq \overline{a}_{12},$$

$$\{\underline{b}_2 - \sum_{j=2}^{n} a_{2j}x_j\}/x_1 \leq \{\overline{b}_1 - \sum_{\substack{j=1\\j\neq 2}}^{n} a_{1j}x_j\}/x_2, \qquad \{\underline{b}_2 - \sum_{j=2}^{n} a_{2j}x_j\}/x_1 \leq \overline{a}_{12},$$

$$\underline{a}_{12} \leq \{\overline{b}_1 - \sum_{\substack{j=1\\j\neq 2}}^{n} a_{1j}x_j\}/x_2, \qquad \underline{a}_{12} \leq \{\overline{b}_2 - \sum_{j=2}^{n} a_{2j}x_j\}/x_1,$$

inequalities without a_{12}, a_{21}

which are equivalent to

$$\exists\, a_{ij} \in \mathbb{R} \quad (i,j = 1,\ldots,n,\ (i,j) \neq (1,2),\ (i,j) \neq (2,1)):$$

$$\{\underline{b}_1 - \sum_{\substack{j=1\\j\neq 2}}^{n} a_{1j}x_j\}x_1 \leq \{\overline{b}_2 - \sum_{j=2}^{n} a_{2j}x_j\}x_2, \qquad \underline{b}_1 - \sum_{\substack{j=1\\j\neq 2}}^{n} a_{1j}x_j \leq \overline{a}_{12}x_2,$$

$$\{\underline{b}_2 - \sum_{j=2}^{n} a_{2j}x_j\}x_2 \leq \{\overline{b}_1 - \sum_{\substack{j=1\\j\neq 2}}^{n} a_{1j}x_j\}x_1, \qquad \underline{b}_2 - \sum_{j=2}^{n} a_{2j}x_j \leq \overline{a}_{12}x_1,$$

$$\underline{a}_{12}x_2 \leq \overline{b}_1 - \sum_{\substack{j=1\\j\neq 2}}^{n} a_{1j}x_j, \qquad \underline{a}_{12}x_1 \leq \overline{b}_2 - \sum_{j=2}^{n} a_{2j}x_j,$$

inequalities without a_{12}, a_{21}.

By inspecting the particular cases $x_1 = 0$ and $x_2 = 0$, respectively, one can see that $x \in S_{\text{sym}}$ remains equivalent to this latter set of inequalities provided that x is restricted to O_1. We thus have described $S_{\text{sym}} \cap O_1$ by a set of inequalities which no longer contain b_1, a_{12} and a_{21}. This process of eliminating successively the parameters $b_i \in [b]_i$, $a_{ij} \in [a]_{ij}$ is the famous Fourier–Motzkin elimination process as presented, e.g., by Schrijver (1986). It finally results in a set of inequalities which contain the components of x at most quadratically, and only the bounds of $[b]_i$ and $[a]_{ij}$. It describes $S_{\text{sym}} \cap O_1$ completely and characterizes halfspaces and sets whose boundaries are quadrics. Analogous statements can be made for the remaining orthants. We refer to Alefeld et al. (1999) for a more general description of the elimination process.

3 The Solution Set S_P

We want to describe now – without proof – several features of the set S_P from (5). We first list some equivalences of the Oettli–Prager Theorem whose proofs can be found in the literature listed below.

Theorem 1 *For a real $n \times n$ interval matrix $[A] = [\check{A} - \Delta, \check{A} + \Delta]$ with $O \leq \Delta \in \mathbb{R}^{n \times n}$ and for a real interval vector $[b] = [\check{b} - \delta, \check{b} + \delta]$ with n components and with $0 \leq \delta \in \mathbb{R}^n$ the following properties are equivalent:*

a) $x \in S$;

b) $[b] \cap [A]x \neq \emptyset$;   *(Beeck 1972)*

c) $0 \in [b] - [A]x$;   *(Beeck 1972)*

d) $|\check{b} - \check{A}x| \leq \Delta \cdot |x| + \delta$;   *(Oettli and Prager 1964)*

e) $\exists D \in \mathbb{R}^{n \times n} : |D| \leq I \wedge \check{b} - \check{A}x = D(\Delta|x| + \delta)$;   *(Rohn 1984)*

f) $\underline{b}_i - \sum\limits_{j=1}^{n} a_{ij}^+ x_j \leq 0 \leq \overline{b}_i - \sum\limits_{j=1}^{n} a_{ij}^- x_j$, $i = 1, \dots, n$,

where a_{ij}^-, a_{ij}^+ are defined by $[a]_{ij} = \begin{cases} [a_{ij}^-, a_{ij}^+] & \text{if } x_j \geq 0 \\ [a_{ij}^+, a_{ij}^-] & \text{if } x_j < 0 \end{cases}$.

(Hartfiel 1980)

In b), c) real interval arithmetic has to be used as introduced, e.g., by Alefeld and Herzberger (1983). It is the representation in f) which can be derived directly by means of the Fourier–Motzkin elimination process. Therefore, in the general case, i.e., if $\mathcal{P}$ means no restriction, the Hartfiel description of S fits into our way of describing S_P. If $\mathcal{P}$ means 'A is symmetric' the second and the last three inequalities of the last equivalence in Section 2 indicate that those in f) for S reappear in those for S_{sym}. This expresses the trivial property $S_{\text{sym}} \subseteq S$.

Part a) of the following theorem is a direct consequence of Theorem 1 f) while part b) summarizes the remarks on S_{sym} in the Sections 1 and 2.

Theorem 2 *Let $[A]$ be a real $n \times n$ interval matrix and let $[b]$ be a real interval vector with n components.*

15

a) In each orthant of $\mathbb{R}^n$ the solution set S can be represented as intersection of finitely many halfspaces.

b) In each orthant O_i of $\mathbb{R}^n$ the symmetric solution set S_{sym} can be represented as the intersection of the solution set $S \cap O_i$ and sets with quadrics as boundaries.

Theorem 2 b) holds analogously for persymmetric matrices and skew–symmetric matrices, respectively. For details see Alefeld et al. (1997). It should be noted that the inequalities for these solution sets remain fixed if the orthant is fixed. For Hankel and Toeplitz matrices Alefeld et al. (1999) showed that the inequalities may change within an orthant. Thus for Hankel matrices from

$$[A] = \begin{pmatrix} 0 & [s] & [d] \\ [s] & [d] & 0 \\ [d] & 0 & 0 \end{pmatrix} \qquad ([s], [d] \text{ given real intervals})$$

and for right–hand sides b from some given interval vector $[b]$ the elimination process reveals such a change in O_1 depending on $x \in C \cap O_1$ and $x \in O_1 \backslash C$, respectively, where C denotes the cone $x_1 x_3 - x_2^2 \geq 0$. For details see again Alefeld et al. (1999). It is an open question what is going on in the general case of Toeplitz or Hankel matrices with perturbations. It is also unknown up to now whether there is a bound for the degree of the algebraic inequalities needed to describe the solution set S_{Toep} and S_{Hank}, respectively.

The example

$$[A] = \begin{pmatrix} [1,2] & 0 & 0 \\ [-4,-2] & [1,2] & 0 \\ [-8,-4] & [-4,-2] & [1,2] \end{pmatrix}, \quad [b] = \begin{pmatrix} 1 \\ 0 \\ 0 \end{pmatrix}$$

shows that S_{Toep} can be described by the inequalities

$$\frac{1}{2} \leq x_1 \leq 4, \quad 2x_1^2 \leq x_2 \leq 4x_1^2, \quad 4x_1^3 \leq x_1 x_3 - x_2^2 \leq 8x_1^3$$

which reveals that $S_{\text{Toep}} \subseteq O_1$ with its boundary partly contained in the two algebraic surfaces

$$x_1 x_3 - x_2^2 - 4x_1^3 = 0, \quad x_1 x_3 - x_2^2 - 8x_1^3 = 0.$$

These surfaces are of order three which means, in particular, that Theorem 2 can no longer hold for S_{Toep} and S_{Hank}, respectively.

All particular solution sets S_P we considered up to now can be transformed into systems of linear equations $Ax = b$ with

$$\left. \begin{aligned} a_{ij} &:= -a_{ij,0} + \sum_{k=1}^{p} a_{ij,k} f_k \\ b_i &:= b_{i,0} + \sum_{k=1}^{p} b_{i,k} f_k \end{aligned} \right\} \qquad i,j = 1, \ldots, n, \qquad (10)$$

16

where $p \in \mathbb{N}_0$, where f_k varies in given intervals $[f]_k$ and where $a_{ij,k}$, $b_{i,k}$ are appropriate coefficients. For the solution set of any linear system subject to (10) it was shown by Alefeld et al. (1998) that it is semialgebraic, i.e., it is a finite union of subsets each of which is defined by a finite system of polynomial equations $P_r(x_1, \ldots, x_n) = 0$ and inequalities of the type $P_s(x_1, \ldots, x_n) > 0$ and $P_t(x_1, \ldots, x_n) \geq 0$ for some polynomials P_r, P_s, P_t.

4 The Eigenpair Set E_P

In order to describe E_P from (1) we first omit any restriction $\mathcal{P}$, i.e., we consider E from (7). Since $Ax = \lambda x$ is equivalent to $(A - \lambda I)x = 0$ we can apply the Oettli–Prager Theorem on the matrix $A - \lambda I$ (assuming that λ is a fixed real number for the moment) and on the right–hand side $b = 0$. This was already done by Deif (1991) ending up with

Theorem 3 *Let* $[A] = [\check{A} - \Delta, \check{A} + \Delta]$ *be a real* $n \times n$ *interval matrix with* $O \leq \Delta \in \mathbb{R}^{n \times n}$. *Then*

$$(x, \lambda) \in E \iff |\check{A}x - \lambda x| \leq \Delta \cdot |x| \text{ and } x \neq 0.$$

In particular, the boundary of E consists of parts of hyperplanes and quadrics. A shortened analogue of Theorem 1 reads

Theorem 4 *For a real* $n \times n$ *interval matrix* $[A] = [\check{A} - \Delta, \check{A} + \Delta]$ *with* $O \leq \Delta \in \mathbb{R}^{n \times n}$ *and for* $x \in \mathbb{R}^n \backslash \{0\}$, $\lambda \in \mathbb{R}$ *the following properties are equivalent:*

a) $(x, \lambda) \in E$;

b) $[A]x \cap \{\lambda x\} \neq \emptyset$;

c) $|\check{A}x - \lambda x| \leq \Delta \cdot |x|$;

d) $\displaystyle\sum_{j=1}^{n} a_{ij}^{-} x_j \leq \lambda x_i \leq \sum_{j=1}^{n} a_{ij}^{+} x_j, \quad i = 1, \ldots, n,$

where a_{ij}^{-} *and* a_{ij}^{+} *are defined by* $[a]_{ij} = \begin{cases} [a_{ij}^{-}, a_{ij}^{+}] & \text{if } x_j \geq 0 \\ [a_{ij}^{+}, a_{ij}^{-}] & \text{if } x_j < 0 \end{cases}$.

Here, d) can form the starting point for describing the eigenvalue set if the entries of $A \in [A]$ are subject to dependencies. Thus the following analogue of Theorem 2 can again be seen from the Fourier–Motzkin elimination process.

Theorem 5 *Let* $[A]$ *be a real* $n \times n$ *interval matrix.*

a) *In each orthant of* $\mathbb{R}^{n+1}$ *the eigenpair set* E *can be represented as intersection of finitely many sets whose boundaries are hyperplanes or quadrics.*

b) In each orthant of $\mathbb{R}^{n+1}$ the symmetric eigenpair set E_{sym} can be represented as intersection of finitely many sets whose boundaries are described by algebraic equations of order ≤ 3.

An analogous statement holds for skew–symmetric and persymmetric matrices. We will illustrate Theorem 5 by the following example:

Let $[A] = \begin{pmatrix} 1 & [-1,1] \\ [-1,1] & 1 \end{pmatrix}$. Then the eigenpair set E is completely described by the inequalities

$$|1 - \lambda| \cdot |x_1| \leq |x_2|, \qquad |1 - \lambda| \cdot |x_2| \leq |x_1| \tag{11}$$

Thus if $\lambda = 1$ then any vector $(x_1, x_2, 1)^T \neq (0, 0, 1)^T$ belongs to E, i.e., E contains the plain $\lambda = 1$ punctured at $(0, 0, 1)^T$. For $\lambda \neq 1$ we have $x_1 \cdot x_2 \neq 0$, and (11) is equivalent to

$$1 - \left|\frac{x_2}{x_1}\right| \leq \lambda \leq 1 + \left|\frac{x_2}{x_1}\right|, \qquad 1 - \left|\frac{x_1}{x_2}\right| \leq \lambda \leq 1 + \left|\frac{x_1}{x_2}\right|,$$

whence

$$1 - \min\left\{\left|\frac{x_2}{x_1}\right|, \left|\frac{x_1}{x_2}\right|\right\} \leq \lambda \leq 1 + \min\left\{\left|\frac{x_2}{x_1}\right|, \left|\frac{x_1}{x_2}\right|\right\}.$$

The symmetric eigenpair set $E_{\text{sym}} \subseteq E$ consists of the plane $\lambda = 1$ punctured at $(x_1, x_2, \lambda)^T = (0, 0, 1)^T$ and of the vectors $(x_1, x_2, \lambda)^T$ satisfying $\lambda \neq 1$, $0 \leq \lambda \leq 2$, $|x_1| = |x_2| > 0$. In order to get an impression of the situation we visualize the intersection $E \cap P_1$ and $E_{\text{sym}} \cap P_1$ in Fig. 1, where P_α denotes the plane $x_2 = \alpha$. We obtain

$$E \cap P_1 = \{ (x_1, 1, 1)^T \mid x_1 \in \mathbb{R} \}$$

$$\cup \left\{ (x_1, 1, \lambda)^T \;\middle|\; 1 - \min\{\tfrac{1}{|x_1|}, |x_1|\} \leq \lambda \leq 1 + \min\{\tfrac{1}{|x_1|}, |x_1|\}, \; x_1 \in \mathbb{R}\backslash\{0\} \right\}$$

and

$$E_{\text{sym}} \cap P_1 = \{ (x_1, 1, 1)^T \mid x_1 \in \mathbb{R} \} \cup \begin{pmatrix} 1 \\ 1 \\ [0,2] \end{pmatrix} \cup \begin{pmatrix} -1 \\ 1 \\ [0,2] \end{pmatrix}.$$

As one can see at once from (11) the maximal domain $[0, 2]$ for λ is attained for $|x_1| = |x_2| \neq 0$ and nowhere else. For $\alpha \neq 0$ the intersections $E \cap P_\alpha$ and $E_{\text{sym}} \cap P_\alpha$ look similar as for $\alpha = 1$. For $\alpha = 0$ they reduce to

$$E \cap P_0 = E_{\text{sym}} \cap P_0 = \{ (x_1, 0, 1)^T \mid x_1 \in \mathbb{R}\backslash\{0\} \}$$

i.e., to a punctured straight line.

18

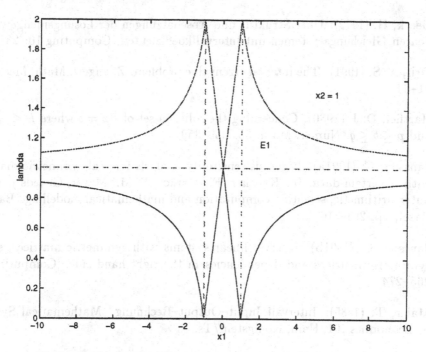

Fig. 1. $E1 := E \cap P_1$ and $E_{\text{sym}} \cap P_1$ (dashed) for the example.

There is no difficulty to apply the ideas of Section 3 to the eigenvalue problem if $A \in [A]$ is subject to more general dependencies.

References

Alefeld, G., Herzberger, J. (1983): Introduction to interval computations. Academic Press, New York

Alefeld, G., Kreinovich, V., Mayer, G. (1997): On the shape of the symmetric, persymmetric, and skew–symmetric solution set. SIAM J. Matrix Anal. Appl. 18: 693–705

Alefeld, G., Kreinovich, V., Mayer, G. (1998): The shape of the solution set of linear interval equations with dependent coefficients. Math. Nachr. 192: 23–26

Alefeld, G., Kreinovich, V., Mayer, G. (1999): On the solution set of particular classes of linear systems. Submitted for publication.

Alefeld, G., Mayer, G. (1995): On the symmetric and unsymmetric solution set of interval systems. SIAM J. Matrix Anal. Appl. 16: 1223–1240

Beeck, H. (1972): Über Struktur und Abschätzungen der Lösungsmenge von linearen Gleichungssystemen mit Intervallkoeffizienten. Computing 10: 231–244

Deif, A. S. (1991): The interval eigenvalue problem. Z. angew. Math. Mech. 71: 61–64

Hartfiel, D. J. (1980): Concerning the solution set of $Ax = b$ where $P \leq A \leq Q$ and $p \leq b \leq q$. Numer. Math. 35: 355–359

Jansson, C. (1991a): Rigorous sensitivity analysis for real symmetric matrices with uncertain data. In: Kaucher, E., Markov; S. M., Mayer, G. (eds.): Computer arithmetic, scientific computation and mathematical modelling. Baltzer, Basel, pp. 293–316

Jansson, C. (1991b): Interval linear systems with symmetric matrices, skew–symmetric matrices and dependencies in the right hand side. Computing 46: 265–274

Maier, T. (1985): Intervall–Input–Output–Rechnung. Mathematical Systems in Economics 101, Hain, Königstein/Ts.

Oettli, W., Prager, W. (1964): Compatibility of approximate solution of linear equations with given error bounds for coefficients and right–hand sides. Numer. Math. 6: 405–409

Rohn, J. (1984): Interval linear systems. Freiburger Intervall–Berichte 84/7: 33–58

Rump, S. M. (1994): Verification methods for dense and sparse systems of equations. In: Herzberger, J. (ed.): Topics in validated computations. Elsevier, Amsterdam, pp. 63–135

Schrijver, A. (1986): Theory of linear and integer programming. Wiley, New York

Wilkinson, J. H. (1963): Rounding errors in algebraic processes. Prentice–Hall, Englewood Cliffs, New Jersey

Symbolic-Numeric Algorithms for Polynomials:
some recent results

Robert M. Corless

1 Introduction

One of the most powerful recent ideas for the solution of systems of multivariate polynomials, namely the conversion of such systems to eigenproblems for commuting families of sparse matrices, will be shown by example.

The facts that the matrices are sparse and commute have theoretical and practical consequences: their sparsity allows efficiency, while their commutativity promotes numerical stability.

Implications for validated computing will be discussed.

2 Motivation and a selection from previous work

In this short tutorial paper we look at converting systems of polynomial equations to eigenvalue problems for commuting families of matrices, and why we should do so. See (Corless et al., 2000) for a concise review of recent work in this area, which is termed "SNAP" for Symbolic-Numeric Algorithms for Polynomials, or "Numerical Nonlinear Algebra". For a detailed discussion of the algorithm used to convert systems of polynomial equations with zero–dimensional varieties (that is, a finite number of solutions) to eigenvalue problems for commuting families of sparse matrices, see for example (Corless, 1996), (Stetter, 1993), (Möller and Stetter, 1995), and (Stetter, 1996). For a deeper and more detailed introduction to these results, see (Emiris and Mourrain, 1999), (Mourrain, 1998), and (Mourrain, 1999).

3 The main idea, via a simple example

To keep the discussion concrete, we consider here a simple example problem, one used in (Li et al. 1989) to show the effectiveness of a particular trick for homotopy methods (the trick was called the 'cheater's homotopy' in that paper). The example is the following system of two polynomial equations in two variables, x

and y:

$$x^3y^2 + c_1x^3y + y^2 + c_2x + c_3 = 0$$
$$c_4x^4y^2 - x^2y + y + c_5 = 0. \qquad (1)$$

The symbols c_k, $k = 1(1)5$, represent parameters, which can take values in $\mathbb{C}$. Generically, as noted in (Li et al., 1989), there are 10 complex roots of this system—by "generically" one means for almost all sets of values of the parameters; for some exceptional values (which are of course interesting) the number of roots may be different.

The classical computer algebra approach to problems like this is to convert this set of polynomial equations into a more convenient set of equations, namely a Gröbner basis, via the Buchberger algorithm (Cox et al., 1992). For convenience, we summarize in the following subsection the results and notation that we use in this paper.

Definitions and properties.

A *Gröbner basis* is a basis of a polynomial ideal that satisfies the properties listed below, with respect to a given monomial ordering. We write a monomial $x_1^{e_1} x_2^{e_2} \cdots x_s^{e_s}$ simply as x^e. The only monomial orderings that we use here are *lexicographic* order, where $x^e > x^f$ if the integer vector $e - f$ has its leftmost nonzero entry positive; and *graded reverse lexicographic* order, where $x^e > x^f$ if the total degree of x^e is larger than the total degree of x^f, that is $e_1 + e_2 + \cdots e_s > f_1 + f_2 + \cdots f_s$, or if the total degree of x^e is the same as the total degree of x^f and the integer vector $e - f$ has its rightmost nonzero entry negative. A graded reverse lexicographic ordering is called a "tdeg" ordering by Maple.

Given an ordering, a set of polynomials G is a Gröbner basis if and only if ideal membership can be decided by multivariate division; that is, a polynomial p is in the ideal generated by G if the remainder on division by G is zero. Note that this remainder is unique if and only if G is a Gröbner basis. An excellent description of multivariate division can be found in (Cox et. al, 1992, *pp.* 60–66).

The *normal set* of G is the set of monomials that cannot be divided by the leading monomial of any polynomial in G. If G generates a zero-dimensional ideal, then the cardinality of the normal set will be finite, and vice-versa. Moreover, the number of elements of the normal set is precisely the number of points in the variety. We typically write the normal set as $[t_1, t_2, \ldots, t_N]$, and by convention we write the t_i in increasing monomial order. Every remainder on division by G can be written as a linear combination of elements of the normal set.

A given set of polynomials can be converted into a Gröbner basis by the Buchberger algorithm. Computation of a lexicographically ordered Gröbner basis (hereafter lex-order basis) can be expensive: the cost to find a lex-order basis is in the worst case doubly exponential in the number of variables, but grevlex-order bases are often "only" singly-exponential cost to compute (particularly for zero-dimensional varieties). See (von zur Gathen and Gerhard, 1999) for an introductory discussion of the complexity of computation of Gröbner bases, but with many more details than presented here.

When it can be computed, and if there are only a finite number of zeros of G, a lex-order basis is in a "triangular" shape: the first polynomial in G, say

g_1, is a polynomial in only one variable, say x_1; then there is a polynomial in G that contains x_1 and only one more variable, say x_2, and so on. Therefore in principle one can solve the first polynomial for all possible values of x_1, and then use the next polynomial to identify all corresponding values of x_2, and so on. This process is known to be numerically unstable.

Gröbner bases are discontinuous with respect to changes in the problem. A simple example is

$$f_1 = \varepsilon x^2 + y^2 - 1$$
$$f_2 = xy - a, \tag{2}$$

which has as its grevlex-order Gröbner basis $G_\varepsilon = \{f_1, f_2, f_3\}$ where

$$f_3 = \varepsilon a x + y^3 - y. \tag{3}$$

However, the computed grevlex-order Gröbner basis of $(y^2 - 1, xy - a)$ is $(x - ya, y^2 - 1)$, not $\lim_{\varepsilon \to 0} G_\varepsilon$. This discontinuity is the cause of fundamental difficulties with computation of Gröbner bases of polynomials with floating-point or interval coefficients.

Multivariate division by a Gröbner basis *is* continuous—but ill-conditioned. For example, division of x^2 by G_ε above gives a unique remainder

$$r = \frac{1 - y^2}{\varepsilon}. \tag{4}$$

It is easy to see by linearity that the derivative of the coefficients of the remainder on division of p by G, with respect to changes in the coefficient of x^e of p, is just the corresponding coefficients of the remainder on division by of x^e itself by G. With the example above, we see that these derivatives can be unbounded. Hence we conclude that the *problem* of multivariate division is ill-conditioned.
Back to the Li-Sauer-Yorke example.

The computation of a lex-order Gröbner basis does not work on this example, without specifying the parameters c_k first. If you try, then you simply run out of computer memory.

This is not the fault of Buchberger's algorithm: it is the fault of the problem. The answer, the lexicographic ordered Gröbner basis itself, contains polynomials of degree 10 with coefficients that are *rational functions* of the c_k of exponentially large length. That is, if we embedded this example into a family of problems with parameters c_k, $k = 1(1)n$, then the length of the coefficients of the Gröbner basis grows at least exponentially with n.

However, not all is lost: one can get, in a few seconds computation on a computer with a modest amount of memory, a grevlex-ordered Gröbner basis for this system. This ordering is not an elimination ordering and does not produce a Gröbner basis with a triangular shape (and therefore at first glance the result seems much less useful). A grevlex-ordered Gröbner basis is usually much smaller than a lex order basis. Even so, the grevlex basis for this example contains some 550 terms in its representation, and hence isn't all that pleasant to look at; but it can be worked with. In particular, one can compute the dual matrix representations of the action of multiplication by x and by y in the ring $\mathbb{C}[x, y]/I$ where I is the ideal generated by G. This is accomplished by computing the remainders on multivariate division of x multiplied by each t_i by G (see for example (Corless, 1996)). In this case, the process takes only a few seconds and generates two 10 by 10 matrices (each containing about 500 terms, again), from which we can find the roots of the original systems in milliseconds, by the following technique.

To find the roots of (1), we just find the eigenvalues and eigenvectors of the matrices. A proof that this is so can be found in (Stetter, 1996), for example. We can compute eigenvalues easily, for any numerical value of the parameters c_k, $k = 1(1)5$, by any convenient eigenvalue package: after all, 10 by 10 matrices are 'trivial'. A program to generate the matrix corresponding to the x values of the roots costs only 232 storage allocations + 233 assignments + 106 subscripts + 282 multiplications + 259 additions + 4 divisions, for any given set of c_k, $k = 1(1)5$. This compares with roughly 35,000 flops for the eigenvalue computation itself (computing eigenvectors also).

These matrices are very sparse, and for efficiency this should be taken advantage of.

Note that for simple roots, the eigenvectors are of the form $[t_1, t_2, \ldots, t_N]$ where the monomials t_i are the entries in the normal set for G. For grevlex ordered Gröbner bases, this normal set usually contains each of x_1, x_2, ..., x_s, and thus the roots of the polynomial system can simply be read off from the eigenvectors. Multiple roots are more complicated; see (Möller and Stetter, 1995) or (Corless et al., 1997) for more details.

A Maple worksheet detailing this calculation is shown in the appendix. This worksheet is the help file for the module NormalSet, which is available at the web page http://www.apmaths.uwo.ca/~rcorless for use with Maple 6.

An alternative approach

An alternative approach would be to specialize the values of the c_k first, and then compute a lexicographic order Gröbner basis of the resulting purely numerical polynomial system. This takes about 8 seconds of computation on a ThinkPad 600E PII 233Mhz. The result, for $c_k = k/10$, is

$$640000\, y^{10} + 12768000\, y^8 + 9680000\, y^7 + 25033600\, y^6 + 17708000\, y^5$$
$$+10951920\, y^4 + 3966200\, y^3 + 890804\, y^2 + 125910\, y + 8405$$

which can then, with modest effort more or less equivalent to the solution of a 10 by 10 eigenvalue problem, be solved for x and y: we solve the degree 10 polynomial in y (which has coefficients of only modest size) for the y-values of the roots, and then use the x equation (not shown here, but which, as is usually the case, is linear in x) to determine the corresponding x value.

This alternative approach is quite attractive if one wishes to solve the system for one or only a few sets of values of the parameters; most people are willing to pay a fair amount of computer time for a relatively straightforward method (in terms of human mental time) for finding roots.

However, it will be obvious to the reader that in some cases this approach is potentially numerically unstable—and to combat the instability with brute force will require large precision for the representations of the polynomial coefficients, and likewise for the iterative computation of the roots, and thus perhaps more cost in general than we are willing to bear.

Finally, consider the common situation where there are many values of the parameters for which we wish to find the roots of (1)—for example, if we wished to plot a section of the surface defined by $z = x(c_1, c_2, 0.3, 0.4, 0.5)$. In this case, it would clearly pay us to do a little pre-processing first, to get an efficient method for computing the roots given the values of the parameters. Solving a 10 by 10 matrix problem numerically every time is thousands of times faster than finding a lex order Gröbner basis every time, even without considering the difficulty of numerical instability.

As another more general example, to demonstrate the common occurrence of numerical instability in the lex-order approach, consider using it on the set of polynomial equations given by the eigenvalue problem $Ax = \lambda x$, together with the normalization $\|x\|^2 = 1$, considered as a set of polynomial equations in the unknowns $x_1, x_2, \ldots, x_n$, and λ. It is clear that a lexicographic order Gröbner basis will contain the characteristic (or minimal) polynomial for λ, and thus this approach is at least as numerically unstable in general as the characteristic polynomial approach: sometimes worth the cost, but usually not (Wilkinson, 1984).

And of course numerical procedures to solve eigenvalue problems directly are quite well-developed, and relatively inexpensive. But for this problem we have even more structure to use, and this is also helpful.

4   Eigenproblems for commuting families

Converting a multivariate polynomial system, say in the variables x_k, $k = 1(1)s$, into an eigenproblem by the methods described in the references in fact gives s commuting matrices, not just one matrix; moreover, the matrices are usually very sparse.

The matrices commute because they are the *dual representation* of the action of multiplication by the variables x_k in a commutative ring of polynomials; if the coefficients of the matrices are computed exactly, then the matrices will commute exactly.

Eigenproblems for commuting families of matrices are interesting, because such families must have common invariant subspaces (see (Gohberg et al., 1986) or (Horn and Johnson, 1985)). This means that the family may be brought simultaneously to Schur form, by a single unitary matrix V. This fact is exploited in (Corless et al., 1997) to construct an algorithm to solve multivariate polynomial systems with multiple roots.

5 Open problems for validated computing

First, the topic of interval Gröbner bases has been very little studied; indeed, until a referee pointed out the interesting paper (Jäger and Ratz, 1995) I was not aware of any work at all. Second, while it seems clear that the approach demonstrated in this present paper is more numerically stable than that of computation of lex-order bases and then solving them, it is not clear how *much* more stable it is. There is the problem of the computation of the (possibly discontinuous with respect to changes in the coefficients of the input) grevlex-order basis. Here, of course, the problem contained only integer coefficients and symbolic parameters, and thus the calculation was exact—but the resulting Gröbner basis is not a so-called *comprehensive* Gröbner basis (Weispfenning, 1992) and therefore evaluating G at certain values of the parameters (specializing G) may lead to incorrect results. Then, there is the question of the possible ill-condition of the numerical multivariate polynomial division problem (again, in our example, solved exactly); then, there is the question of the numerical stability of the evaluation of the entries in the commuting families of matrices (here, rational functions of the parameters are to be evaluated, and of course there is the certainty of rounding errors here). Finally, some eigenproblems are themselves ill-conditioned. It would be very useful to have a validated study of examples of these problems solved by the methods of this paper.

The problem of the computation of eigenvalues of *nearly*-commuting families of matrices in an interval context is also interesting. During the discussion at Dagstuhl at this talk, it was pointed out by S. Rump and A. Neumaier that some progress might be made if the entries in the matrix family were constructed out of parameters (such as the c_k in the example given here) and the values of the parameters could be enclosed. It was pointed out by H. J. Stetter that point methods could be used to approximate the answers, and that this approximation might be used as the basis for a subsequent enclosure by validated methods. It would be interesting to see these questions investigated further.

6 Concluding Remarks

This tutorial paper has discussed, by a concrete example, some recent developments in the use of eigenproblems for commuting families of matrices for the solution of systems of multivariate equations. The theory and algorithms involved are developed elsewhere, by many people; this paper is intended mostly as a pointer to an emerging and interesting area. For more details, consult the references of (Corless et al., 2000).

Acknowledgements

I would like to thank the organizers of this Dagstuhl seminar, and in particular Götz Alefeld, for the opportunity to have many stimulating discussions and to meet so many new colleagues. This work was supported by the National Science and Engineering Research Council of Canada, and by the Ontario Research Centre for Computer Algebra.

References

Corless, R. M. (1996). Gröbner bases and matrix eigenproblems. *Sigsam Bulletin: Communications in Computer Algebra*, 30(4):26–32.

Corless, R. M., Gianni, P. M., and Trager, B. M. (1997). A reordered Schur factorization method for zero-dimensional polynomial systems with multiple roots. In *Proceedings of the ACM-SIGSAM International Symposium on Symbolic and Algebraic Computation*, Association for Computing Machinery, New York pp. 133–140.

Corless, R. M., Kaltofen, E., and Watt, S. M. (2000). *Symbolic-Numeric Algorithms for Polynomials*. Springer-Verlag, to appear.

Cox, D., Little, J., and O'Shea, D. (1992). *Ideals, Varieties, and Algorithms*. Springer-Verlag, New York.

Emiris, I. Z. and Mourrain, B. (1999). Matrices in elimination theory. *J. Symb. Comp.*, 28(1 & 2):3–44.

von zur Gathen, J. and Gerhardt, J. (1999). *Modern Computer Algebra*. Cambridge University Press.

Gohberg, I., Lancaster, P., and Rodman, L. (1986). *Invariant Subspaces of Matrices with Applications*. Canadian Mathematical Society, Toronto.

Horn, R. and Johnson, C. (1985). *Matrix Analysis*. Cambridge University Press.

Jäger, C. and Ratz, D. (1995). A combined method for enclosing all solutions of nonlinear systems of polynomial equations. *Reliable Computing*, 1:41–64.

Li, T. Y., Sauer, T., and Yorke, J. A. (1989). The cheater's homotopy. *SIAM J. Num. Anal.*, 26(5):1241–1251.

Möller, H. M. and Stetter, H. J. (1995). Multivariate polynomial equations with multiple zeros solved by matrix eigenproblems. *Numer. Math.*, 70(3):311–330.

Mourrain, B. (1998). Computing isolated polynomial roots by matrix methods. *J. Symb. Comp.*, 26(6):715–738.

Mourrain, B. (1999). An introduction to linear algebra methods for solving polynomial equations. In Lipitakis, E. A., editor, *HERCMA, pp.* 179–200.

Stetter, H. J. (1993). Multivariate polynomial equations as matrix eigenproblems. In *Contributions to Numerical Mathematics*, volume 2 of *Series in Applicable Analysis, pp.* 355–371. World Scientific.

Stetter, H. J. (1996). Matrix eigenproblems are at the heart of polynomial system solving. *Sigsam Bulletin: Communications in Computer Algebra*, 30(4):22–25.

Weispfenning, V. (1992). Comprehensive Gröbner bases. *J. Symb. Comp.*, 14:1–29.

Wilkinson, J. H. (1984). *The Perfidious Polynomial*, In: MAA Studies, volume 24, *pp.* 1–28. Mathematical Association of America.

A Normal Sets and Multiplication Matrices in Maple

Function: NormalSet:-SetBasis — compute the normal set of a 0-dim Groebner basis

Function: NormalSet:-MulMatrix — compute a multiplication matrix from a normal set

Calling Sequences:
```
> with(NormalSet);
> ns, rv := SetBasis(gb, termorder):
```

This assigns the normal set to "ns" and the reversed table to "v".

You must make sure that the ordering **termorder** is the same as that used to compute the Groebner basis gb. Only the case of zero-dimensional varieties (finite-dimensional normal sets) is handled.

Once "ns" and "rv" are computed, you can compute the multiplication matrices corresponding to each variable by using MulMatrix.
```
> mx := MulMatrix(x, ns, rv, gb, termorder):
> my := MulMatrix(y, ns, rv, gb, termorder):
```

Output:

The output of **SetBasis** is an expression sequence with two elements. The first element,
```
> ns;
```
is a list containing the elements of the normal set of the Groebner basis. It will usually look like $[1, t_2, t_3, ..., t_n]$, where the t_k are the terms of the normal set. The second element, rv, contains exactly the same information, but reversed in a table, for efficient access by the routine **MulMatrix**. You will not normally want to look at rv.

The output of **MulMatrix** is the multiplication matrix corresponding to the variable that was input. The eigenvalues of the multiplication matrices give the roots of the original polynomial system.

Examples:

Example 1: A Lagrange multiplier problem.

First load in the routines.

```
> with(NormalSet);
```
$$[MulMatrix, SetBasis]$$
```
> f := x^3 + 2*x*y*z - z^2:
> g := x^2 + y^2 + z^2 - 1:
> F := f + lambda*g;
```
$$F := x^3 + 2xyz - z^2 + \lambda(x^2 + y^2 + z^2 - 1)$$
```
> gF := convert(linalg[grad](F, [lambda, x, y, z]),set);
```
$$gF := \{x^2 + y^2 + z^2 - 1,\, 3x^2 + 2yz + 2\lambda x,\, 2xz + 2\lambda y,\, 2xy - 2z + 2\lambda z\}$$
```
> with(Groebner):
> gb := gbasis(gF, tdeg(lambda,x,y,z)):
```

Now call SetBasis to get the normal set of this Groebner basis.

```
> ns, rv := SetBasis(gb, tdeg(lambda,x,y,z)):
> ns;
```
$$[1,\, z,\, y,\, x,\, \lambda,\, z^2,\, yz,\, y^2,\, xz,\, \lambda z,\, \lambda^2,\, z^3]$$

We don't wish to look at these 12 by 12 matrices here, but we will verify that the matrices commute.

```
> M[x] := MulMatrix(x, ns, rv, gb, tdeg(lambda,x,y,z)):
> M[y] := MulMatrix(y, ns, rv, gb, tdeg(lambda,x,y,z)):
> M[z] := MulMatrix(z, ns, rv, gb, tdeg(lambda,x,y,z)):
> M[lambda] := MulMatrix(lambda, ns, rv, gb, tdeg(lambda,x,y,z)):
> LinearAlgebra:-Norm( M[x].M[y]-M[y].M[x], infinity);
```
$$0$$

Example 2: A geometric intersection problem.

```
> restart:
> with(NormalSet);
```
$$[MulMatrix, SetBasis]$$
```
> f[1] := 3*x[1]^2*x[2] + 9*x[1]^2+2*x[1]*x[2]+5*x[1]
> + x[2] - 3;
```
$$f_1 := 3x_1^2 x_2 + 9x_1^2 + 2x_1 x_2 + 5x_1 + x_2 - 3$$
```
> f[2] := 2*x[1]^3*x[2] + 6*x[1]^3 - 2*x[1]^2 - x[1]*x[2]
> - 3*x[1] - x[2] + 3;
```
$$f_2 := 2x_1^3 x_2 + 6x_1^3 - 2x_1^2 - x_1 x_2 - 3x_1 - x_2 + 3$$
```
> f[3] := x[1]^3*x[2] + 3*x[1]^3 + x[1]^2*x[2] + 2*x[1]^2;
```

$$f_3 := x_1{}^3 x_2 + 3 x_1{}^3 + x_1{}^2 x_2 + 2 x_1{}^2$$

```
> with(Groebner):
> gb := gbasis( [ f[1], f[2], f[3] ], tdeg(x[1],x[2])):
> ns, rv := SetBasis(gb, tdeg(x[1],x[2])):
> ns;
```

$$[1, x_2, x_1]$$

```
> M[1] := MulMatrix(x[1], ns, rv, gb, tdeg(x[1],x[2]));
```

$$M_1 := \begin{bmatrix} 0 & 0 & 1 \\ -3 & 1 & -1 \\ 3 & -1 & \dfrac{3}{2} \end{bmatrix}$$

```
> M[2] := MulMatrix(x[2], ns, rv, gb, tdeg(x[1],x[2]));
```

$$M_2 := \begin{bmatrix} 0 & 1 & 0 \\ \dfrac{3}{2} & \dfrac{5}{2} & 4 \\ -3 & 1 & -1 \end{bmatrix}$$

We find that this problem has only simple roots and so we may use eigenvectors to find out everything.

```
> M[1] . M[2] - M[2] . M[1] ;
```

$$\begin{bmatrix} 0 & 0 & 0 \\ 0 & 0 & 0 \\ 0 & 0 & 0 \end{bmatrix}$$

```
> with(LinearAlgebra):
> e[1],V[1] := Eigenvectors(M[1]);
```

$$e_1, V_1 := \begin{bmatrix} 0 \\ \dfrac{5}{4} + \dfrac{1}{4}\sqrt{65} \\ \dfrac{5}{4} - \dfrac{1}{4}\sqrt{65} \end{bmatrix}, \begin{bmatrix} 1 & 1 & 1 \\ 3 & -\dfrac{3}{4} - \dfrac{1}{4}\sqrt{65} & -\dfrac{3}{4} + \dfrac{1}{4}\sqrt{65} \\ 0 & \dfrac{5}{4} + \dfrac{1}{4}\sqrt{65} & \dfrac{5}{4} - \dfrac{1}{4}\sqrt{65} \end{bmatrix}$$

Since the eigenvectors are [1, x[2], x[1]], we can simply read off the roots. But we can also show that the matrices are simultaneously diagonalized, as below.

```
> P := JordanForm(M[1],output='Q');
```

$$P := \begin{bmatrix} \dfrac{-1}{5} & \dfrac{3}{325}(5 + \sqrt{65})\sqrt{65} & \dfrac{3}{325}(-5 + \sqrt{65})\sqrt{65} \\ \dfrac{-3}{5} & \dfrac{3}{26}\sqrt{65} + \dfrac{3}{10} & -\dfrac{3}{26}\sqrt{65} + \dfrac{3}{10} \\ 0 & -\dfrac{6}{65}\sqrt{65} & \dfrac{6}{65}\sqrt{65} \end{bmatrix}$$

```
> j[1] := map(radnormal, P^(-1) . M[1] . P );
```

$$
j_1 := \begin{bmatrix} 0 & 0 & 0 \\ 0 & \dfrac{5}{4} - \dfrac{1}{4}\sqrt{65} & 0 \\ 0 & 0 & \dfrac{5}{4} + \dfrac{1}{4}\sqrt{65} \end{bmatrix}
$$

```
>  j[2] := map(radnormal, P^(-1) . M[2] . P );
```

$$
j_2 := \begin{bmatrix} 3 & 0 & 0 \\ 0 & -\dfrac{3}{4} + \dfrac{1}{4}\sqrt{65} & 0 \\ 0 & 0 & -\dfrac{3}{4} - \dfrac{1}{4}\sqrt{65} \end{bmatrix}
$$

```
>  answers := seq( {x[1]=j[1][i,i],x[2]=j[2][i,i]}, i=1..3);
```

$$
answers := \{x_1 = 0,\ x_2 = 3\},\ \{x_1 = \frac{5}{4} - \frac{1}{4}\sqrt{65},\ x_2 = -\frac{3}{4} + \frac{1}{4}\sqrt{65}\},
$$
$$
\{x_1 = \frac{5}{4} + \frac{1}{4}\sqrt{65},\ x_2 = -\frac{3}{4} - \frac{1}{4}\sqrt{65}\}
$$

So one root is x=0 and y=3. The others are almost as uncomplicated.

```
>  seq(simplify(subs(answers[i], [f[1],f[2],f[3]])),i=1..3);
```

$$
[0, 0, 0],\ [0, 0, 0],\ [0, 0, 0]
$$

Example 3. A problem with free parameters. This is taken from (Li et al., 1989). This problem cannot be done by a pure lexicographic ordered Groebner basis (the answer is just too complicated to be of any use even if we could calculate it in a reasonable length of time). But the approach here works in under a minute on a 486, with only a small amount of memory.

```
>  restart:
>  LSY[1] := x^3*y^2+c1*x^3*y+y^2+c2*x+c3;
```

$$
LSY_1 := x^3 y^2 + c1\, x^3 y + y^2 + c2\, x + c3
$$

```
>  LSY[2] := c4*x^4*y^2-x^2*y+y+c5;
```

$$
LSY_2 := c4\, x^4 y^2 - x^2 y + y + c5
$$

```
>  with(Groebner):
>  with(NormalSet);
```

$$
[MulMatrix,\ SetBasis]
$$

The following step does not succeed if you ask for a plex order basis, but with a tdeg ordering it takes only a few seconds.

```
>  gb := gbasis({LSY[1],LSY[2]},tdeg(x,y)):
>  ns, rv := SetBasis(gb, tdeg(x,y)):
>  ns;
```

$$
[1,\ y,\ x,\ y^2,\ x\,y,\ x^2,\ y^3,\ x\,y^2,\ x^2\,y,\ x^3]
$$

```
>  nops(ns);
```

```
>   M[x] := MulMatrix(x, ns, rv, gb, tdeg(x,y)):
```

The entries of these sparse matrices are rational functions in the parameters c1 through c5. For example,

```
>   M[x][10,3];
```

$$(c3^2 \, c4 \, c1^2 \, c5 + c3^2 \, c4^2 \, c2^2 - c3^2 \, c4^2 \, c2 \, c1^2 + 2 \, c3 \, c1^2 \, c5^2 + c3 \, c1^3 \, c5 + 2 \, c3 \, c4^2 \, c2^2 \, c1^2$$
$$+ \, c3 \, c2^2 + c4 \, c2^3 \, c1 + 2 \, c1^2 \, c4 \, c2^2 \, c5 + 2 \, c3 \, c2 \, c5 \, c1 + c4 \, c1 \, c3^3 + c4 \, c1^3 \, c2^2$$
$$+ \, 2 \, c3 \, c4^3 \, c2^3 \, c1 + c4^3 \, c3^4 \, c1 + c1^3 \, c4^3 \, c3^3 + 2 \, c4^2 \, c1 \, c3^2 \, c5 \, c2)/(c1 \, (c3 \, c4^3 \, c2^3$$
$$+ \, c3 \, c4^2 \, c2 \, c1^2 \, c5 + c1 \, c4^2 \, c3 \, c2^2 + c4^2 \, c5 \, c1^4 \, c2 + c1 \, c4 \, c2^2 \, c5 + 2 \, c2 \, c5 \, c1^2$$
$$+ \, c4^3 \, c2^3 \, c1^2 + c4^2 \, c1^3 \, c2^2 + c4 \, c2^3 + c5^2 \, c1^3 + c1 \, c2^2))$$

```
>   M[y] := MulMatrix(y, ns, rv, gb, tdeg(x,y)):
```

At this point, one may insert numerical values for c1 through c5 and find eigenvalues of a generic (convex random) combination of these two matrices, cluster any multiple roots, and use the Schur vectors to find the roots of the system. We see that there are generically 10 roots. We take random values for the parameters below, as an example, ignoring the possibility of multiple roots.

```
>   c1 := rand()/10.^12:
>   c2 := rand()/10.^12:
>   c3 := rand()/10.^12:
>   c4 := rand()/10.^12:
>   c5 := rand()/10.^12:
>   Digits := trunc(evalhf(Digits)):
>   with(LinearAlgebra):
>   xvals, V := Eigenvectors(evalf(M[x])):
```

We may read off the corresponding y-values of the roots from the known structure of ns. Since the 2nd element of ns is y, the 2nd element of each eigenvector will be the y-value of the root (if each eigenvector is normalized so that its first entry is 1).

```
>   yvals := [seq(V[2,i]/V[1,i],i=1..10)]:
```

Substitute the computed values of x and y into the original equations, to see how nearly the computed quantities satisfy the original equations. To know how accurate our computed x and y are, we need more than just these residuals; we should look at how perturbations in these polynomials affect the roots. We do not do that here.

```
>   LSY[1];
>   LSY[2];
```

$$x^3 \, y^2 + .9148251370 \, x^3 \, y + y^2 + .2813148623 \, x + .4541007452$$

$$.7690344101 \, x^4 \, y^2 - x^2 \, y + y + .08070311130$$

```
>   residuals :=
>   [seq(subs(x=xvals[i],y=yvals[i],{LSY[1],LSY[2]}),i=1..10)];
```

$residuals := [\{-.193\,10^{-11} - .14\,10^{-12}\,I, .86\,10^{-12} + .143\,10^{-11}\,I\},$
$\{-.281\,10^{-11} + .112\,10^{-11}\,I, -.188\,10^{-11} - .64\,10^{-12}\,I\},$
$\{-.86\,10^{-12} - .4\,10^{-14}\,I, -.3810\,10^{-12} + .286\,10^{-12}\,I\},$
$\{-.14\,10^{-12} - .102\,10^{-12}\,I, -.782\,10^{-13} - .105\,10^{-12}\,I\},$
$\{.304\,10^{-11} - .3889\,10^{-11}\,I, -.192\,10^{-11} + .2\,10^{-12}\,I\},$
$\{.324\,10^{-11} + .2805\,10^{-11}\,I, -.88\,10^{-12} - .3\,10^{-12}\,I\},$
$\{-.1342\,10^{-10} + .87\,10^{-11}\,I, .318\,10^{-11} - .1537\,10^{-10}\,I\},$
$\{.428\,10^{-11} + .441\,10^{-11}\,I, -.1092\,10^{-10} - .12\,10^{-11}\,I\},$
$\{-.9\,10^{-13} + .18\,10^{-12}\,I, -.22\,10^{-12} + .1327\,10^{-11}\,I\},$
$\{-.69\,10^{-12} - .141\,10^{-12}\,I, -.14\,10^{-12} + .5\,10^{-13}\,I\}]$

Alternatively one can use tdeg bases to explore the bifurcation behaviour of these equations; one can quite easily determine parameter sets that force double, triple, or even quadruple roots.

Symbolic-Numeric QD-algorithms with applications in Function theory and Linear algebra

Annie Cuyt

1. Introduction

The univariate qd-algorithm is very useful for the determination of poles of meromorphic functions and eigenvalues of certain tridiagonal matrices. Both applications are linked to the theory of orthogonal polynomials, in particular the formally orthogonal Hadamard polynomials.

When looking for the pole curves of multivariate functions, or for the eigenvalue curves of some parameterized tridiagonal matrices, the qd-algorithm has to be generalized in order to deal with multivariate data. Indeed, the univariate algorithm only involves number manipulations and these multivariate problems require the manipulation of expressions.

For the computation of the poles of a multivariate meromorphic function a symbolic generalization of the qd-algorithm implemented in floating-point polynomial arithmetic seems to be possible. For the parameterized eigenvalue problem a homogeneous version implemented in exact polynomial arithmetic can be used.

In Section 2 we summarize the univariate prerequisites for the material. The multivariate pole detection problem and the floating-point polynomial qd-algorithm are treated in Section 3 while the parameterized eigenvalue problem is solved in Section 4 using the homogeneous version of the qd-algorithm.

2. The univariate floating-point qd-algorithm.

Let us define a linear functional c from the vector space $\mathbb{C}[z]$ to $\mathbb{C}$ by

$$c(z^j) = c_j \qquad j = 0, 1, \ldots$$

Hence c is completely determined by the sequence $\{c_j\}_{j \in \mathbb{N}}$ which is called the sequence of moments of the functional c (in the sequel $c_j = 0$ when $j < 0$). With the c_j we also associate the Hankel determinants

$$H_m^{(n)} = \begin{vmatrix} c_n & \cdots & c_{n+m-1} \\ \vdots & \ddots & c_{n+m} \\ & & \vdots \\ c_{n+m-1} & \cdots & c_{n+2m-2} \end{vmatrix} \qquad H_0^{(n)} = 1$$

The functional c is called s-normal if

$$H_m^{(n)} \neq 0 \qquad n \geq 0 \qquad m = 0, \ldots, s$$

From now on we shall assume that this is the case.

With the sequence $\{c_j\}_{j \in N}$ we can also set up the qd-scheme where subscripts denote columns and superscripts downward sloping diagonals [Henrici 1974]. Its start columns are given by

$$e_0^{(n)} = 0 \qquad n = 1, 2, \ldots$$
$$q_1^{(n)} = \frac{c_{n+1}}{c_n} \qquad n = 0, 1, \ldots$$

and the rhombus rules for continuation of the scheme by

$$e_m^{(n)} = q_m^{(n+1)} - q_m^{(n)} + e_{m-1}^{(n+1)} \qquad m = 1, 2 \ldots \quad n = 0, 1 \ldots$$
$$q_{m+1}^{(n)} = \frac{e_m^{(n+1)}}{e_m^{(n)}} q_m^{(n+1)} \qquad m = 1, 2 \ldots \quad n = 0, 1, \ldots$$

In what follows we need the next lemma.

LEMMA 1:
If the functional c is s-normal, then the values $q_m^{(n)}$ and $e_m^{(n)}$ exist for $m = 1, \ldots, s$ and $n \geq 0$ and they are given by

$$q_m^{(n)} = \frac{H_m^{(n+1)} H_{m-1}^{(n)}}{H_m^{(n)} H_{m-1}^{(n+1)}}$$

$$e_m^{(n)} = \frac{H_{m+1}^{(n)} H_{m-1}^{(n+1)}}{H_m^{(n)} H_m^{(n+1)}}$$

The following properties of the qd-algorithm can be found in [Henrici 1974].

2.1. Meromorphic functions and root finding.

THEOREM 1 [Henrici 1974, pp. 612–613]:
Let

$$\sum_{j=0}^{\infty} c_j z^j$$

be the Taylor series at $z = 0$ of a function $f(z)$ meromorphic in the disk $B(0, R) = \{z : |z| < R\}$ and let the poles z_i of f in $B(0, R)$ be numbered such that

$$z_0 = 0 < |z_1| \leq |z_2| \leq \ldots < R$$

each pole occuring as many times in the sequence $\{z_i\}_{i \in N}$ as indicated by its order. If f is s-normal for some integer $s > 0$, then the qd-scheme associated with f has the following properties (put $z_{s+1} = \infty$ if f has only s poles):
(a) for each m with $0 < m \leq s$ and $|z_{m-1}| < |z_m| < |z_{m+1}|$,

$$\lim_{n \to \infty} q_m^{(n)} = z_m^{-1}$$

36

(b) for each m with $0 < m \le s$ and $|z_m| < |z_{m+1}|$,

$$\lim_{n\to\infty} e_m^{(n)} = 0$$

Any index m such that the strict inequality

$$|z_m| < |z_{m+1}|$$

holds, is called a critical index. It is clear that the critical indices of a function do not depend on the order in which the poles of equal modulus are numbered. The theorem above states that if m is a critical index and f is m-normal, then

$$\lim_{n\to\infty} e_m^{(n)} = 0$$

Thus the qd-table of a meromorphic function is divided into subtables by those e-columns tending to zero. This property motivated Rutishauser [Henrici 1974, p. 614] to apply the rhombus rules satisfied by the q- and e-values, namely

$$q_m^{(n+1)} e_m^{(n+1)} = e_m^{(n)} q_{m+1}^{(n)}$$
$$e_{m-1}^{(n+1)} + q_m^{(n+1)} = q_m^{(n)} + e_m^{(n)}$$

in their progressive form [Henrici 1974]: when computing the q-values from the top down rather than from left to right, one avoids divisions by possibly small e-values that can inflate rounding errors. Other reformulations can be found in [Von Matt 1997] and [Fernando et al. 1994]. Any q-column corresponding to a simple pole of isolated modulus is flanked by such e-columns and converges to the reciprocal of the corresponding pole. If a subtable contains $j > 1$ columns of q-values, the presence of j poles of equal modulus is indicated. In [Henrici 1974] it is also explained how to determine these poles if $j > 1$.

THEOREM 2 [[Henrici 1974, p. 642]:
Let m and $m + j$ with $j > 1$ be two consecutive critical indices and let f be $(m + j)$-normal. Let the polynomials $\rho_k^{(n)}$ be defined by

$$\rho_0^{(n)}(z) = 1$$
$$\rho_{k+1}^{(n)}(z) = z\rho_k^{(n+1)}(z) - q_{m+k+1}^{(n)}\rho_k^{(n)}(z) \qquad n \ge 0 \qquad k = 0, 1, \ldots, j-1$$

Then there exists a subsequence $\{n(\ell)\}_{\ell\in N}$ such that

$$\lim_{\ell\to\infty} \rho_j^{(n(\ell))}(z) = (z - z_{m+1}^{-1}) \cdots (z - z_{m+j}^{-1})$$

The polynomials $\rho_k^{(n)}$ are closely related to the formally orthogonal Hadamard polynomials which will be discussed in the next section. From the above theorems the qd-scheme seems to be an ingenious tool for determining, under certain conditions, the poles of a meromorphic function f .

2.2. Hadamard polynomials and eigenvalue problems.

With the sequence $\{c_j\}_{j\in N}$ we can also associate the Hadamard polynomials

$$p_m^{(n)}(z) = \frac{H_m^{(n)}(z)}{H_m^{(n)}} \qquad m \geq 0, n \geq 0$$

where

$$H_m^{(n)}(z) = \begin{vmatrix} c_n & \cdots & c_{n+m-1} & c_{n+m} \\ \vdots & \ddots & & \\ & & \vdots & \vdots \\ c_{n+m-1} & \cdots & & c_{n+2m-1} \\ 1 & \cdots & z^{m-1} & z^m \end{vmatrix} \qquad H_0^{(n)}(z) = 1$$

These monic polynomials are formally orthogonal with respect to the linear functional c because they satisfy [Brezinski 1980, pp. 40–41]

$$c\left(z^i p_m^{(n)}(z)\right) = 0 \qquad i = 0, \ldots, m-1$$

In [Henrici 1974, pp. 634–636] it was shown that

$$p_m^{(n)}(z) = \det(zI - A_m^{(n)})$$

where $A_m^{(n)}$ denotes the matrix

$$A_m^{(n)} = \begin{pmatrix} q_1^{(n)} + e_0^{(n)} & q_1^{(n)} e_1^{(n)} & & & 0 \\ 1 & q_2^{(n)} + e_1^{(n)} & q_2^{(n)} e_2^{(n)} & & \\ & \ddots & \ddots & \ddots & \\ & & 1 & q_{m-1}^{(n)} + e_{m-2}^{(n)} & q_{m-1}^{(n)} e_{m-1}^{(n)} \\ 0 & & & 1 & q_m^{(n)} + e_{m-1}^{(n)} \end{pmatrix}$$

Hence the zeros of the Hadamard polynomials are the eigenvalues of the matrix $A_m^{(n)}$, or equivalently of the matrix $B_m^{(n)}$ where

$$B_m^{(n)} = \begin{pmatrix} q_1^{(n)} + e_0^{(n)} & -q_1^{(n)} & & & 0 \\ -e_1^{(n)} & q_2^{(n)} + e_1^{(n)} & -q_2^{(n)} & & \\ & \ddots & \ddots & \ddots & \\ & & -e_{m-2}^{(n)} & q_{m-1}^{(n)} + e_{m-2}^{(n)} & -q_{m-1}^{(n)} \\ 0 & & & -e_{m-1}^{(n)} & q_m^{(n)} + e_{m-1}^{(n)} \end{pmatrix}$$

The next theorem tells us that the qd-algorithm can be an ingenious way to compute the eigenvalues of such tridiagonal matrices.

THEOREM 3 [Henrici 1974, pp. 634–636]:

Let the functional c be m-normal and let the eigenvalues z_i of $B_m^{(n)}$ be numbered such that

$$|z_1| \geq |z_2| \geq \ldots \geq |z_m| \geq 0 = |z_{m+1}|$$

each eigenvalue occuring as many times in this sequence as indicated by its multiplicity. Then the qd-scheme associated with the sequence $\{c_j\}_{j \in N}$ has the following properties:

(a) for each k with $0 < k \leq m$ and $|z_k| > |z_{k+1}|$, it holds that

$$\lim_{n \to \infty} e_k^{(n)} = 0$$

(b) for each k with $0 < k \leq m$ and $|z_{k-1}| > |z_k| > |z_{k+1}|$, it holds that

$$\lim_{n \to \infty} q_k^{(n)} = z_k$$

(c) for each k and $j > 1$ such that $0 < k < k+j \leq m$ and $|z_{k-1}| > |z_k| = \ldots = |z_{k+j-1}| > |z_{k+j}|$, it holds that for the polynomials $\pi_i^{(n)}$ defined by

$$\pi_0^{(n)}(z) = 1$$

$$\pi_{i+1}^{(n)}(z) = z\pi_i^{(n+1)}(z) - q_{k+i+1}^{(n)}\pi_i^{(n)}(z) \qquad n \geq 0 \qquad i = 0, 1, \ldots, j-1$$

there exists a subsequence $\{\pi_j^{(n_\ell)}\}_{\ell \in N}$ such that

$$\lim_{\ell \to \infty} \pi_j^{(n_\ell)}(z) = (z - z_{k+1}) \ldots (z - z_{k+j})$$

3. Multivariate pole detection.

Let a multidimensional table $\{c_{i_1,\ldots,i_k}\}_{i_j \in N}$ be given and let us introduce the notations

$$\vec{i} = (i_1, \ldots, i_k) \qquad |\vec{i}| = i_1 + \ldots + i_k$$

$$\vec{x} = (x_1, \ldots x_k) \qquad \vec{x}^{\vec{i}} = x_1^{i_1} \ldots x_k^{i_k}$$

$$c_{\vec{i}} = c_{i_1 \ldots i_k}$$

For a formal Taylor series expansion

$$f(\vec{x}) = \sum_{\vec{i} \in N^k} c_{\vec{i}} \, \vec{x}^{\vec{i}}$$

we must specify in which order we shall deal with the index tuples $\vec{i}$ in N^k. Let us enumerate them as

$$f(\vec{x}) = \sum_{\ell=0}^{\infty} c_{\vec{i}(\ell)} \, \vec{x}^{\vec{i}(\ell)}$$

with the only requirement for the enumeration that it satisfies the inclusion property. By this we mean that when a point is added, every point in the polyrectangular subset emanating from the origin with the added point as its furthermost corner, must already be enumerated previously. As a consequence, always $\vec{i}(0) = (0,\dots,0)$. Let us also introduce the monomials

$$C_{m,n}(\vec{x}) = c_{\vec{i}(n)-\vec{i}(m)}\ \vec{x}^{\,\vec{i}(n)-\vec{i}(m)} \qquad i_j(n) \geq i_j(m) \quad j = 1,\dots,k$$

which are comparable to the univariate terms $c_{n-m}z^{n-m}$. In what follows we consider functions $f(\vec{x})$ which are meromorphic in a polydisc $B(0,R_{\vec{i}}) = \{\vec{x} : |x_{i_j}| < R_{i_j}\}$, meaning that there exists a polynomial

$$D(\vec{x}) = \sum_{\vec{i} \in M \subseteq N^2} r_{\vec{i}}\,\vec{x}^{\,\vec{i}} = \sum_{\ell=0}^{m} r_{\vec{i}(\ell)}\,\vec{x}^{\,\vec{i}(\ell)} \qquad r_{\vec{i}(0)}r_{\vec{i}(m)} \neq 0 \qquad (1)$$

indexed by the finite set M such that $(fD)(\vec{x})$ is analytic in the polydisc above. The index tuples belonging to M are determined by the powers of the variables x_{i_j} occurring in the polynomial $D(\vec{x})$. As a consequence of the inclusion property satisfied by the enumeration, one must allow for some of the coefficients $r_{\vec{i}}$ to be zero.

3.1. The general order qd-algorithm.

Let us introduce help entries $g_{0,m}^{(n)}$ by:

$$g_{0,m}^{(n)} = \sum_{k=0}^{n} C_{m,k}(\vec{x}) - \sum_{k=0}^{n} C_{m-1,k}(\vec{x})$$

$$g_{m,r}^{(n)} = \frac{g_{m-1,r}^{(n)}g_{m-1,m}^{(n+1)} - g_{m-1,r}^{(n+1)}g_{m-1,m}^{(n)}}{g_{m-1,m}^{(n+1)} - g_{m-1,m}^{(n)}} \qquad r = m+1, m+2, \dots$$

keeping in mind that $c_{\vec{i}} = 0$ if some $i_j < 0$. The general order multivariate qd-algorithm is then defined by [Cuyt 1988]:

$$Q_1^{(n)}(\vec{x}) = \frac{C_{0,n+1}(\vec{x})}{C_{0,n}(\vec{x})} \frac{g_{0,1}^{(n+1)}}{g_{0,1}^{(n+1)} - g_{0,1}^{(n+2)}}$$

$$Q_m^{(n+1)}(\vec{x}) = \frac{E_{m-1}^{(n+2)}(\vec{x})Q_{m-1}^{(n+2)}(\vec{x})}{E_{m-1}^{(n+1)}(\vec{x})} \frac{g_{m-2,m-1}^{(n+m-1)} - g_{m-2,m-1}^{(n+m)}}{g_{m-2,m-1}^{(n+m-1)}} \frac{g_{m-1,m}^{(n+m)}}{g_{m-1,m}^{(n+m)} - g_{m-1,m}^{(n+m+1)}} \qquad (2)$$

$$m \geq 2$$

$$E_m^{(n+1)}(\vec{x}) + 1 = \frac{g_{m-1,m}^{(n+m)} - g_{m-1,m}^{(n+m+1)}}{g_{m-1,m}^{(n+m)}} \left(Q_m^{(n+2)}(\vec{x}) + 1\right) \qquad m \geq 1 \qquad (3)$$

If we arrange the values $Q_m^{(n)}(\vec{x})$ and $E_m^{(n)}(\vec{x})$ as in the univariate case, where subscripts indicate columns and superscripts indicate downward sloping diagonals, then (2) links the elements in the rhombus

$$
\begin{array}{ccc}
 & E_{m-1}^{(n+1)}(\vec{x}) & \\
Q_{m-1}^{(n+2)}(\vec{x}) & & Q_m^{(n+1)}(\vec{x}) \\
 & E_{m-1}^{(n+2)}(\vec{x}) &
\end{array}
$$

and (3) links two elements on an upward sloping diagonal

$$
\begin{array}{cc}
 & E_m^{(n+1)}(\vec{x}) \\
Q_m^{(n+2)}(\vec{x}) &
\end{array}
$$

In analogy with the univariate case it is also possible to give explicit determinant formulas for the multivariate Q- and E-functions. Let us define the determinants

$$
H_{0,m}^{(n)}(\vec{x}) = \begin{vmatrix} C_{0,n}(\vec{x}) & \cdots & C_{m-1,n}(\vec{x}) \\ \vdots & & \vdots \\ C_{0,n+m-1}(\vec{x}) & \cdots & C_{m-1,n+m-1}(\vec{x}) \end{vmatrix} \qquad H_{0,0}^{(n)} = 0
$$

$$
H_{1,m}^{(n)}(\vec{x}) = \begin{vmatrix} 1 & \cdots & 1 \\ C_{0,n}(\vec{x}) & \cdots & C_{m,n}(\vec{x}) \\ \vdots & & \vdots \\ C_{0,n+m-1}(\vec{x}) & \cdots & C_{m,n+m-1}(\vec{x}) \end{vmatrix} \qquad H_{1,-1}^{(n)} = 0
$$

$$
H_{2,m}^{(n)}(\vec{x}) = \begin{vmatrix} \sum_{k=0}^{n-1} C_{0,k}(\vec{x}) & \cdots & \sum_{k=0}^{n-1} C_{m,k}(\vec{x}) \\ C_{0,n}(\vec{x}) & \cdots & C_{m,n}(\vec{x}) \\ \vdots & & \vdots \\ C_{0,n+m-1}(\vec{x}) & \cdots & C_{m,n+m-1}(\vec{x}) \end{vmatrix} \qquad H_{2,-1}^{(n)} = 0
$$

$$
H_{3,m}^{(n)}(\vec{x}) = \begin{vmatrix} 1 & \cdots & 1 \\ \sum_{k=0}^{n-1} C_{0,k}(\vec{x}) & \cdots & \sum_{k=0}^{n-1} C_{m,k}(\vec{x}) \\ C_{0,n}(\vec{x}) & \cdots & C_{m,n}(\vec{x}) \\ \vdots & & \vdots \\ C_{0,n+m-2}(\vec{x}) & \cdots & C_{m,n+m-2}(\vec{x}) \end{vmatrix} \qquad T_{3,-1}^{(n)} = 0
$$

and the polynomial

$$
\hat{H}_{1,m}^{(n)}(\vec{x}) = \frac{H_{1,m}^{(n)}(\vec{x})}{\vec{x}^{\sigma}} \qquad \sigma = \sum_{j=1}^{m} \left(\vec{i}(n+j) - \vec{i}(j) \right)
$$

41

By factoring out $\vec{x}^{\sigma}$, we obtain a polynomial that satisfies

$$\hat{H}_{1,m}^{(n)}(\vec{0}) \neq 0$$

By means of recurrence relations for these determinants we can prove the following lemma [Cuyt 1988].

LEMMA 2:
For well-defined $Q_m^{(n)}(\vec{x})$ and $E_m^{(n)}(\vec{x})$ the following determinant formulas hold:

$$-Q_m^{(n)}(\vec{x}) = \frac{H_{0,m}^{(n+m)} H_{1,m-1}^{(n+m-1)} H_{3,m}^{(n+m)}}{H_{0,m}^{(n+m-1)} H_{1,m}^{(n+m)} H_{3,m-1}^{(n+m)}}(\vec{x})$$

$$-E_m^{(n)}(\vec{x}) = \frac{H_{0,m+1}^{(n+m)} H_{1,m-1}^{(n+m)} H_{3,m}^{(n+m+1)}}{H_{0,m}^{(n+m)} H_{1,m}^{(n+m+1)} H_{3,m}^{(n+m)}}(\vec{x})$$

3.2. Application to pole detection.

For the chosen enumeration that satisfies the inclusion property, we also define the k functions

$$\nu_j(n) = \max\{i_j(\ell) \mid 0 \leq \ell \leq n\} \qquad j = 1, \ldots, k$$

Let m zeros $\vec{x}(h)$ of $D(\vec{x})$, specified by (1), be given in $B(0, R_{\vec{i}})$:

$$D(\vec{x}(h)) = 0 \qquad h = 1, \ldots, m \tag{4}$$

The next theorems are formulated for the so-called "simple pole" case where no information on derivatives at pole points $\vec{x}(h)$ is used. It is of course true that the following results can also be written down for the so-called "multipole" case introduced in [Cuyt 1992].

THEOREM 4:
Let $f(\vec{x})$ be a function which is meromorphic in the polydisc $B(0, R_{\vec{i}}) = \{\vec{x} : |x_j| < R_{i_j}, j = 1, \ldots, k\}$. Let (1) be satisfied and let m zeros $\vec{x}(h)$ of $D(\vec{x})$ in $B(0, R_{\vec{i}})$ be given, satisfying

$$(fD)(\vec{x}(h)) \neq 0 \qquad h = 1, \ldots, m \tag{5a}$$

and

$$\begin{vmatrix} \vec{x}(1)^{\vec{i}(1)} & \cdots & \vec{x}(1)^{\vec{i}(m)} \\ \vdots & & \vdots \\ \vec{x}(m)^{\vec{i}(1)} & \cdots & \vec{x}(m)^{\vec{i}(m)} \end{vmatrix} \neq 0 \tag{5b}$$

Then if $\lim_{n \to \infty} \nu_j(n) = \infty$ for $j = 1, \ldots, k$, one finds

$$\lim_{n \to \infty} \hat{H}_{1,m}^{(n+m)}(\vec{x}) = D(\vec{x})$$

42

From Lemma 2 and Theorem 4 we see that if f is a meromorphic function, then in some column the expressions $Q_m^{(n+1)}(\vec{x})$ contain information on the poles of f, because in that case the factor $\hat{H}_{1,m}^{(n+m)}(\vec{x})$ in the denominator of $Q_m^{(n+1)}(\vec{x})$ converges to the poles of the meromorphic f. This particular factor $\hat{H}_{1,m}^{(n+m)}$ is easily isolated because it is the only one in the denominator of $Q_m^{(n+1)}(\vec{x})$ (except for a constant) that evaluates different from zero in the origin. Let us also define the functions

$$\hat{E}_m^{(n+1)}(\vec{x}) = \frac{H_{0,m+1}^{(n+m)} H_{3,m}^{(n+m+1)}}{H_{0,m}^{(n+m)} H_{3,m}^{(n+m)}}(\vec{x})$$

which contain only some of the factors in $E_m^{(n+1)}(\vec{x})$, namely those factors that do not contain direct pole curve information. Again these specific factors are easily isolated in $E_m^{(n+1)}(\vec{x})$ because they all evaluate to zero in the origin and the remaining factors don't.

THEOREM 5:
Let $f(\vec{x})$ be a function which is meromorphic in the polydisc $B(0, R_{\vec{i}})$, meaning that there exists a polynomial $D(\vec{x})$ such that $(fD)(\vec{x})$ is analytic in the polydisc above. Let the polynomial $D(\vec{x})$ be given by (1) and let the conditions (5) be satisfied. If the first m columns of the general order multivariate qd-scheme are defined, then with the enumeration of $\mathbb{N}^k$ satisfying the same conditions as in Theorem 4, one finds:
(a)

$$\lim_{n \to \infty} \hat{E}_m^{(n+1)}(\vec{x}) = 0$$

in measure in a neighbourhood of the origin
(b)

$$\lim_{n \to \infty} \hat{H}_{1,m}^{(n+m)}(\vec{x}) = D(\vec{x})$$

How the general order multivariate qd-scheme is used as a tool to detect successive factors of the polar singularities of a multivariate function is further detailed in Section 5. We shall discuss how to deal with the situation where several algebraic curves define polar singularities and we shall also indicate which columns have to be inspected. How the general order qd-scheme can be implemented in floating-point polynomial arithmetic instead of symbolically is indicated in [Cuyt et al. 1999].

4. Parameterized eigenvalue problems.

For the multidimensional table $\{c_{i_1,\ldots,i_k}\}_{i_j \in \mathbb{N}}$ we introduce the additional notation

$$\vec{\lambda} = (\lambda_1, \ldots, \lambda_k) \in \mathbb{C}^k \qquad ||\vec{\lambda}||_p = 1$$

where $||.||_p$ is one of the Minkowski-norms on $\mathbb{C}^k$. In the sequel we shall often switch from a cartesian coordinate system to a spherical one (and back) and hence we also introduce

$$\vec{x} = (\lambda_1 z, \ldots, \lambda_k z) \qquad x_1, \ldots, x_k, z \in \mathbb{C} \qquad \|\vec{\lambda}\|_p = 1$$

$$c_j(\vec{\lambda}) = \sum_{|\vec{i}|=j} c_{\vec{i}} \, \vec{\lambda}^{\,\vec{i}}$$

$$C_j(\vec{x}) = \sum_{|\vec{i}|=j} c_{\vec{i}} \, \vec{x}^{\,\vec{i}} \tag{6}$$

$$= \left(\sum_{|\vec{i}|=j} c_{\vec{i}} \, \vec{\lambda}^{\,\vec{i}} \right) z^j$$

$$= c_j(\vec{\lambda}) z^j \tag{7}$$

The expressions (6) and (7) will be used interchangeably for the homogeneous polynomial $C_j(\vec{x})$. We denote by $\mathbb{C}[z]$ the linear space of polynomials in the variable z with complex coefficients, by $\mathbb{C}[\lambda_1, \ldots, \lambda_k]$ the linear space of multivariate polynomials in $\vec{\lambda}$ with complex coefficients, by $\mathbb{C}(\lambda_1, \ldots, \lambda_k)$ the commutative field of rational functions in $\vec{\lambda}$ with complex coefficients, by $\mathbb{C}(\lambda_1, \ldots, \lambda_k)[z]$ the linear space of polynomials in the variable z with coefficients from $\mathbb{C}(\lambda_1, \ldots, \lambda_k)$ and finally by $\mathbb{C}[\lambda_1, \ldots, \lambda_k][z]$ the linear space of polynomials in the variable z with coefficients from $\mathbb{C}[\lambda_1, \ldots, \lambda_k]$.

With the $c_j(\vec{\lambda})$ we can now define a parameterized linear functional Γ acting on the space $\mathbb{C}[z]$ by

$$\Gamma(z^j) = c_j(\vec{\lambda})$$

and parameterized Hankel determinants in $\mathbb{C}[\lambda_1, \ldots, \lambda_k]$,

$$\mathcal{H}_m^{(n)} = \begin{vmatrix} c_n(\vec{\lambda}) & \cdots & c_{n+m-1}(\vec{\lambda}) \\ \vdots & \ddots & c_{n+m}(\vec{\lambda}) \\ & & \vdots \\ c_{n+m-1}(\vec{\lambda}) & \cdots & c_{n+2m-2}(\vec{\lambda}) \end{vmatrix} \qquad \mathcal{H}_0^{(n)} = 1$$

as well as parameterized Hadamard polynomials in $\mathbb{C}[\lambda_1, \ldots, \lambda_k][z]$,

$$\mathcal{P}_m^{(n)}(z) = \frac{\mathcal{H}_m^{(n)}(z)}{h_m^{(n)}(\vec{\lambda})} \qquad m \geq 0 \qquad n \geq 0 \tag{8}$$

where

$$\mathcal{H}_m^{(n)}(z) = \begin{vmatrix} c_n(\vec{\lambda}) & \cdots & c_{n+m-1}(\vec{\lambda}) & c_{n+m}(\vec{\lambda}) \\ \vdots & \ddots & & c_{n+m+1}(\vec{\lambda}) \\ c_{n+m-1}(\vec{\lambda}) & & \cdots & c_{n+2m-1}(\vec{\lambda}) \\ 1 & z & \cdots & z^m \end{vmatrix} \qquad \mathcal{H}_0^{(n)}(z) = 1$$

and where the polynomial $h_m^{(n)}(\vec{\lambda})$ is a polynomial greatest common divisor of the polynomial coefficients of the powers of z in $\mathcal{P}_m^{(n)}(z)$. In this way its polynomial coefficients are relatively prime and hence $\mathcal{P}_m^{(n)}(z)$ is primitive. Note that $\mathcal{P}_m^{(n)}(z)$ belongs to $\mathbb{C}[\lambda_1,\ldots,\lambda_k][z]$ but does not belong to $\mathbb{C}[x_1,\ldots,x_k]$ because the powers in $\vec{\lambda}$ and z do not match. These parameterized Hadamard polynomials $\mathcal{P}_m^{(n)}(z)$ were introduced in [Benouahmane et al. 1999] and satisfy the formal orthogonality conditions

$$\Gamma\left(z^i \mathcal{P}_m^{(n)}(z)\right) = 0 \qquad i = 0,\ldots,m-1$$

We will call the functional Γ s-normal if

$$\mathcal{H}_m^{(n)}(\vec{\lambda}) \not\equiv 0 \qquad n \geq 0 \qquad m = 0,\ldots,s$$

4.1. The homogeneous qd-algorithm.

The homogeneous multivariate qd-algorithm is defined by the starting values

$$E_0^{(n)}(\vec{x}) = 0$$
$$Q_1^{(n)}(\vec{x}) = \frac{C_{n+1}(\vec{x})}{C_n(\vec{x})} \qquad n = 1,2,\ldots$$

and the continuation rules

$$E_m^{(n)}(\vec{x}) = Q_m^{(n+1)}(\vec{x}) - Q_m^{(n)}(\vec{x}) + E_{m-1}^{(n+1)}(\vec{x}) \qquad m = 1,2\ldots \quad n = 0,1\ldots$$
$$Q_{m+1}^{(n)}(\vec{x}) = \frac{E_m^{(n+1)}(\vec{x})Q_m^{(n+1)}(\vec{x})}{E_m^{(n)}(\vec{x})} \qquad m = 1,2,\ldots \quad n = 0,1,\ldots$$

which can be executed symbolically or for one particular $\vec{x}$ numerically. It was shown in [Chaffy 1984, pp. 22–28] that the homogeneous qd-algorithm satisfies

$$Q_m^{(n)}(\vec{x}) = Q_m^{(n)}(\lambda_1 z,\ldots,\lambda_k z)$$
$$= q_m^{(n)}(\vec{\lambda})z \tag{9}$$
$$E_m^{(n)}(\vec{x}) = E_m^{(n)}(\lambda_1 z,\ldots,\lambda_k z)$$
$$= e_m^{(n)}(\vec{\lambda})z \tag{10}$$

where

$$e_0^{(n)}(\vec{\lambda}) = 0 \qquad n = 1,2,\ldots$$
$$q_1^{(n)}(\vec{\lambda}) = \frac{c_{n+1}(\vec{\lambda})}{c_n(\vec{\lambda})} \qquad n = 0,1,\ldots$$
$$e_m^{(n)}(\vec{\lambda}) = q_m^{(n+1)}(\vec{\lambda}) - q_m^{(n)}(\vec{\lambda}) + e_{m-1}^{(n+1)}(\vec{\lambda}) \qquad m = 1,2\ldots \quad n = 0,1\ldots$$
$$q_{m+1}^{(n)}(\vec{\lambda}) = \frac{e_m^{(n+1)}(\vec{\lambda})q_m^{(n+1)}(\vec{\lambda})}{e_m^{(n)}(\vec{\lambda})} \qquad m = 1,2,\ldots \quad n = 0,1,\ldots$$

In other words the homogeneous qd-algorithm can be regarded as a parameterized univariate qd-algorithm. It is then easy to write down the following parameterized version of a result proved in [Cuyt 1994].

45

LEMMA 2:
If the functional Γ is s-normal, then the values $Q_m^{(n)}(\vec{x})$ and $E_m^{(n)}(\vec{x})$ exist for $m = 1, \ldots, s$ and $n \geq 0$ and they are given by

$$Q_m^{(n)}(\vec{x}) = \frac{\mathcal{H}_m^{(n+1)} \mathcal{H}_{m-1}^{(n)}}{\mathcal{H}_m^{(n)} \mathcal{H}_{m-1}^{(n+1)}} z$$

$$E_m^{(n)}(\vec{x}) = \frac{\mathcal{H}_{m+1}^{(n)} \mathcal{H}_{m-1}^{(n+1)}}{\mathcal{H}_m^{(n)} \mathcal{H}_m^{(n+1)}} z$$

From this lemma it is easy to see that $q_m^{(n)}(\vec{\lambda})$ and $e_m^{(n)}(\vec{\lambda})$ introduced in (9) and (10) are rational expressions in $\lambda_1, \ldots, \lambda_k$ and hence that they belong to $\mathbb{C}(\lambda_1, \ldots, \lambda_k)$.

4.2. Application to parameterized eigenvalue problems.

Theorem 6 will generalize Theorem 3 for the computation of the eigenvalues of $B_m^{(n)}(\vec{x})$ given by (to save space we have respectively denoted $Q_m^{(n)}(\vec{x})$ and $E_m^{(n)}(\vec{x})$ in the matrix by $Q_m^{(n)}$ and $E_m^{(n)}$)

$$\begin{pmatrix} (Q_1^{(n)} + E_0^{(n)}) & -Q_1^{(n)} & & & 0 \\ -E_1^{(n)} & (Q_2^{(n)} + E_1^{(n)}) & -Q_2^{(n)} & & \\ & \ddots & \ddots & \ddots & \\ & & -E_{m-2}^{(n)} & (Q_{m-1}^{(n)} + E_{m-2}^{(n)}) & -Q_{m-1}^{(n)} \\ 0 & & & -E_{m-1}^{(n)} & (Q_m^{(n)} + E_{m-1}^{(n)}) \end{pmatrix}$$

$$(11)$$

Let us consider the following rational function, belonging to $\mathbb{C}(\lambda_1, \ldots, \lambda_k)[z]$,

$$r^{(n)}(\vec{\lambda}z) = \frac{1}{\lceil z} + \frac{-q_1^{(n)}(\vec{\lambda})z}{\lceil z} + \frac{-e_1^{(n)}(\vec{\lambda})z}{\lceil z} + \frac{-q_2^{(n)}(\vec{\lambda})z}{\lceil z} + \ldots + \frac{-q_m^{(n)}(\vec{\lambda})z}{\lceil z}$$

$$= \frac{1}{\lceil z} + \frac{-Q_1^{(n)}(\vec{x})}{\lceil z} + \frac{-E_1^{(n)}(\vec{x})}{\lceil z} + \frac{-Q_2^{(n)}(\vec{x})}{\lceil z} + \ldots + \frac{-Q_m^{(n)}(\vec{x})}{\lceil z} \quad (12)$$

The denominator $r_D(\vec{\lambda}z)$ of $r^{(n)}(\vec{\lambda}z)$ is monic of degree m in z and is of the form

$$r_D(\vec{\lambda}z) = \sum_{i=0}^{m} b_i(\vec{\lambda})z^i \qquad b_i(\vec{\lambda}) \in \mathbb{C}(\lambda_1, \ldots, \lambda_k) \qquad b_m(\vec{\lambda}) = 1$$

A special collection of vectors $\vec{\lambda}$ is the following one. If we carefully examine Theorem 3, then we notice that the difference between the cases (b) and (c) stems from the fact that (c) deals with poles of equal modulus whereas (b) doesn't. Since $\mathbb{C}(\lambda_1, \ldots, \lambda_k)[z]$ is a unique factorization domain, $r_D(\vec{\lambda}z)$ can be factored as

$$r_D(\vec{\lambda}z) = \sum_{i=0}^{m} b_i(\vec{\lambda})z^i = \prod_{i=1}^{\ell} \beta_i(\vec{\lambda}z) \qquad \beta_i(\vec{\lambda}z) \in \mathbb{C}(\lambda_1, \ldots, \lambda_k)[z]$$

For some $\vec{\lambda}$ it may happen that a zero $z^*(\vec{\lambda})$ of $\beta_i(\vec{\lambda}z)$ is at the same time a zero of $\beta_j(\vec{\lambda}z)$ with $i \neq j$ and this $\vec{\lambda}$ then gives rise to a pole of higher order than the multiplicity of $z^*(\vec{\lambda})$ as a zero of $\beta_i(\vec{\lambda}z)$ alone. For this $\vec{\lambda}$ the pole $z^*(\vec{\lambda})$ is then at least a multipole and its inverse a degenerate eigenvalue.

If $z^*(\vec{\lambda})$ is a multipole of only $\beta_i(\vec{\lambda}z)$ and not of another factor, then the parameterized interpretation of the $q_m^{(n)}(\vec{\lambda})$- and $e_m^{(n)}(\vec{\lambda})$-values is not disturbed, although $1/z^*(\vec{\lambda})$ is still a degenerate eigenvalue. If it nullifies at least two different factors, then for this $\vec{\lambda}$ the qd-table has to be interpreted differently. Let us collect the vectors $\vec{\lambda}$ that give rise to such an occasional pole of higher order in the set

$$M = \{\vec{\lambda} \in \mathbb{C}^k : \|\vec{\lambda}\| = 1, \ r^{(n)}(\vec{\lambda}z) \text{ has a pole that cancels}$$
$$\text{more than one factor of } r_D(\vec{\lambda}z)\}$$

It is important to see that the set M is a finite set [??].

THEOREM 6:
Let the functional Γ be m-normal and fix $\vec{\lambda}$ on the unit sphere in $\mathbb{C}^k$ excluding M. Let the eigenvalues of $B_m^{(n)}(\vec{x})$ in $\{\vec{\lambda}z : \|\vec{\lambda}\|_p = 1, \vec{\lambda} \notin M, z \in \mathbb{C}\}$ be numbered such that

$$|z_1(\vec{\lambda})| \geq |z_2(\vec{\lambda})| \geq \ldots \geq |z_m(\vec{\lambda})| \geq 0 = |z_{m+1}(\vec{\lambda})|$$

each eigenvalue occuring as many times in this sequence as indicated by its multiplicity. Then the homogeneous qd-scheme associated with the sequence $\{c_j(\vec{\lambda})\}_{j \in \mathbb{N}}$ has the following properties:
(a) for each k with $0 < k \leq m$ and $|z_k(\vec{\lambda})| > |z_{k+1}(\vec{\lambda})|$, it holds that

$$\lim_{n \to \infty} e_k^{(n)}(\vec{\lambda}) = 0$$

(b) for each k with $0 < k \leq m$ and $|z_{k-1}(\vec{\lambda})| > |z_k(\vec{\lambda})| > |z_{k+1}(\vec{\lambda})|$, it holds that

$$\lim_{n \to \infty} q_k^{(n)}(\vec{\lambda}) = z_k(\vec{\lambda})$$

(c) for each k and $j > 1$ such that $0 < k < k+j \leq m$ and $|z_{k-1}(\vec{\lambda})| > |z_k(\vec{\lambda})| = \ldots = |z_{k+j-1}(\vec{\lambda})| > |z_{k+j}(\vec{\lambda})|$, it holds that for the polynomials $\pi_i^{(n)}(\vec{\lambda}, z)$ defined by

$$\pi_0^{(n)}(\vec{\lambda}, z) = 1$$
$$\pi_{i+1}^{(n)}(\vec{\lambda}, z) = z\pi_i^{(n+1)}(\vec{\lambda}, z) - q_{k+i+1}^{(n)}(\vec{\lambda})\pi_i^{(n)}(\vec{\lambda}, z) \qquad n \geq 0, 0 \leq i \leq j - 1$$

there exists a subsequence $\{\pi_j^{(n_\ell)}\}_{\ell \in N}$ such that

$$\lim_{\ell \to \infty} \pi_j^{(n_\ell)}(\vec{\lambda}, z) = (z - z_{k+1}(\vec{\lambda})) \dots (z - z_{k+j}(\vec{\lambda}))$$

For more details we refer to [Benouahmane et al. 2000].

5. Numerical illustration.

Each of the above techniques will be illustrated by means of a numerical example. In order to let the reader fully understand the difference between the two generalizations of the qd-algorithm presented above, we have chosen the examples such that the solution of both the pole detection problem and the parameterized eigenvalue problem is identical, meaning that our example is chosen such that for both problems the solution curves are identical. Nevertheless, the symbolic output of both generalizations of the qd-algorithm will not be identical, because they treat the multivariate data in a different way, as explained in respectively Section 3 and 4.

5.1. Pole detection mechanism.

Since we want to illustrate the techniques with some graphs, we shall now restrict ourselves to the case of two real variables. Consider a meromorphic function with polar singularities, given by

$$f(x_1, x_2) = \frac{\exp(x_1 + 2x_2)}{-5x_1^3 - x_1^2 x_2 - 5x_1 x_2^2 - x_2^3 - 5x_1^2 + 8x_1 x_2 + 7x_2^2 + 40x_1 - 4x_2 - 30}$$

The set M indexing the denominator polynomial $D(x_1, x_2)$ indicates the order in which the coefficients in the Taylor series representation of f are dealt with:

$$M = \{(0,0), (1,0), (0,1), \dots, (3,0), \dots, (0,3)\} = \{\vec{i}(0), \dots, \vec{i}(9)\}$$

Furthermore, the denominator polynomial can be factored as the product of a "straight line" $D_1(x_1, x_2)$ and a "circle" $D_2(x_1, x_2)$:

$$D(x_1, x_2) = D_1(x_1, x_2) D_2(x_1, x_2) = (5 - 5x_1 - x_2)(x_1^2 + 2x_1 + x_2^2 - 2x_2 - 6)$$

For the factors D_1 and D_2 the sets M_1 and M_2 as defined in (1) for D, are respectively given by:

$$M_1 = \{(0,0), (1,0), (0,1)\}$$
$$M_2 = \{(0,0), (1,0), (0,1), (2,0), (0,2)\}$$

From Lemma 3 and Theorem 5 we know that the information on the poles of $f(x_1, x_2)$ is to be found in the denominator of some Q-expressions. We need only look at the factor $H_{1,m}^{(n+m)}$ and more particularly at $\hat{H}_{1,m}^{(n+m)}$. From now on, when we say that a Q-column delivers pole information, we mean that the

$\hat{H}$-function in question has been extracted from its denominator. One cannot know from the factored form of $D(\vec{x})$ in what order one is to expect the factors $D_\ell(\vec{x})$. However Theorem 5 states that a Q-column delivering in its denominator as a limiting value the product of some pole factors is immediately followed by an $\hat{E}$-column tending to zero. The general order multivariate qd-table is divided into subtables by these vanishing $\hat{E}$-columns. The column number of such a vanishing $\hat{E}$-column is called a critical index.

From M_1, M_2 and M we can easily obtain the candidates for critical index:

$$m = \#M_1 - 1 = 2$$
$$m = \#M_2 - 1 = 4$$
$$m = \#M - 1 = 9$$

After inspection of the $\hat{E}$-expressions, it is clear that the true critical indices in this example are $m = 2$ and $m = 9$.

We remark that in a real-life case the degree of the factors $D_\ell(x_1, x_2)$ is not necessarily known in advance and hence that it is very important (much more important than in the univariate case) to have vanishing $\hat{E}$-columns signaling where to look for the pole information. Numerical experiments have also shown that the enumeration of $I\!N^k$ is closely linked to the order in which the pole factors are to be traced by the algorithm. Indeed the sequence of index tuples $\{\vec{i}(\ell)\}_{\ell \in N}$ indicates in which order the algorithm deals with the input Taylor coefficients of $f(\vec{x})$.

We print out the approximants obtained for the factor $D_1(x_1, x_2)$ and for the denominator $D(x_1, x_2) = D_1(x_2, x_2)D_2(x_1, x_2)$ after inputting the first 351 coefficients of $f(x_1, x_2)$, namely $\{c_{i_1 i_2}\}_{0 \leq i_1 + i_2 \leq 25}$:

$$D_1(x_1, x_2) \approx \hat{H}_{1,2}^{(333)}(x_1, x_2)$$
$$= -1.018 + x_1 + 0.2120x_2$$
$$(D_1 D_2)(x_1, x_2) \approx \hat{H}_{1,9}^{(340)}(x_1, x_2)$$
$$= (6.000 - 8.000x_1 + 0.8000x_2 + 1.000x_1^2 - 1.600x_1x_2 - 1.400x_2^2$$
$$+ x_1^3 + 0.2000x_1^2x_2 + 1.000x_1x_2^2 + 0.2000x_2^3)$$

Some of these functions are better understood from their graphs. The zeros of $\hat{H}_{1,2}^{(333)}$ and $\hat{H}_{1,9}^{(340)}$ are plotted in $[-5, 3] \times [-2, 6] \subset I\!R^2$. The surfaces $\hat{E}_2^{(332)}$ and $\hat{E}_9^{(332)}$ are shown in $[-1, 1] \times [-1, 1] \times [-1, 1] \subset I\!R^3$.

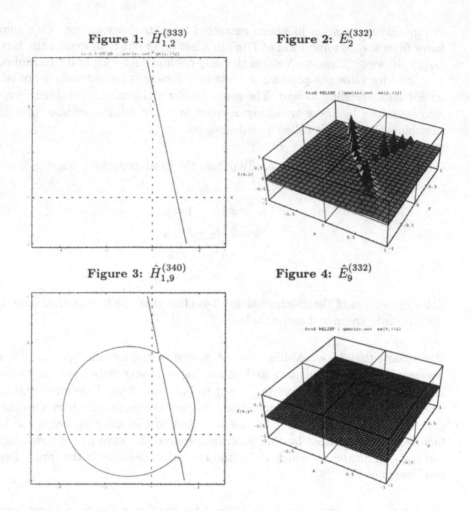

Figure 1: $\hat{H}_{1,2}^{(333)}$

Figure 2: $\hat{E}_2^{(332)}$

Figure 3: $\hat{H}_{1,9}^{(340)}$

Figure 4: $\hat{E}_9^{(332)}$

5.2. Parameterized eigenvalue technique.

We again take the dimension of our multivariate problem $k = 2$ and consider the matrix $B_3^{(1)}(\vec{x})$ of the form (11) with

$$Q_1^{(1)}(x_1, x_2) = \frac{-225x_1^4 + 60x_1^3x_2 - 190x_1^2x_2^2 - 132x_1x_2^3 - 45x_2^4}{(5x_1 + x_2)(5x_1^2 - 8x_1x_2 - 7x_2^2)(x_1^2 + x_2^2)}$$

$$Q_2^{(1)}(x_1, x_2) = \frac{2(5x_1^2 - 8x_1x_2 - 7x_2^2)(11875x_1^3 + 75x_1^2x_2 + 3075x_1x_2^2 - 661x_2^3)}{(725x_1^2 - 40x_1x_2 + 113x_2^2)(225x_1^4 - 60x_1^3x_2 + 190x_1^2x_2^2 + 132x_1x_2^3 + 45x_2^4)}$$

$$Q_3^{(1)}(x_1, x_2) = \frac{15(725x_1^2 - 40x_1x_2 + 113x_2^2)}{11875x_1^3 + 75x_1^2x_2 + 3075x_1x_2^2 - 661x_2^3}$$

$$E_1^{(1)}(x_1, x_2) = \frac{2(5x_1 + x_2)(x_1^2 + x_2^2)(725x_1^2 - 40x_1x_2 + 113x_2^2)}{(5x_1^2 - 8x_1x_2 - 7x_2^2)(225x_1^4 - 60x_1^3x_2 + 190x_1^2x_2^2 + 132x_1x_2^3 + 45x_2^4)}$$

$$E_2^{(1)}(x_1, x_2) = \frac{3375(-225x_1^4 + 60x_1^3x_2 - 190x_1^2x_2^2 - 132x_1x_2^3 - 45x_2^4)}{(725x_1^2 - 40x_1x_2 + 113x_2^2)(11875x_1^3 + 75x_1^2x_2 + 3075x_1x_2^2 - 661x_2^3)}$$

The eigenvalue problem for this 3×3 matrix can easily be solved exactly by hand, or by calling a computer algebra system. But we want to illustrate the use of the symbolic homogeneous qd-algorithm which remains applicable to large-scale problems and delivers an accurate approximation. If we look at the exact eigenvalues of the matrix $B_3^{(1)}(\vec{x})$ defined above, then we find the expressions

$$z_1(x_1, x_2) = \frac{5}{5x_1 + x_2}$$

$$z_2(x_1, x_2) = \frac{-x_1 + x_2 - \sqrt{7x_1^2 - 2x_1x_2 + 7x_2^2}}{x_1^2 + x_2^2}$$

$$z_3(x_1, x_2) = \frac{-x_1 + x_2 + \sqrt{7x_1^2 - 2x_1x_2 + 7x_2^2}}{x_1^2 + x_2^2}$$

To give the reader an idea of what this looks like, we give a parametric plot of the eigenvalues for $\vec{x}$ varying over the unit disk in $\mathbb{R}^2$, in other words for $\vec{x} = e^{i\theta}$ or $x_1 = \cos\theta$ and $x_2 = \sin\theta$ with $\theta \in [0, 2\pi[$. This results in the pictures

Figure 5: $z_1(x_1, x_2)$ **Figure 6:** $z_2(x_1, x_2)$ **Figure 7:** $z_3(x_1, x_2)$

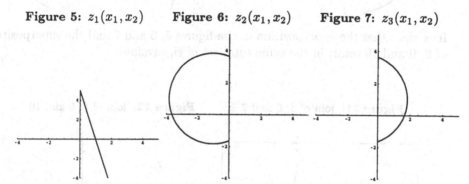

When on the other hand the qd-technique is used, we first switch to polar coordinates. By the homogeneous qd-algorithm the eigenvalues are then being approximated, and more important, being regrouped such that, for each vector $\vec{\lambda}$, and because of the symbolic computation for all vectors $\vec{\lambda}$ at the same time (except for those in M), the expression

$$z_1(\vec{\lambda}) = \lim_{n \to \infty} q_1^{(n)}(\vec{\lambda})$$

delivers the in modulus largest eigenvalues and the expression

$$z_3(\vec{\lambda}) = \lim_{n \to \infty} q_3^{(n)}(\vec{\lambda})$$

delivers the in modulus smallest eigenvalues. When applied to the matrix $B_3^{(1)}$ this results for instance in the approximations

$$z_1(\lambda_1, \lambda_2) \approx q_1^{(9)}(\vec{\lambda})$$

$$z_2(\lambda_1, \lambda_2) \approx q_2^{(7)}(\vec{\lambda})$$

$$z_3(\lambda_1, \lambda_2) \approx q_3^{(5)}(\vec{\lambda}) = \frac{15(177359375\lambda_1^6 + 68062500\lambda_1^5\lambda_2 + 108140625\lambda_1^4\lambda_2^2 - 20047000\lambda_1^3\lambda_2^3 + 25741125\lambda_1^2\lambda_2^4 - 8636220\lambda_1\lambda_2^5 + 2417987\lambda_2^6)}{2689140625\lambda_1^7 + 1469890625\lambda_1^6\lambda_2 + 1996921875\lambda_1^5\lambda_2^2 - 206283125\lambda_1^4\lambda_2^3 + 555975875\lambda_1^3\lambda_2^4 - 222944925\lambda_1^2\lambda_2^5 + 93486145\lambda_1\lambda_2^6 - 21496039\lambda_2^7}$$

For $\vec{\lambda}$ varying over the unit disk in $I\!R^2$ the parameterized approximations for the eigenvalues now look like

Figure 8: $q_1^{(24)}(\lambda_1, \lambda_2)$ **Figure 9:** $q_2^{(22)}(\lambda_1, \lambda_2)$ **Figure 10:** $q_3^{(20)}(\lambda_1, \lambda_2)$

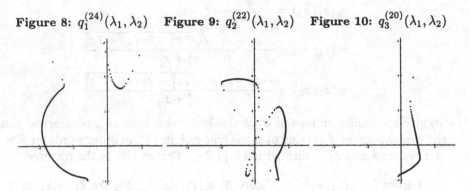

It is clear that the superposition of the figures 5, 6 and 7 and the superposition of 8, 9 and 10 result in the same total set of eigenvalues.

Figure 11: join of 5, 6 and 7   **Figure 12:** join of 8, 9 and 10

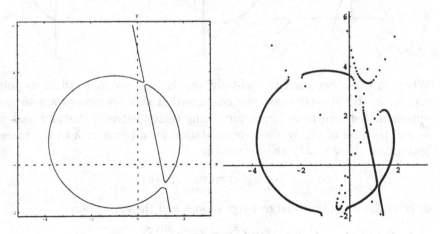

The difference between the two approaches lies in the fact that in the second approach the eigenvalues are automatically ordered in modulus, and that it is therefore an approximation method, not an explicit solution method. Remains the problem of the problematic vectors in M. For our example

$$M = \{\vec{\lambda}_1^* = (0.0843271, 0.996438), \vec{\lambda}_2^* = (0.817216, -0.576331)\}$$

as shown in Figure 13.

Figure 13: set M of problematic parameters

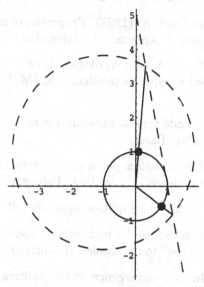

If we carefully examine the expression $e_1^{(23)}(\vec{\lambda})$ then we notice that for $\vec{\lambda} \neq \vec{\lambda}_2^*$, the value $e_1^{(23)}(\vec{\lambda})$ tends to zero. For instance, again restricting ourselves to the unit circle in $\mathbb{R}^2$,

$$e_1^{(23)}(1,0) \approx 1.3 \times 10^{-7}$$

On the other hand for $\vec{\lambda} = \vec{\lambda}_2^*$,

$$e_1^{(23)}(\vec{\lambda}_2^*) \approx -0.00572$$

Hence $q_1^{(24)}(\vec{\lambda})$ delivers the largest eigenvalues except for $\vec{\lambda} = \vec{\lambda}_2^*$ where a double pole occurs in $r(\vec{\lambda}z)$. Examining $e_2^{(21)}(\vec{\lambda})$ shows that this expression tends to zero for $\vec{\lambda} \neq \vec{\lambda}_1^*$. But careful: for $\vec{\lambda} = \vec{\lambda}_2^*$, the expression $q_2^{(22)}(\vec{\lambda})$ must be interpreted differently. If $\vec{\lambda} \notin M$, then $q_2^{(22)}(\vec{\lambda})$ delivers the second eigenvalue. For $\vec{\lambda} = \vec{\lambda}_2^*$ Theorem 1(c) must be applied with $k = 1$ and $j = 2$. Continuing our investigation reveals that $e_2^{(21)}(\vec{\lambda}_1^*)$ does not tend to zero,

$$e_2^{(21)}(1,0) \approx -2.3 \times 10^{-6}$$
$$e_2^{(21)}(\vec{\lambda}_1^*) \approx -0.00252$$

while $e_3^{(19)}(\vec{\lambda}_1^*)$ does. Hence the expression $q_3^{(20)}(\vec{\lambda})$ must now be interpreted differently for $\vec{\lambda} = \vec{\lambda}_1^*$ and for all other $\vec{\lambda}$. For $\vec{\lambda} \neq \vec{\lambda}_1^*$ the expression $q_3^{(20)}(\vec{\lambda})$ delivers the smallest eigenvalues. For $\vec{\lambda} = \vec{\lambda}_1^*$ Theorem 1(c) must be applied with $k = 2$ and $j = 2$.

References

Benouahmane, B. and Cuyt, A. (1999): Properties of multivariate homogeneous orthogonal polynomials. J. Approx. Th. submitted.

Benouahmane, B., Cuyt, A. and Verdonk, B. (2000): On the solution of parameterized (smallest) eigenvalue problems. SIAM J. Matrix Anal. & Applcs. submitted.

Brezinski, C. (1980): Padé type approximation and general orthogonal polynomials. Birkhauser Verlag, Basel.

Chaffy, C. (1984): Interpolation polynomiale et rationnelle d'une fonction de plusieurs variables complexes. Thèse. Inst. Polytech., Grenoble.

Cuyt, A. (1988): A multivariate qd-like algorithm. BIT 28: 98–112.

Cuyt, A. (1992): Extension of "A multivariate convergence theorem of the de Montessus de Ballore type" to multipoles. J. Comput. Appl. Math.41: 323–330.

Cuyt, A. (1994): On the convergence of the multivariate "homogeneous" qd-algorithm. BIT 34: 535–545.

Cuyt, A. and Verdonk, B. (1999): Extending the qd-algorithm to tackle multivariate problems. In: Papamichael, N., Ruscheweyh, S., and Saff, B. (eds.): Proceedings of the Third CMFT Conference "Computational methods and function theory 1997", Nicosia, Cyprus. World Scientific, pp. 135–159.

Fernando, K. and Parlett, B. (1994): Accurate singular values and differential qd algorithms. Numer. Math. 67: 191–229.

Henrici, P. (1974): Applied and computational complex analysis I. John Wiley, New York.

Von Matt, U. (1997): The orthogonal QD-algorithm. SIAM J. Sci. Statist. Comput. 18: 1163–1186.

---

* Research Director FWO-Vlaanderen

Isoefficiency and the Parallel Descartes Method

Thomas Decker and Werner Krandick

1 Introduction

The efficiency of a parallel algorithm with input x on $P \geq 1$ processors is defined as $E(x, P) = \frac{T(x,1)}{PT(x,P)}$ where $T(x, P)$ denotes the time it takes to perform the computation using P processors and $T(x, 1)$ is the sequential execution time. The efficiency of many parallel algorithms decreases when the number of processors increases and the sequential execution time is fixed; likewise, the efficiency increases when the sequential computing time increases and the number of processors is fixed. The term scalability refers to this change of efficiency (Sahni & Thanvantri, 1996). Intuitively, a parallel algorithm is scalable if it stays efficient when the number of processors and the sequential execution time are both increased.

One of the concepts that quantify this intuition is the notion of isoefficiency introduced by Kumar et al. (1988; Grama et al., 1993); isoefficiency relates the number of processors and the sequential computing time necessary to maintain a given efficiency. The concept was never rigorously defined but, nevertheless, used in the analysis of a number of algorithms (Kumar & Singh, 1991; Gupta & Kumar, 1993; Gupta et al., 1997; Mahapatra & Dutt, 1997; Yang & Lin, 1997). In these applications, isoefficiency is characterized as follows. If e is a given efficiency and $I_e(P)$ is the isoefficiency function then, for all x and P, $T(x, 1) = I_e(P)$ implies $E(x, P) = e$. This property does not suffice to define $I_e(P)$ since there may be inputs that have the same sequential computing time but different efficiencies; also, there may be inputs that have different sequential computing times but the same efficiency. Both situations arise in parallel search algorithms. In the Descartes method for real root isolation, for example, the efficiency depends strongly on the shape of the search tree—and not merely on the sequential computing time.

Still, the scalability of such algorithms can be discussed in a meaningful way. We ask for the minimal sequential time that guarantees a given efficiency when the algorithm is executed on more than one processor. If X is the set of inputs we define

$$I_e(P) = \inf\{T(x, 1) \mid (\forall y \in X)\, T(y, 1) \geq T(x, 1) \Rightarrow E(y, P) \geq e\} \ .$$

If the set on the right-hand side is empty, and only then, $I_e(P)$ is undefined. This is the case precisely when, for any sequential computing time, there is a computation that takes at least that much time and does not achieve efficiency e on P processors. In this case one may say that the algorithm does not scale for efficiency e. Whenever $I_e(P)$ is defined any sequential computation taking longer than $I_e(P)$ is guaranteed to run at an efficiency of at least e on P processors,

$$T(x, 1) > I_e(P) \text{ implies } E(x, P) \geq e \ . \tag{1}$$

In this case, the growth rate of $I_e(P)$ in P quantifies the scaling behavior of the algorithm at efficiency e. If, for some efficiency e, the isoefficiency function $I_e(P)$ grows at most polynomially in P, the algorithm is efficient parallel in the sense of Kruskal et al. (1990, Theorem 3.2).

The Descartes method serves to isolate polynomial real roots. Given a univariate squarefree polynomial with integer coefficients the method finds disjoint isolating intervals for all the real roots of the polynomial, that is, intervals that contain exactly one root each. Using sixty-four 600-MHz DEC-Alpha processors of the Cray T3E, the parallel Descartes method isolates the real roots of random polynomials of degree 5000 in 9.8 seconds on the average; on one processor the average is 449 seconds.

The original sequential method devised by Collins and Akritas (1976; Krandick, 1995) uses exact integer arithmetic. A validated version of the method using hardware interval arithmetic was discussed by Johnson and Krandick (1997). That version may fail for polynomials of high degree or with very close roots due to either insufficient precision or exponent overflow. A parallel version of the Descartes method developed by Decker and Krandick (1999) uses arbitrary precision floating point interval arithmetic and a full word for the exponent. By repeating any failing computation with a higher precision the method will ultimately succeed.

We analyze the isoefficiency function $I_e(P)$ of the parallel Descartes method using the dominance terminology due to Collins (1974): If f and g are real-valued functions defined on a common domain S, we say that f *is dominated by* g, and write $f \preceq g$, in case there is a positive real number c such that $f(x) \leq cg(x)$ for all elements $x \in S$. If $f \preceq g$ and $g \preceq f$, then we say that f and g are *codominant* and write $f \sim g$.

We carry out our analysis assuming a fixed precision. That is, we analyze one run of the method on an interval polynomial of a given precision—no matter whether the precision is sufficient or not. We present a class of inputs on which the isoefficiency is codominant with P^2. In Theorems 20 and 21 we assume that we obtain the input polynomial by embedding each coefficient of a given squarefree integral polynomial into an interval with floating point endpoints. Since the precision is fixed the coefficient size of the embedded integral polynomial is bounded. We show that, relative to any set of such input polynomials,

$$I_e(P) \preceq P^3 (\log P)^2 .$$

Here, e may be any efficiency less than 1. By contrast, the other known parallelizations of the Descartes method (Collins et al., 1992; Schreiner et al., 2000) do not scale for any efficiency.

The parallel Descartes method has three levels of parallelism. At the top level the method performs a tree search. The corresponding isoefficiency analysis in Section 3 is carried out in general terms. The middle level of the Descartes method selects one of two parallel algorithms; Section 4 gives an isoefficiency analysis of such poly-algorithms. Section 5 describes the parallel Descartes method; its isoefficiency is analyzed in Section 6.

2 The isoefficiency function

Definition 1. *Let a parallel algorithm be given and let X be a set of allowed inputs. For every $x \in X$ and $P \in \mathbb{N}$ let $T(x, P)$ denote the time the algorithm takes to process* *x on P processors. Let*

$$E(x, P) := \frac{T(x, 1)}{P\,T(x, P)}$$

be the efficiency *of this computation. For* $e \in [0, 1)$, *let*

$$S_e(P) := \{T(x, 1) \mid (\forall y \in X)\ T(y, 1) \geq T(x, 1) \Rightarrow E(y, P) \geq e\}$$

be the set of all sequential computing times that guarantee efficiency e. *If* $S_e(P)$ *is non-empty, the* isoefficiency function *is defined by*

$$I_e(P) := \inf S_e(P);$$

if $S_e(P)$ *is empty,* $I_e(P)$ *is undefined.*

Theorem 1. *The isoefficiency function is monotonic in* e, *that is,*

$$e_1 < e_2 \ \text{implies} \ I_{e_1}(P) \leq I_{e_2}(P).$$

for all efficiencies e_1, e_2 *and any number of processors* P.

Proof. Immediate from $S_{e_2}(P) \subseteq S_{e_1}(P)$. □

The next theorem shows that an upper bound for the parallel execution time yields an upper bound for the isoefficiency. Clearly, an upper bound for the parallel execution time immediately provides a lower bound for the efficiency. Intuitively, an underestimate of the efficiency results in an overestimate of the problem size needed to obtain a given efficiency.

Theorem 2. *Let the functions* $\underline{E} : X \times I\!N \to (0, 1]$, $\overline{E} : X \times I\!N \to (0, 1]$ *bracket the true efficiency* E,

$$\underline{E}(x, P) \leq E(x, P) \leq \overline{E}(x, P)$$

for all $x \in X$ *and* $P \in I\!N$. *For all efficiencies* $e \in (0, 1)$ *let*

$$\underline{S}_e(P) := \{T(x, 1) \mid (\forall y \in X)\ T(y, 1) \geq T(x, 1) \Rightarrow \underline{E}(y, P) \geq e\} \ \text{and}$$
$$\overline{S}_e(P) := \{T(x, 1) \mid (\forall y \in X)\ T(y, 1) \geq T(x, 1) \Rightarrow \overline{E}(y, P) \geq e\} \ ,$$

and let

$$\underline{I}_e(P) := \inf \overline{S}_e(P) \ , \ \text{and} \ \overline{I}_e(P) := \inf \underline{S}_e(P)$$

whenever the respective set is non-empty. Then,

$$\underline{I}_e(P) \leq I_e(P) \leq \overline{I}_e(P)$$

for all numbers of processors P *and all efficiencies* e.

Proof. Immediate from $\underline{S}_e(P) \subseteq S_e(P) \subseteq \overline{S}_e(P)$. □

Definition 2. *A set of real numbers is* discrete *if there exists a positive lower bound for the absolute value of the difference of any two distinct elements.*

In any computation model where the durations of the basic operations are multiples of a fixed unit time, the set of computing times is discrete.

Theorem 3. *If the computing times are discrete,*

(i) $I_e(P) \in S_e(P)$, and
(ii) $T(x, 1) \geq I_e(P)$ implies $E(x, P) \geq e$.

Proof. By discreteness, $\inf S_e(P) = \min S_e(P)$, so assertion (i) holds. If $T(x, 1) = I_e(P)$, assertion (i) implies $T(x, 1) \in S_e(P)$, so $E(x, P) \geq e$. If $T(x, 1) > I_e(P)$, apply the characteristic property (1) of the isoefficiency function. □

Definition 3. *An efficiency function has the* speed-up *property, if, for all $x \in X$,*

$$P_1 < P_2 \text{ implies } E(x, P_1) > E(x, P_2).$$

Theorem 4. *If the efficiency function has the speed-up property,*

$$P_1 < P_2 \text{ implies } I_e(P_1) \leq I_e(P_2)$$

for all efficiencies e and all numbers of processors P_1 and P_2.

Proof. By contradiction. Let $P_1 < P_2$ be natural numbers and suppose $I_e(P_1) > I_e(P_2)$. Then $S_e(P_2) \not\subset S_e(P_1)$, so there exists $x \in X$ such that $T(x, 1) \in S_e(P_2) - S_e(P_1)$. Hence, there exists $y \in X$ such that $T(y, 1) \geq T(x, 1)$ and $E(y, P_1) < e$ and $E(y, P_2) \geq e$. In particular, $E(y, P_1) < E(y, P_2)$—contradicting the speed-up property of E. □

The sequential execution time can be inferred from the efficiency if the following property holds.

Definition 4. *An efficiency function has the* size-up *property if, for all $x, y \in X$ and all $P > 1$,*

$$T(x, 1) < T(y, 1) \text{ if and only if } E(x, P) < E(y, P).$$

Theorem 5. *If an efficiency function E has the size-up property, the following implications hold for all $x \in X$ and $P \in \mathbb{N}$.*

(i) $E(x, P) = e$ implies $T(x, 1) = I_e(P)$, and
(ii) $E(x, P) \geq e$ implies $T(x, 1) \geq I_e(P)$.

Proof. To show (i) and a part of (ii) let x be such that $E(x, P) = e$. Due to the size-up property, $T(x, 1) \in S_e(P)$. Since all inputs with smaller sequential computing times have a smaller efficiency, $T(x, 1)$ is the smallest element of $S_e(P)$ and thus $T(x, 1) = I_e(P)$. To show the remainder of (ii) let x be such that $E(x, P) > e$. Again, $T(x, 1) \in S_e(P)$, and so $I_e(P) \leq T(x, 1)$. □

Theorem 6. *If the computing times are discrete and the efficiency function has the size-up property,*

$$E(x, P) \leq e \text{ implies } T(x, 1) \leq I_e(P).$$

Proof. Theorem 5(i) and Theorem 3(ii). □

Figure 1 shows how isoefficiency curves can be obtained by experiment.

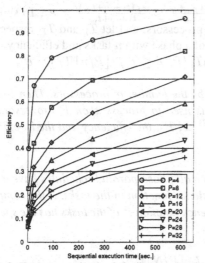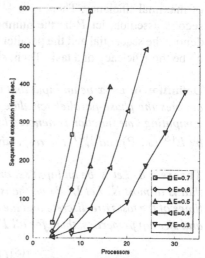

Fig. 1. Each curve in the left diagram is obtained with a different number of processors. For a given number of processors, certain input polynomials (Mignotte polynomials) of degrees $100, 200, \ldots, 500$ are processed one-by-one; the efficiency of each parallel computation and the computing time of the corresponding sequential computation form the coordinates of a point on the curve; consecutive points are connected by straight line segments. The right diagram shows isoefficiency curves. The curve corresponding to a given efficiency is obtained by intersecting the curves in the left diagram with the appropriate horizontal line; each intersection point yields a number of processors and a sequential execution time—the coordinates of a point in the right diagram. The points are connected by straight line segments. In this example, the speed-up and the size-up property both hold.

3 Isoefficiency of tree searches

We consider schedules that can be used to perform a parallel breadth-first search in a tree. One processes the levels of the tree sequentially and the nodes on each level in parallel.

Definition 5. *Let P be a number of processors and let n tasks be given, $n \in \{1, \ldots, P\}$. Let each task be parallelizable and let the computing time functions be identical. A parallel phase is a schedule that assigns to each task a set of $\lfloor P/n \rfloor$ processors such that each processor is assigned to at most one task. The computing time of a phase is equal to the computing time of any of its tasks. A phase-parallel schedule $s = (n_1, \ldots, n_m)$ is a sequence of m parallel phases, where phase i processes $n_i \leq P$ tasks. The width of a phase-parallel schedule is the maximum number of tasks in a phase. The total number of tasks in a phase-parallel schedule s is denoted by $|s|_1$. The computing time of a phase-parallel computation is the sum of the computing times of its phases.*

Theorem 7. *If a phase-parallel computation has efficiency e it contains a phase that has at least efficiency e. If a phase has efficiency e' all of its tasks have at least efficiency e'.*

Proof. Let a phase-parallel computation $(n_1, \ldots, n_m)$ have efficiency e. Let t_i denote the parallel execution time of phase i, and let e_i be the efficiency. Let j be a phase of maximal efficiency. Then, $e = \frac{e_1 t_1 + \cdots + e_m t_m}{t_1 + \cdots + t_m} \leq \frac{e_j (t_1 + \cdots + t_m)}{t_1 + \cdots + t_m} = e_j$. To show the second assertion, let P be the number of processors, and let T_1 and T_P respectively denote the sequential and the parallel time of a phase with n tasks and efficiency e'. Let e'' be the efficiency of a task. Then, $e' = nT_1/PT_P \leq T_1/(\lfloor P/n \rfloor T_P) = e''$. □

Definition 6. *Let x be an input and let P be the number of processors. Then $S(x, P)$ denotes the phase-parallel schedule that is used to process x on P processors. The computing time function is denoted by $T^{(S)}(x, P)$, the efficiency and the isoefficiency by $E^{(S)}(x, P)$ and $I_e^{(S)}(P)$, respectively.*

Theorem 8. *Let X be an input set on which all phase-parallel schedules have a width that is at most N. Let $I_e(P)$ be the isoefficiency function of the tasks. If the computing times of the tasks are discrete and the efficiency function E of the tasks has the speed-up and size-up properties, then, for all $P \geq N$,*

$$I_e^{(S)}(P) \geq I_e(\lfloor P/N \rfloor).$$

Proof. Since the task computing times are discrete the schedule computing times $T^{(S)}$ are also discrete. By Theorem 3, there is an input x such that $I_e^{(S)}(P) = T^{(S)}(x, 1)$ and $E^{(S)}(x, 1) \geq e$. If $T(x, 1)$ is the corresponding sequential time of a task we have $T^{(S)}(x, 1) \geq T(x, 1)$ since the schedule has at least one task. The most efficient task computation uses $P' \geq \lfloor P/N \rfloor$ processors and, by Theorem 7, has efficiency $e' \geq e$; that is, $E(x, P') = e'$. Since E has the size-up property, $T(x, 1) = I_{e'}(P')$ (Theorem 5(i)). Since the isoefficiency is monotonic in e, $I_{e'}(P') \geq I_e(P')$. By the speed-up property of E and Theorem 4, $I_e(P') \geq I_e(\lfloor P/N \rfloor)$. In total, $I_e^{(S)}(P) = T^{(S)}(x, 1) \geq T(x, 1) \geq I_e(\lfloor P/N \rfloor)$. □

Theorem 9. *Let X be an input set on which all phase-parallel schedules have at most N tasks. Let $I_e(P)$ be the isoefficiency function of the tasks. If the computing times of the tasks are discrete and the efficiency function E of the tasks has the speed-up and size-up properties, then*

$$I_e^{(S)}(P) \leq N I_e(P).$$

Proof. By discreteness and Theorem 3 there is an input x with $I_e^{(S)}(P) = T^{(S)}(x, 1)$. The least efficient task computation uses $P' \leq P$ processors and has at most efficiency e, that is $E(x, P') \leq e$. Since E has the size-up property, Theorem 6 yields $T(x, 1) \leq I_e(P')$. By the speed-up property of E and Theorem 4, $I_e(P') \leq I_e(P)$. Hence, $I_e^{(S)}(P) = T^{(S)}(x, 1) \leq N T(x, 1) \leq N I_e(P') \leq N I_e(P)$. □

4 Isoefficiency of poly-algorithms

Some algorithms can be parallelized in more than one way. If one can predict how efficient each parallel algorithm will be on any given input, one will execute the most efficient algorithm for each input. The following theorem analyzes the isoefficiency of the resulting poly-algorithm.

Theorem 10. *Let A and B be parallel algorithms that operate on the same set of inputs and that have the same sequential computing time $T(x, 1)$ for every input x. Let $E^{(A)}$ and $E^{(B)}$ respectively denote the efficiencies of algorithms A and B, and let $I_e^{(A)}$ and $I_e^{(B)}$ be the isoefficiencies. Consider the poly-algorithm C that, for input x, applies algorithm A if $E^{(A)}(x, P) > E^{(B)}(x, P)$ and algorithm B otherwise. Let $E^{(C)}$ be the efficiency and $I_e^{(C)}$ the isoefficiency of algorithm C. Then,*

(i) $I_e^{(C)}(P) \leq \min\left(I_e^{(A)}(P), I_e^{(B)}(P)\right)$.

(ii) If $E^{(A)}$ and $E^{(B)}$ have the speed-up property, then so does $E^{(C)}$.

(iii) If $E^{(A)}$ and $E^{(B)}$ have the size-up property, then so does $E^{(C)}$; if, in addition, the computing times are discrete, $I_e^{(C)}(P) = \min\left(I_e^{(A)}(P), I_e^{(B)}(P)\right)$.

Proof. Assertion (i) holds since $S_e^{(C)}(P) \supseteq S_e^{(A)}(P) \cup S_e^{(B)}(P)$ implies $I_e^{(C)}(P) =$ inf $S_e^{(C)}(P) \leq$ inf $\left(S_e^{(A)}(P) \cup S_e^{(B)}(P)\right) = \min\left(\text{inf } S_e^{(A)}(P), \text{inf } S_e^{(B)}(P)\right) =$ $\min\left(I_e^{(A)}(P), I_e^{(B)}(P)\right)$. Assertion (ii) follows from Definition 3. To show assertion (iii) consider inputs x, y with $T(x, 1) < T(y, 1)$. If C applies A for x and B for y we have $E^{(C)}(x, P) = E^{(A)}(x, P) < E^{(A)}(y, P) \leq E^{(B)}(y, P) = E^{(C)}(y, P)$; the other cases are similar. Likewise, $E^{(C)}(x, P) < E^{(C)}(y, P)$ implies $T(x, 1) < T(y, 1)$. To show the second part of the assertion let $T(x, 1) \in S_e^{(C)}(P)$. If algorithm A is executed for x we have $E^{(A)}(x, P) = E^{(C)}(x, P) \geq e$ and so, by Theorem 5(ii), $T(x, 1) \geq I_e^{(A)}(P)$ and hence, by Theorem 3, $T(x, 1) \in S_e^{(A)}(P)$. If B is executed for x one obtains instead $T(x, 1) \in S_e^{(B)}(P)$. So, $S_e^{(C)}(P) \subseteq S_e^{(A)}(P) \cup S_e^{(B)}(P)$, and hence $I_e^{(C)}(P) \geq \min\left(I_e^{(A)}, I_e^{(B)}\right)$. □

5   The parallel Descartes method

The parallel Descartes method uses parallelism on three levels as shown in Figure 2. The top level is a breadth-first search in a binary tree that uses the phase-parallel schedules of Section 3. On the middle level, the method chooses between two ways of scheduling three tasks; such a poly-algorithm is analyzed in Section 4. The bottom level performs an operation called *Taylor shift*. The intermediate results of this operation and their dependencies form a graph called *pyramid-DAG* that is scheduled using a method called *pie-mapping* (Decker & Krandick, 1999).

6   Analysis of the Descartes method

We analyze the Descartes method from the bottom up. The computing times of Taylor shift and node computation are discrete and the efficiency functions have the size-up and speed-up properties. The isoefficiency of Taylor shift and node computations can thus be analyzed by solving equations. The isoefficiency of the Descartes method is then obtained from the isoefficiency of the node computations.

The letters T, E, I respectively refer to computing time, efficiency, and isoefficiency of the parallel Taylor shift; the corresponding functions for the node computations are denoted by $T^{(v)}, E^{(v)}, I^{(v)}$; for the complete Descartes method we use

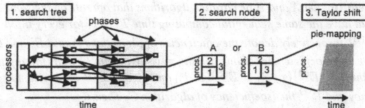

Fig. 2. The parallel Descartes method uses hierarchical load balancing. The levels of the search tree are processed in parallel phases. The search nodes schedule three Taylor shifts according to scheme A or B; method A performs two shifts in parallel on $\lfloor P/2 \rfloor$ processors each and then one on P processors; if 3 divides P method B allocates $P/3$ processors to one shift and executes the other two sequentially on the remaining processors. Each parallel Taylor shift uses pie-mapping.

$T^{(\mathcal{D})}, E^{(\mathcal{D})}, I^{(\mathcal{D})}$; by adding overbars and underbars we denote the corresponding upper and lower bounds.

We analyze the computing time of the parallel Taylor shift using a variant of the LogP-model (Culler et al., 1996). The four parameters L, o, g, and P describe system properties. Sending m floating point numbers across the network takes Lm time units. Processors are blocked for o cycles while sending or receiving messages; no other computations can be done during this time. The parameter g is used to model the bandwidth of the network, and P, as before, is the number of processors. Furthermore, we assume that two floating point numbers can be added in time t.

6.1 Isoefficiency of the Taylor shift

We show that the isoefficiency function of pie-mapping is codominant with P^2. Kumar et al. (1994) propose a different method ("cyclic-checkerboard mapping") for the pyramid DAG and try to show that its isoefficiency function is dominated by $P\sqrt{P}$, but their analysis is defective. In fact, we showed that cyclic-checkerboard mapping is less efficient than pie-mapping. The granularity of the parallel Taylor shift is determined by a parameter $s \in \mathbb{N}$ that is related to pie-mapping. Here, we consider only odd values of s; a realistic choice is $s = 3$; higher values of s mean a finer granularity. Since the computing time of the Taylor shift depends only on the degree n of the input polynomial A we write $T(n, P)$ instead of $T(A, P)$.

Theorem 11. *(Decker & Krandick, 1999) If a polynomial of degree $n - 1$ is Taylor-shifted using floating point arithmetic and pie-mapping on $P > 1$ processors, the parallel execution time is $T(n, P) = T_{calc}(n, P) + T_{com}(n, P)$, where T_{com} is the communication overhead and T_{calc} the computing time. We have $\underline{T}_{com}(n, P) :=$ $(sP - 1)o \le T_{com}(n, P) \le (sP - 1)(4o + nL/(sP)) =: \overline{T}_{com}(n, P)$, and*

$$T_{calc}(n, P) = \frac{1}{2}\left(((s^2 + 1)P + s - 1)\left(\frac{n}{sP}\right)^2 - \frac{n}{P}\left(\frac{n}{sP} + 1\right)\right)t.$$

The sequential execution time is $T(n, 1) = (tn(n - 1))/2$.

Clearly, the sets $\{T(n, 1) \mid n \in \mathbb{N}\}$, $\{\underline{T}(n, 1)\}$, and $\{\overline{T}(n, 1)\}$ are discrete.

Definition 7. *For all n and P let*

$$\underline{T}(n, P) := T_{calc}(n, P) + \underline{T}_{com}(n, P), \quad \overline{E}(n, P) := T(n, 1)/(P\underline{T}(n, P)),$$
$$\overline{T}(n, P) := T_{calc}(n, P) + \overline{T}_{com}(n, P), \quad \underline{E}(n, P) := T(n, 1)/(P\overline{T}(n, P)),$$

and define $\underline{I}_e$ and $\overline{I}_e$ as in Theorem 2.

Theorem 12. *For all natural numbers n and P and all $e \in (0, 1)$,*

$$\underline{T}(n, P) \leq T(n, P) \leq \overline{T}(n, P),$$
$$\underline{E}(n, P) \leq E(n, P) \leq \overline{E}(n, P),$$
$$\underline{I}_e(n, P) \leq I_e(n, P) \leq \overline{I}_e(n, P).$$

Furthermore, $\underline{E}$ and $\overline{E}$ have the speed-up and the size-up property.

Theorem 13. *Let $\overline{e}(s, P) := \frac{s^2 P}{s^2 P + P - 1}$; then, for any $e \in (0, \overline{e}(s, P))$, the values $I_e(P)$, $\underline{I}_e(P)$, and $\overline{I}_e(P)$ are defined. Let $\overline{e}(s) := \frac{s^2}{s^2 + 1}$; then, for any $e \in (0, \overline{e}(s))$, the functions $I_e(P)$, $\underline{I}_e(P)$, and $\overline{I}_e(P)$ are defined for all P and codominant with P^2.*

Proof. The first assertion holds since $\overline{E}$ and $\underline{E}$ grow monotonically in n and since $\lim_{n \to \infty} E(n, P) = \lim_{n \to \infty} \underline{E}(n, P) = \lim_{n \to \infty} \overline{E}(n, P) = \overline{e}(s, P)$. For the second assertion note that $\lim_{P \to \infty} \overline{e}(s, P) = \overline{e}(s)$. For the last assertion it suffices to show $\overline{I}_e(P) \preceq P^2$ and $\underline{I}_e \succeq P^2$. Let $e \in (0, \overline{e}(s))$. Then $\underline{E}(n, P) \geq e$ is equivalent to

$$0 \geq \underbrace{\left(s^2 P + P - 1 - \frac{s^2 P}{e} \right) \frac{t}{2}}_{=: A} n^2 + \underbrace{\left(\frac{1 - e}{e} \frac{s^2 P t}{2} + (sP - 1)sLP \right)}_{=: B} n + \underbrace{4 (sP - 1) o s^2 P^2}_{=: C}.$$

Since $e < \overline{e}(s)$, A is negative; hence,

$$\underline{E}(n, P) \geq e \quad \text{if and only if} \quad n \geq \left\lceil -\frac{B}{2A} + \sqrt{\frac{B^2}{4A^2} - \frac{C}{A}} \right\rceil =: n_0,$$

and thus $\overline{I}_e(P) = T(n_0, 1)$. Since $\frac{B}{A} \preceq P$ and $\frac{C}{A} \preceq P^2$ we have $n_0 \preceq P$. Now, $T(n, 1) \sim n^2$ implies $\overline{I}_e(P) \preceq P^2$. The proof of $\underline{I}_e \succeq P^2$ is similar. $\square$

6.2 Isoefficiency of the node computations

We analyze the poly-algorithm that chooses between methods A and B of Figure 2 to schedule three Taylor shifts. We start with method B; this method is used only if P is divisible by 3.

Definition 8. *For all $n \in \mathbb{N}$ and all P divisible by 3 let $T^{(B)}(n, P)$ and $E^{(B)}(n, P)$ be the parallel computing time and the efficiency of method B. Define $\underline{T}^{(B)}(n, P) := 2\underline{T}\left(n, \frac{2}{3}P\right)$ and $\overline{E}^{(B)}(n, P) := \frac{3 T(n, 1)}{T^{(B)}(n, P)}$. For $e \in (0, 1)$ let $I_e^{(B)}(P)$ be the isoefficiency; define $\underline{I}_e^{(B)}(P)$ as in Theorem 2. Define $\overline{T}^{(B)}$, $\underline{E}^{(B)}$, and $\overline{I}_e^{(B)}$ analogously.*

Theorem 14. *For all n and P, and all $e \in (0, 1)$,*

$$\underline{T}^{(B)}(n, P) \leq T^{(B)}(n, P) \leq \overline{T}^{(B)}(n, P),$$
$$\underline{E}^{(B)}(n, P) \leq E^{(B)}(n, P) \leq \overline{E}^{(B)}(n, P),$$
$$\underline{I}_e^{(B)}(n, P) \leq I_e^{(B)}(n, P) \leq \overline{I}_e^{(B)}(n, P).$$

Further, $\underline{E}^{(B)}$ and $\overline{E}^{(B)}$ have the speed-up and the size-up property. If $e \in (0, \overline{e}(s))$ the functions $I_e(P)$, $\overline{I}_e^{(B)}(P)$, and $\underline{I}_e^{(B)}(P)$ are defined for all P and are codominant with P^2.

Proof. Since $\overline{E}(n, P)$ is monotone decreasing in P the Taylor shift on $P/3$ processors is more efficient than the two other shifts. Therefore, $\underline{T}^{(B)}(n, P) \le T^{(B)}(n, P)$. Now, $\overline{E}^{(B)}(n, P) = \dfrac{3\,T(n, 1)}{P \cdot \left(2\,\underline{T}\left(n, \frac{2}{3}P\right)\right)} = \dfrac{T(n, 1)}{\frac{2}{3}P\,\underline{T}\left(n, \frac{2}{3}P\right)} = \overline{E}\left(n, \frac{2}{3}P\right)$, so, by Theorem 12, $\overline{E}^{(B)}$ has the size-up and the speed-up property. Finally, the function $\underline{I}_e^{(B)}(P) = $
$\inf \left\{ T^{(B)}(n, 1) \mid (\forall m \in \mathbb{N})\; T^{(B)}(m, 1) \ge T^{(B)}(n, 1) \Rightarrow \overline{E}^{(B)}(m, P) \ge e \right\}$
$= \inf \left\{ 3\,T(n, 1) \mid (\forall m \in \mathbb{N})\; T(m, 1) \ge T(n, 1) \Rightarrow \overline{E}\left(m, \frac{2}{3}P\right) \ge e \right\} = 3\,\underline{I}_e\left(\frac{2}{3}P\right)$
is codominant with P^2 (Theorem 13). The remaining proofs are analogous. Note that, for every $e < \overline{e}(s)$, $I_e^{(v)}(P)$ exists for every P since $I_e(P)$ exists and, on P processors, a search node allows higher efficiencies than a single Taylor shift. □

Definition 9. *For method A, define for all n and all $P > 1$, $T^{(A)}$, $E^{(A)}$, $I_e^{(A)}$, etc. in analogy to Definition 8 but let $\underline{T}^{(A)}(n, P) := \underline{T}\left(n, \lfloor \frac{P}{2} \rfloor\right) + \underline{T}(n, P)$.*

Theorem 15. *The assertions of Theorem 14 hold with $^{(B)}$ replaced by $^{(A)}$.*

Proof. Here, the assertions need to be checked by computation. For example, the proof that $\underline{I}_e^{(A)}(P) \sim P^2$ is analogous to the proof of Theorem 13. □

Definition 10. *For the node computations define $T^{(v)}$, $E^{(v)}$, $I_e^{(v)}$, etc. in analogy to Definitions 8 and 9.*

Theorem 16. *The bounds for the efficiency of the node computations, $\underline{E}^{(v)}(n, P)$ and $\overline{E}^{(v)}(n, P)$, have the size-up and the speed-up property. The bounds for the isoefficiency, $\underline{I}_e^{(v)}(P)$ and $\overline{I}_e^{(v)}(P)$, are codominant with P^2.*

Proof. By Theorem 10 from Theorems 14 and 15. □

6.3 Isoefficiency of the Descartes method

Figure 3 shows that the isoefficiency depends on the shape of the search trees; we start by considering search trees of bounded width or with a bounded number of nodes. In our analysis we denote by $\mathcal{D}(A, P)$ the phase-parallel schedule that is used for isolating the real roots of a polynomial A on P processors. Bounds for the node computation times yield bounds for the total isolation time $T^{(\mathcal{D})}(A, P)$ and thus, as before, bounds for the efficiency $E^{(\mathcal{D})}(A, P)$ and the isoefficiency $I_e^{(\mathcal{D})}(P)$. Again, we denote these bounds by underbars and overbars. Throughout we assume that the set of computing times is discrete. As before, the isoefficiency functions are defined for all P whenever $e < \overline{e}(s)$.

Theorem 17. *Let $I_e^{(\mathcal{D})}$ be the isoefficiency of the Descartes method for the set of polynomials that generate search trees of width at most N. Then $P^2 \preceq I_e^{(\mathcal{D})}(P)$.*

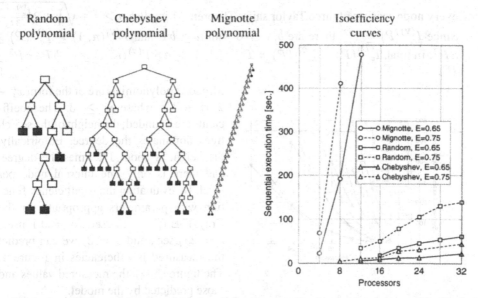

Fig. 3. Search trees arising from isolating the real roots of polynomials of degree 20. Shaded nodes correspond to isolating intervals. The isoefficiency curves for Mignotte, random, and Chebyshev polynomials are shown at efficiencies .65 and .75. The isoefficiency depends on the shape of the search trees.

Proof. The width of $\mathcal{D}(A, P)$ is bounded by the width of the search tree of A. Assume $P > 2N$ for an analysis of the asymptotic behavior of $I_e^{(\mathcal{D})}(P)$. Since $\overline{E}^{(\vee)}$ has the size-up and the speed-up property, $\underline{I}_e^{(\vee)}(\lfloor P/N \rfloor) \leq I_e^{(\mathcal{D})}(P)$ (Theorem 8). So, $P^2 \sim (P/N)^2 \sim (\lfloor P/N \rfloor)^2 \sim \underline{I}_e^{(\vee)}(\lfloor P/N \rfloor) \leq I_e^{(\mathcal{D})}(P) \leq I_e^{(\mathcal{D})}(P)$. $\qquad \square$

Theorem 18. *Let $I_e^{(\mathcal{D})}$ be the isoefficiency for polynomials that generate search trees with at most N search nodes. Then $I_e^{(\mathcal{D})}(P) \sim P^2$.*

Proof. Since $\underline{E}^{(\vee)}$ has the size-up and the speed-up property, $\overline{I}_e^{(\mathcal{D})}(P) \leq N\overline{I}_e^{(\vee)}(P)$ (Theorem 9). So, $I_e^{(\mathcal{D})}(P) \leq \overline{I}_e^{(\mathcal{D})}(P) \leq N\overline{I}_e^{(\vee)}(P) \sim NP^2 \sim P^2$. Since the width of search trees with at most N nodes is bounded by N, Theorem 17 yields $I_e^{(\mathcal{D})}(P) \succeq P^2$. $\qquad \square$

The polynomials $x^n - 2$, $n \geq 2$, have bounded coefficients and search trees with at most 3 nodes; on this input set the Descartes method has quadratic isoefficiency.

Theorem 19. *Let $I_e^{(\mathcal{D})}(P)$ be the isoefficiency for polynomials whose search trees have only one internal node per level and have heights that dominate the degrees. Then $P^3 \preceq I_e^{(\mathcal{D})}(P)$.*

Proof. Let X be a class of such polynomials. Then there is an $a > 0$ such that, for all $A \in X$, $a \deg A \leq |\mathcal{D}(A, 1)|_1$. So, $T^{(\mathcal{D})}(A, 1) = |\mathcal{D}(A, 1)|_1 T^{(\vee)}(\deg A, 1) \geq a \deg A\, T^{(\vee)}(\deg A, 1)$ for all $A \in X$. Let $A \in X$ such that $\underline{I}_e^{(\mathcal{D})}(P) = T^{(\mathcal{D})}(A, 1)$ and let $n = \deg A$. All nodes have the same efficiency, so $\overline{E}^{(\vee)}(n, P) = \overline{E}^{(\mathcal{D})}(A, P) \geq e$. Since $\overline{E}^{(\vee)}$ has the size-up property, Theorem 5 yields $T^{(\vee)}(n, 1) \geq \underline{I}_e^{(\vee)}(P)$. Since

every node consists of three Taylor shifts Theorem 11 implies $n \geq \frac{1}{2} + \sqrt{\frac{1}{4} + 2\frac{\underline{I}_e^{(v)}(P)}{3\,t}}$.
Since $\underline{I}_e^{(v)}(P) \succeq P^2$, there are $b, c > 0$ with $n \geq b\,P$ and $T^{(v)}(n, 1) \geq \underline{I}_e^{(v)}(P) \geq c\,P^2$. In total, $I_e^{(\mathcal{D})}(P) \geq \underline{I}_e^{(\mathcal{D})}(P) = T^{(\mathcal{D})}(A, 1) \geq a\,n\,T^{(v)}(n, 1) \geq a \cdot b\,P \cdot c\,P^2 \succeq P^3$. □

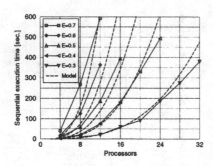

Mignotte polynomials are of the form $x^n - 2(5x - 1)^2$ where $n \geq 3$. The coefficients are bounded; the height of the search trees dominates the degree. Empirically, the height is about 1.2 times the degree, and there is only one internal node per level. By evaluating the isoefficiency function with parameters appropriate for the Cray T3E ($L = 20\,\mu\text{sec.}$, $o = 0.1\,\text{msec}$, $t = 25\,\mu\text{sec}$, and $s = 3$) we can predict the measured isoefficiencies in Figure 1. The figure shows the measured values and those predicted by the model.

Theorem 20. *Assume that the real roots of a squarefree integral polynomial are isolated using the Descartes method (floating point interval arithmetic using a fixed precision and a fixed exponent field). If the polynomial has degree n, the number of search nodes that are processed is dominated by $n\,(\log n)^2$.*

Proof. Since only finitely many floating point numbers are available, the set of integers that can be embedded into floating point intervals is bounded. Hence, the coefficients of the input polynomial A are bounded by some constant D. Let Υ be the search tree generated by A. Let K be the depth of Υ. For $1 \leq k \leq K$, let $h(k)$ denote the number of nodes on level k. Since, according to Krandick (1995, Theorem 48), $k\,h(k) \preceq n\log(n^2 D) \preceq n\log n$ and $K \preceq n\log(n^2 D) \preceq n\log n$, we can bound the number of nodes of Υ. Let H_K denote the K-th harmonic number. Then there is a $C > 0$ such that $\sum_{k=1}^{K} h(k) \leq \sum_{k=1}^{K} \frac{C\,n\log(n)}{k} \leq C\,n\,\log(n)\,H_K \preceq n\log(n)\,\log(n\,\log(n)) \preceq n\,(\log n)^2$. □

Theorem 21. *The isoefficiency function $I_e^{(\mathcal{D})}(P)$ of the Descartes method is dominated by $P^3\,(\log P)^2$.*

Proof. Let P be a number of processors and let A be a polynomial of degree n such that $\overline{I}_e^{(\mathcal{D})}(P) = T^{(\mathcal{D})}(A, 1)$. Since the root of the search tree of A is processed by P processors its efficiency is at most e, that is, $\underline{E}^{(v)}(n, P) \leq e$. Since $E^{(v)}$ has the size-up property, Theorem 6 yields $T^{(v)}(n, 1) \leq \overline{I}_e^{(v)}(P)$. Thus, $n \leq \frac{1}{2} + \sqrt{\frac{1}{4} + 2\frac{\overline{I}_e^{(v)}(P)}{3\,t}}$. By Theorem 16, $\overline{I}_e^{(v)} \preceq P^2$, so there are positive numbers a and b such that $n \leq a\,P$ and $T^{(v)}(n, 1) \leq b\,P^2$. By Theorem 20 there are positive numbers c, c' such that $|\mathcal{D}(A, 1)|_1 \leq c\,n\,(\log(n))^2 \leq c\,a\,P\,(\log(a\,P))^2 \leq c'\,P\,(\log P)^2$. In total, $I_e^{(\mathcal{D})}(P) \leq \overline{I}_e^{(\mathcal{D})}(P) = T^{(\mathcal{D})}(A, 1) = |\mathcal{D}(A, 1)|_1\,T^{(v)}(n, 1) \leq c'\,P\,(\log P)^2 \cdot b\,P^2 \preceq P^3\,(\log P)^2$. □

7 Acknowledgements

We thank the anonymous referee for suggesting an improved definition of isoefficiency.

References

Collins, G. E. (1974). The computing time of the Euclidean algorithm. *SIAM Journal on Computing, 3*(1), 1-10.

Collins, G. E., & Akritas, A. G. (1976). Polynomial real root isolation using Descartes' rule of signs. In R. D. Jenks (Ed.), *Proceedings of the 1976 ACM Symposium on Symbolic and Algebraic Computation* (pp. 272–275). ACM.

Collins, G. E., Johnson, J. R., & Küchlin, W. (1992). Parallel real root isolation using the coefficient sign variation method. In R. E. Zippel (Ed.), *Computer Algebra and Parallelism.*, LNCS 584, pp. 71-87. Springer-Verlag.

Culler, D. E., Karp, R. M., Patterson, D., Sahay, A., Santos, E. E., Schauser, K. E., Subramonian, R., & von Eicken, T. (1996). LogP: A practical model of parallel computation. *Communications of the ACM, 39*(11), 78–85.

Decker, T., & Krandick, W. (1999). Parallel real root isolation using the Descartes method. In P. Banerjee, V. K. Prasanna, & B. P. Sinha (Eds.), *High Performance Computing - HIPC'99*, LNCS 1745, pp. 261–268. Springer-Verlag.

Grama, A. Y., Gupta, A., & Kumar, V. (1993). Isoefficiency: Measuring the scalability of parallel algorithms and architectures. *IEEE Parallel and Distributed Technology, 1*(3), 12-21.

Gupta, A., Karypis, G., & Kumar, V. (1997). Highly scalable parallel algorithms for sparse matrix factorization. *IEEE Transactions on Parallel and Distributed Systems, 8*(5), 502–520.

Gupta, A., & Kumar, V. (1993). The scalability of FFT on parallel computers. *IEEE Transactions on Parallel and Distributed Systems, 4*(8), 922–932.

Johnson, J. R., & Krandick, W. (1997). Polynomial real root isolation using approximate arithmetic. In W. Küchlin (Ed.), *International Symposium on Symbolic and Algebraic Computation* (pp. 225–232). ACM.

Krandick, W. (1995). Isolierung reeller Nullstellen von Polynomen. In J. Herzberger (Ed.), *Wissenschaftliches Rechnen* (pp. 105–154). Akademie Verlag, Berlin.

Kruskal, C. P., Rudolph, L., & Snir, M. (1990). A complexity theory of efficient parallel algorithms. *Theoretical Computer Science, 71*(1), 95–132.

Kumar, V., Grama, A., Gupta, A., & Karypis, G. (1994). *Introduction to Parallel Computing: Design and Analysis of Algorithms.* Redwood City, CA, USA: Benjamin/Cummings.

Kumar, V., Nageshwara Rao, V., & Ramesh, K. (1988). Parallel depth first search on the ring architecture. In D. H. Bailey (Ed.), *Proceedings of the 1988 International Conference on Parallel Processing* (Vol. III, pp. 128–132). The Pennsylvania State University Press.

Kumar, V., & Singh, V. (1991). Scalability of parallel algorithms for the all-pairs shortest-path problem. *Journal of Parallel and Distributed Computing, 13*, 124–138.

Mahapatra, N. R., & Dutt, S. (1997). Scalable global and local hashing strategies for duplicate pruning in parallel A* graph search. *IEEE Transactions on Parallel and Distributed Systems, 8*(7), 738–756.

Sahni, S., & Thanvantri, V. (1996). Performance metrics: Keeping the focus on runtime. *IEEE Parallel and Distributed Technology, 4*(1), 43–56.

Schreiner, W., Mittermaier, C., & Winkler, F. (2000). On solving a problem in algebraic geometry by cluster computing. In A. Bode, T. Ludwig, W. Karl, & R. Wismüller (Eds.), *Euro-Par 2000 Parallel Processing*, LNCS 1900, pp. 1196–1200. Springer-Verlag.

Yang, T.-R., & Lin, H.-X. (1997). Isoefficiency analysis of CGLS algorithm for parallel least squares problems. In B. Hertzberger & P. Sloot (Eds.), *High-Performance Computing and Networking*, LNCS 1225, pp. 452–461. Springer-Verlag.

Matrix Methods for Solving Algebraic Systems

Ioannis Z. Emiris

1 Introduction

The problem of computing all common zeros of a system of polynomials is of fundamental importance in a wide variety of scientific and engineering applications. This article surveys efficient methods based on the sparse resultant for computing all *isolated* solutions of an arbitrary system of n polynomials in n unknowns. In particular, we construct matrix formulae which yield nontrivial multiples of the resultant thus reducing root-finding to the eigendecomposition of a square matrix.

Our methods can exploit structure of the polynomials as well as that of the resulting matrices. This is an advantage as compared to other algebraic methods, such as Gröbner bases and characteristic sets. All approaches have complexity exponential in n, but Gröbner bases suffer in the worst case by a quadratic exponent, whereas for matrix-based methods the exponent is linear. Moreover, they are discontinuous with respect to perturbations in the input coefficients, unlike resultant matrix methods in general. Of course, Gröbner bases provide a complete description of arbitrary algebraic systems and have been well developed, including public domain stand-alone implementations or as part of standard computer algebra systems. There is also a number of numerical methods for solving algebraic systems, but their enumeration goes beyond this article's scope.

The next section describes briefly the main steps in the relatively young theory of sparse elimination, which aspires to generalize the results of its mature counterpart, classical elimination. Section 3 presents the construction of sparse resultant matrices. Section 4 reduces solution of arbitrary algebraic systems to numerical linear algebra, avoiding any issues of convergence. Our techniques find their natural application in problems arising in a variety of fields, including problems expressed in terms of geometric and kinematic constraints in robotics, vision and computer-aided modelling. We describe in detail problems from structural biology, in section 5.

The emphasis is placed on recent and new implementations, described in each respective section, with pointers to where they can be found. They have been ported on several architectures, including Sun, DEC, Linux and Iris platforms. Previous work and open questions are mentioned in the corresponding sections.

2 Sparse elimination

Sparse elimination generalizes several results of classical elimination theory on multivariate polynomial systems of arbitrary degree by considering their structure. This leads to stronger algebraic and combinatorial results in general (Gelfand et al. 1994), (Sturmfels 1994), (Cox et al. 1998); the reader may consult these references for details and proofs. Assume that the number of variables is n; roots in $(\mathbb{C}^*)^n$ are called *toric*. We use x^e to denote the monomial $x_1^{e_1} \cdots x_n^{e_n}$, where $e = (e_1, \ldots, e_n) \in \mathbb{Z}^n$. Let the input *Laurent* polynomials be

$$f_1, \ldots, f_n \in \mathbb{Q}[x_1^{\pm 1}, \ldots, x_n^{\pm 1}]. \tag{1}$$

The discussion applies to arbitrary coefficient fields and roots in the torus of their algebraic closure. Let *support* $\mathcal{A}_i = \{a_{i1}, \ldots, a_{im_i}\} \subset \mathbb{Z}^n$ denote the set of exponent vectors corresponding to monomials in f_i with nonzero coefficients: $f_i = \sum_{a_{ij} \in \mathcal{A}_i} c_{ij} x^{a_{ij}}$, for $c_{ij} \neq 0$. The *Newton polytope* $Q_i \subset \mathbb{R}^n$ of f_i is the convex hull of support $\mathcal{A}_i$. Function sys_Maple() of package spares (see the next section) implements both operations. For arbitrary sets A and $B \subset \mathbb{R}^n$, their *Minkowski sum* is $A + B = \{a + b \,|\, a \in A, b \in B\}$.

Definition 2.1 *Given convex polytopes $A_1, \ldots, A_n, A_k' \subset \mathbb{R}^n$, mixed volume, is the unique real-valued function $MV(A_1, \ldots, A_n)$, invariant under permutations, such that, $MV(A_1, \ldots, \mu A_k + \rho A_k', \ldots, A_n) = \mu \, MV(A_1, \ldots, A_k, \ldots, A_n) + \rho MV(A_1, \ldots, A_k', \ldots, A_n)$, for $\mu, \rho \in \mathbb{R}_{\geq 0}$, and, $MV(A_1, \ldots, A_n) = n! \, \mathrm{Vol}(A_1)$, when $A_1 = \cdots = A_n$, where $\mathrm{Vol}(\cdot)$ denotes standard euclidean volume in $\mathbb{R}^n$.*

If the polytopes have integer vertices, their mixed volume takes integer values. We are now ready to state a generalization of Bernstein's theorem (Gelfand et al. 1994), (Cox et al. 1998):

Theorem 2.2 *Given system (1), the cardinality of common isolated zeros in $(\mathbb{C}^*)^n$, counting multiplicities, is bounded by $MV(Q_1, \ldots, Q_n)$. Equality holds when certain coefficients are generic.*

Newton polytopes model the polynomials' structure and provide a "sparse" counterpart of total degree. Similarly for mixed volume and Bézout's bound (simply the product of all total degrees), the former being usually significantly smaller for systems encountered in engineering applications. The generalization to *stable volume* provides a bound for non-toric roots.

Ι The algorithm by Emiris and Canny (1995) has resulted to program mixvol:

Input: supports of n polynomials in n variables
Output: mixed volume and mixed cells
Language: C
Availability: www.inria.fr/saga/emiris/soft_alg.html

It is also available as part of the INRIA library ALP: www.inria.fr/saga/logiciels/ALP. Program mixvol enumerates all *mixed cells* in the subdivision of $Q_1 + \cdots + Q_n$, thus identifying the integer points comprising a monomial basis of the quotient ring of the ideal defined by the input polynomials. Mixed cells also correspond to start systems (with immediate solution) for a *sparse homotopy* of the original system's roots. Important work in these areas has been done by T.Y. Li and his collaborators (Gao et al. 1999).

The *resultant* of a polynomial system of $n+1$ polynomials in n variables with indeterminate coefficients is a polynomial in these indeterminates, whose vanishing provides a necessary and sufficient condition for the existence of common roots of the system. Different resultants exist depending on the space of the roots we wish to characterize, namely projective, affine or toric. *Sparse* or *toric resultants* express the existence of toric roots. Let

$$f_0, \ldots, f_n \in \mathbb{Q}[x_1^{\pm 1}, \ldots, x_n^{\pm 1}], \tag{2}$$

with f_i corresponding to generic point $c_i = (c_{i1}, \ldots, c_{im_i})$ in the space of polynomials with support $\mathcal{A}_i$. This space is identified with projective space $\mathbb{P}^{m_i-1}$. Then system (2) can be thought of as point $c = (c_0, \ldots, c_n)$. Let Z denote the Zariski closure, in the product of projective spaces, of the set of all c such that the system has a solution in $(\mathbb{C}^*)^n$. Z is an irreducible variety.

Definition 2.3 *The* sparse resultant $R = R(\mathcal{A}_0, \ldots, \mathcal{A}_n)$ *of system (2) is a polynomial in* $\mathbb{Z}[c]$. *If* $codim(Z) = 1$ *then R is the defining irreducible polynomial of the hypersurface Z. If* $codim(Z) > 1$ *then* $R = 1$.

The resultant is homogeneous in the coefficients of each polynomial. If $MV_{-i} = MV(Q_0, \ldots, Q_{i-1}, Q_{i+1}, \ldots, Q_n)$, then the degree of R in the coefficients of f_i is $\deg_{f_i} R = MV_{-i}$. deg R will stand for the total degree.

3 Matrix formulae

Different means of expressing a resultant are possible (Cox et al. 1998), (Emiris and Mourrain 1999). Ideally, we wish to express it as a matrix determinant, or a divisor of such a determinant where the quotient is a nontrivial extraneous factor. This section discusses matrix formulae for the sparse resultant, which exploit the monomial structure of the Newton polytopes. These are *sparse resultant*, or *Newton, matrices*. We restrict ourselves to Sylvester-type matrices which generalize the coefficient matrix of a linear system and Sylvester's matrix of two univariate equations.

There are two main approaches to construct a well-defined, square, generically nonsingular matrix M, such that $R|\det M$. The rows of M will always be indexed by the product of a monomial with an input polynomial. The entries of a row are coefficients of that product, each corresponding to the monomial indexing the respective column. The degree of $\det M$ in the coefficients of f_i, equal to the number of rows with coefficients of f_i, is greater or equal to $\deg_{f_i} R$. Obviously, the smallest possible matrix has dimension deg R.

The first approach, introduced by Canny and Emiris in 1993, relies on a *mixed subdivision* of the Minkowski sum of the Newton polytopes $Q = Q_0 + \cdots + Q_n$

(Canny and Pedersen 1993), (Sturmfels 1994), (Canny and Emiris 2000). The algorithm uses a subset of $(Q + \delta) \cap \mathbb{Z}^n$ to index the rows and columns of M. $\delta \in \mathbb{Q}^n$ must be *sufficiently generic* so that every integer point lies in the relative interior of a unique n-dimensional cell of the mixed subdivision of $Q + \delta$. In addition, δ is small enough so that this cell is among those that had the point on their boundary. Clearly, the dimension of the resulting matrix is at most equal to the number of points in $(Q + \delta) \cap \mathbb{Z}^n$. This construction allows us to pick any one polynomial so that it corresponds to exactly $\deg_{f_i} R$ rows.

The greedy version of Canny and Pedersen (1993) uses a minimal point set and is the algorithm implemented by function spares() in the Maple package of the same name. It is also included as function spresultant() in Maple package multires developed at INRIA (www.inria.fr/saga/logiciels/multires.html):

Input: $n + 1$ polynomials in n variables, an arbitrary number of parameters
Output: sparse resultant matrix in the parameters
Language: Maple
Availability: www.inria.fr/saga/emiris/soft_alg.html

For instance, spares([f0,f1,f2],[x1,x2]) constructs the sparse resultant matrix of the 3 polynomials by eliminating variables x1, x2. The function also expresses the polynomial coefficients in terms of any indeterminates other than x1, x2. Optional arguments may specify vector δ and the subdivision of Q.

The second algorithm, by Emiris and Canny (1995), is *incremental* and yields usually smaller matrices and, in any case, of dimension no larger than the cardinality of $(Q + \delta) \cap \mathbb{Z}^n$. We have observed that in most cases of systems with dimension bounded by 10 the algorithm gives a matrix at most 4 times the optimal. The selection of integer points corresponding to monomials multiplying the row polynomials uses a vector $v \in (\mathbb{Q}^*)^n$. In those cases where a minimum matrix of Sylvester type provably exists, the incremental algorithm produces this matrix. These are precisely the systems for which v can be deterministically specified; otherwise, a random v can be used.

The algorithm proceeds by constructing candidate rectangular matrices in the input coefficients. Given such a matrix with the coefficients specialized to generic values, the algorithm verifies whether its rank is complete using modular arithmetic. If so, any square nonsingular submatrix can be returned as M; otherwise, new rows (and columns) are added to the candidate. This is the first part of program far, also integrated in the ALP library. The entire far has:

Input: $n + 1$ polynomials in n variables to be eliminated, one in the coefficient field
Output: sparse resultant matrix and a superset of the common roots
Language: C
Availability: www.inria.fr/saga/emiris/soft_alg.html

For instance, commands "far -nco trial input" and "far -nco -ms 0 trial input" construct a sparse resultant matrix, where file input.exps contains the supports and file input.coef contains vector v and the MV_{-i}, if known (otherwise the program computes them by calling mixvol and writes them in file temp_all_mvs). In the first case, we assumed file trial.msum exists and contains all needed integer points for matrix construction. In the second example this file is created and filled in by far. A number of command line options exists, including "-iw trial.indx" to store the matrix definition in file trial.indx in order to be used by subsequent executions.

Sparse resultant matrices, including the candidates constructed by the incremental algorithm, are characterized by a structure that generalizes the Toeplitz structure and has been called *quasi-Toeplitz* (Emiris and Pan 1997). An open implementation problem is to exploit this structure in verifying full rank, aspiring to match the asymptotic acceleration of almost one order of magnitude. Another open question concerns exploiting quasi-Toeplitz structure for accelerating the solution of an eigenproblem.

4 Algebraic solving by linear algebra

To solve the well-constrained system (1) by the resultant method we define an overconstrained system and apply the resultant matrix construction. Matrix M need only be computed once for all systems with the same supports. So this step can be carried out offline, while the matrix operations to approximate all isolated roots for each coefficient specialization are online.

We present two ways of defining an overconstrained system. The first method *adds an extra polynomial* f_0 to the given system (thus defining a well-studied object, the u-resultant). The constant term is a new indeterminate:

$$f_0 = x_0 + c_{01}x_1 + \cdots + c_{0n}x_n \in (\mathbb{Q}[x_0])[x_1^{\pm 1}, \ldots, x_n^{\pm 1}].$$

Coefficients c_{0j} are usually random. M describes the multiplication map for f_0 in the coordinate ring of the ideal defined by (1). An alternative way to obtain an overconstrained system is by *hiding* one of the variables in the coefficient field and consider (after modifying notation to unify the subsequent discussion) system:

$$f_0, \ldots, f_n \in (\mathbb{Q}[x_0])[x_1^{\pm 1}, \ldots, x_n^{\pm 1}].$$

M is a matrix polynomial in x_0, and may not be linear.

In both cases, the idea is that when x_0 is equal to the respective coordinate of a common root, then the resultant and, hence, the matrix determinant vanish. An important issue concerns the degeneracy of the input coefficients. This may result in the trivial vanishing of the sparse resultant or of $\det M$ when there is an infinite number of common roots (in the torus or at toric infinity) or simply due to the matrix constructed. An infinitesimal perturbation has recently been proposed by D'Andrea and Emiris (2000), which respects the structure of Newton polytopes and is computed at minimal extra cost.

The perturbed determinant is a polynomial in the perturbation variable, whose leading coefficient is guaranteed to be nonzero. The trailing *nonzero* coefficient is always a multiple of a generalized resultant, in the sense that it vanishes when x_0 takes its values at the system's roots. This is a univariate polynomial in x_0, hence univariate equation solving yields these coordinates. Moreover, the u-resultant allows us to recover all coordinates via polynomial factorization. The perturbed matrix can be obtained by package **spares**, provided that local variable **PERT_DEGEN_COEFS** is appropriately set, as explained in the package's documentation. An open problem concerns the combination of this perturbation with the matrix operations described below.

4.1 Eigenproblems

This section describes the online matrix solver of **far**. Most of the computation is numeric, yet the method has global convergence and avoids issues related to the starting point of iterative methods. We use double precision floating point arithmetic and the LAPACK library because it implements state-of-the-art algorithms, offering the choice of a tradeoff between speed and accuracy, and provides efficient ways for computing estimates on the condition numbers and error bounds.

A basic property of resultant matrices is that right vector multiplication expresses evaluation of the row polynomials. Specifically, multiplying by a column vector containing the values of column monomials q at some $\alpha \in \mathbb{C}^n$ produces the values of the row polynomials $\alpha^p f_{i_p}(\alpha)$, where integer point (or, equivalently, monomial) p indexes a row. Letting $\mathcal{E}$ be the monomial set indexing the matrix rows and columns,

$$M(x_0) \begin{bmatrix} \vdots \\ \alpha^q \\ \vdots \end{bmatrix} = \begin{bmatrix} \vdots \\ \alpha^p f_{i_p}(x_0, \alpha) \\ \vdots \end{bmatrix}, \quad q, p \in \mathcal{E}, i_p \in \{0, \ldots, n\}.$$

Computationally it is preferable to have to deal with as small a matrix as possible. To this end we partition M into four blocks M_{ij} so that the upper left submatrix M_{11} is of maximal possible dimension under the following conditions: it must be square, independent of x_0, and well-conditioned relative to some user-defined threshold. **far** first concentrates all constant columns to the left and within these columns permutes all zero rows to the bottom; both operations could be implemented offline. To specify M_{11} according to the above conditions, an LU decomposition with column pivoting is applied, though an SVD (or QR decomposition) might be preferable.

Once M_{11} is specified, let $A(x_0) = M_{22}(x_0) - M_{21}(x_0) M_{11}^{-1} M_{12}(x_0)$. To avoid computing M_{11}^{-1}, we use its decomposition to solve $M_{11} X = M_{12}$ and compute $A = M_{22} - M_{21} X$. The routine used depends on $\kappa(M_{11})$, with the slower but more accurate function **dgesvx** called when $\kappa(M_{11})$ is beyond some threshold.

If $(\alpha_0, \alpha) \in \mathbb{C}^{n+1}$ is a common root with $\alpha \in \mathbb{C}^n$, then $\det M(\alpha_0) = 0 \Rightarrow \det A(\alpha_0) = 0$. Let point (or monomial) set $\mathcal{B} \subset \mathcal{E}$ index matrix A. For any vector $v' = [\cdots \alpha^q \cdots]^T$, where q ranges over $\mathcal{B}$, $A(\alpha_0) v' = 0$. Moreover,

$$\begin{bmatrix} M_{11} & M_{12}(\alpha_0) \\ 0 & A(\alpha_0) \end{bmatrix} \begin{bmatrix} v \\ v' \end{bmatrix} = \begin{bmatrix} 0 \\ 0 \end{bmatrix} \Rightarrow M_{11} v + M_{12}(\alpha_0) v' = 0,$$

determines v once v' has been computed. Vectors v and v' together contain the values of every monomial in $\mathcal{E}$ evaluated at α.

It can be shown that $\mathcal{E}$ affinely spans $\mathbb{Z}^n$ and an affinely independent subset can be computed in polynomial time. Given v, v' and these points, we can compute the coordinates of α. If all independent points are in $\mathcal{B}$ then v' suffices. To find the vector entries that will allow us to recover the root coordinates it is sufficient to search for pairs of entries corresponding to q_1, q_2 such that $q_1 - q_2 = (0, \ldots, 0, 1, 0, \ldots, 0)$. This lets us compute the i-th coordinate, if the unit appears at the i-th position, by taking ratios of the vector entries.

To reduce the problem to an eigendecomposition, let r be the dimension of $A(x_0)$, and $d \geq 1$ the highest degree of x_0 in any entry. We wish to find x_0:

$$A(x_0) = x_0^d A_d + x_0^{d-1} A_{d-1} + \cdots + x_0 A_1 + A_0$$

becomes singular. These are the eigenvalues of the *matrix polynomial*. Furthermore, for every eigenvalue λ, there is a basis of the kernel of $A(\lambda)$ defined by the *right eigenvectors* of the matrix polynomial. If A_d is nonsingular then the eigenvalues and right eigenvectors of $A(x_0)$ are the eigenvalues and right eigenvectors of *monic* matrix polynomial $A_d^{-1} A(x_0)$. This is always the case when adding an extra linear polynomial, since $d = 1$ and $A_1 = I$ is the $r \times r$ identity matrix; then $A(x_0) = -A_1(-A_1^{-1} A_0 - x_0 I)$. Generally, the *companion matrix* of a monic matrix polynomial is a square matrix C of dimension rd:

$$C = \begin{bmatrix} 0 & I & \cdots & 0 \\ \vdots & & \ddots & \\ 0 & 0 & \cdots & I \\ -A_d^{-1} A_0 & -A_d^{-1} A_1 & \cdots & -A_d^{-1} A_{d-1} \end{bmatrix}$$

The eigenvalues of C are precisely the eigenvalues λ of $A_d^{-1} A(x_0)$, whereas its right eigenvector equals the concatenation of $v_1, \ldots, v_d$: v_1 is a right eigenvector of $A_d^{-1} A(x_0)$ and $v_i = \lambda^{i-1} v_1$, for $i = 2, \ldots, d$. There is an iterative and a direct algorithm in LAPACK for solving this eigenproblem, respectively implemented in hsein and trevc. Experimental evidence points to the former as being faster on large problems. Further, an iterative solver could eventually exploit the fact that we are only interested in real eigenvalues and eigenvectors.

We now address the question of a singular A_d. The following *rank balancing* transformation is used in far also to improve the conditioning of the leading matrix: If matrix polynomial $A(x_0)$ is not singular for all x_0, then there exists a transformation $x_0 \mapsto (t_1 y + t_2)/(t_3 y + t_4)$ for some $t_i \in \mathbb{Z}$, that produces a new matrix polynomial of the same degree and with nonsingular leading coefficient. The new matrix polynomial has coefficients of the same rank, for sufficiently generic t_i. We have observed that for matrices of dimension larger than 200, at least two or three quadruples should be tried since a lower condition number by two or three orders of magnitude is sometimes achieved. The asymptotic as well as practical complexity of this stage is dominated by the eigendecomposition.

If a matrix polynomial with invertible leading matrix is found, then the eigendecomposition of the corresponding companion matrix is undertaken. If

A_d is ill-conditioned for all linear rank balancing transformations, then we build the matrix pencil and call the *generalized eigendecomposition* routine **dgegv** to solve $C_1 x + C_0$. The latter returns pairs (α, β) such that matrix $C_1 \alpha + C_0 \beta$ is singular with an associated right eigenvector. For $\beta \neq 0$ the eigenvalue is α/β, while for $\beta = 0$ we may or may not wish to discard the eigenvalue. $\alpha = \beta = 0$ occurs if and only if the pencil is identically zero within machine precision.

If the x_0-root coordinates have all unit *geometric multiplicity* and $A(x_0)$ is not identically singular, then we have reduced root-finding to an eigenproblem and some evaluations to eliminate extraneous eigenvectors and eigenvalues. The complexity lies in $O^*\left(2^{O(n)}(\deg R)^3 d\right)$, where polylogarithmic terms are ignored.

These operations are all implemented in the solver of far: Command "far trial input" constructs the sparse resultant matrix and solves the system with supports in input.exps, vector v, mixed volumes MV_{-i} and (integer) coefficients in input.coef, and integer points in trial.msum. Command line option "-ir trial.indx" reads the matrix definition from file trial.indx and option -a tells the program to set $A(x_0) = M(x_0)$, thus avoiding the decomposition of M_{11} and any related numerical errors. Other options control the condition number thresholds, printing of various information, and verification of results.

The section concludes with accuracy issues, irrespective of whether A_d is regular or not. Since there is no restriction in picking which variable to hide, it is enough that one of the original $n + 1$ variables have unit geometric multiplicity. If none can be found, we can specialize the hidden variable to each of the eigenvalues and solve every one of the resulting subsystems. Other numerical and algebraic remedies are under study, including the aforementioned perturbed determinant. Still, there is an open implementation problem in verifying the multiplicity and solving in such cases. Clustering neighbouring eigenvalues and computing the error on the average value will help handling such cases. Lastly, self-validating methods should be considered to handle ill-conditioned matrix polynomials, in particular in the presence of defective eigenvalues.

5 Molecular conformations

A relatively new branch of computational biology has been emerging as an effort to apply successful paradigms and techniques from geometry and robot kinematics to predicting the structure of molecules and embedding them in euclidean space. This section examines the problem of computing all 3-dimensional *conformations* of a molecule described by certain geometric characteristics.

Energy minima can be approximated by allowing only the dihedral angles to vary, while considering bond lengths and bond angles as rigid. We consider cyclic molecules of six atoms to illustrate our approach and show that the corresponding algebraic formulation conforms to our model of sparseness. An in-depth study of cyclic molecules has been presented by Emiris and Mourrain (1999). Direct geometric analysis yields a 3×3 polynomial system

$$f_i = \beta_{i1} + \beta_{i2}x_j^2 + \beta_{i3}x_k^2 + \beta_{i4}x_j^2 x_k^2 + \beta_{i5}x_j x_k = 0, \quad i \in \{1, 2, 3\},$$

for $\{i, j, k\} = \{1, 2, 3\}$. The β_{ij} are functions of known parameters. The system has a Bézout bound of 64 and mixed volume 16.

The first instance tried is a synthetic example for which β_{ij} is the j-th entry of $(-9, -1, -1, 3, 8)$ for all i. The symmetry of the problem is bound to produce root coordinates of high multiplicity, so we add $f_0 = x_0 + c_{01}x_1 - c_{02}x_2 + c_{03}x_3$ with randomly selected c_{0j}. The 3-fold mixed volumes are $16, 12, 12, 12$ hence $\deg R = 52$. M is regular and has dimension 86, with 30 rows corresponding to f_0. The entire 56×56 constant submatrix is relatively well-conditioned. In the 30×30 matrix polynomial, matrix A_1 is numerically singular; random transformations fail to improve significantly its conditioning. The generalized eigenproblem routine produces 12 complex solutions, 3 infinite real solutions and 15 finite real roots. The absolute value of the four polynomials on the candidate values lies in $[0.6 \cdot 10^{-9}, 0.3 \cdot 10^{-3}]$ for values that approximate true solutions and in $[7.0, 3.0 \cdot 10^{20}]$ for spurious answers. Our program computes the true roots to at least 5 digits, the true roots being $\pm(1, 1, 1)$, $\pm(5, -1, -1)$, $\pm(-1, 5, -1)$, $\pm(-1, -1, 5)$. The average CPU time of the online phase on a SUN SPARC 20 with clock rate 60MHz and 32MB of memory is 0.4 seconds.

Usually noise enters in the process that produces the coefficients. To model this phenomenon, we consider the *cyclohexane* which has equal inter-atomic distances and equal bond angles. We randomly perturb these values by about 10% to obtain

$$\beta = \begin{bmatrix} -310 & 959 & 774 & 1313 & 1389 \\ -365 & 755 & 917 & 1269 & 1451 \\ -413 & 837 & 838 & 1352 & 1655 \end{bmatrix}.$$

We defined an overconstrained system by hiding variable x_3. M has dimension 16 and is quadratic in x_3, whereas the 2-fold mixed volumes are all 4 and $\deg R = 12$. The monic quadratic polynomial reduces to a 32×32 companion matrix on which the standard eigendecomposition is applied. After rejecting false candidates each solution contains at least 8 correct digits. CPU time is 0.2 seconds on average for the online phase.

Last is an instance where the input parameters are sufficiently generic to produce 16 real roots. Let β_{ij} be the j-th entry of $(-13, -1, -1, -1, 24)$. We hide x_3 and obtain $\dim M = 16$, whereas $\deg R = 12$, and the companion matrix has dimension 32. There are 16 real roots. Four of them correspond to eigenvalues of unit geometric multiplicity, while the rest form four groups, each corresponding to a triple eigenvalue. For the latter the eigenvectors give us no valid information, so we recover the values of x_1, x_2 by looking at the other solutions and by relying on symmetry. The computed roots are correct to at least 7 decimal digits. The average CPU time is 0.2 seconds.

An equivalent approach to obtaining the same algebraic system may be based on *distance geometry*. A *distance matrix* is a square, real symmetric matrix, with zero diagonal. It can encode all inter-atomic distances by associating its rows and columns to atoms. When the entries are equal to a scalar multiple of the corresponding squared pairwise distance, the matrix is said to be *embeddable* in $\mathbb{R}^3$. Necessary and sufficient conditions for such matrices to be embeddable are known in terms of the eigenvalues and rank.

The main interest of this approach lies in large molecules. We have examined it in relation with experimental data that determine intervals in which the unknown distances lie. Optimization methods have been developed and applied successfully to molecules with a few hundreds of atoms (Havel et al. 1997). Ours are direct linear algebra techniques which are, for now, in a preliminary stage. We apply results from distance matrix theory and structured matrix perturbations to reduce the rank of the interval matrix respecting the experimental bounds. The Matlab code developed in collaboration with Nikitopoulos (1999) can handle molecules with up to 30 atoms.

References

Canny, J., Emiris, I.Z. (2000): A subdivision-based algorithm for the sparse resultant. J. ACM. 47:417–451

Canny, J., Pedersen, P. (1993): An algorithm for the Newton resultant. Technical Report 1394, Computer Science Department, Cornell University, Ithaca, New York

Cox, D., Little, J., O'Shea, D. (1998): Using algebraic geometry. Springer, New York (Graduate Texts in Mathematics, vol. 185)

D'Andrea, C., Emiris, I.Z. (2000): Solving sparse degenerate algebraic systems. Submitted for publication. `ftp://ftp-sop.inria.fr/saga/emiris/publis/DAEm00.ps.gz`

Emiris, I.Z., Canny, J.F. (1995): Efficient incremental algorithms for the sparse resultant and the mixed volume. J. Symb. Comput. 20:117–149

Emiris, I.Z., Mourrain, B. (1999): Computer algebra methods for studying and computing molecular conformations. Algorithmica, Special issue on algorithms for computational biology. 25:372–402

Emiris, I.Z., Pan, V.Y. (1997): The structure of sparse resultant matrices. In: Proceedings of the ACM-SIGSAM International Symposium on Symbolic and Algebraic Computation, ISSAC '97, Maui, Hawaii. Association of Computing Machinery, New York, pp. 189–196

Gao, T., Li, T.Y., Wang, X. (1999): Finding isolated zeros of polynomial systems in C^n with stable mixed volumes. J. Symb. Comput. 28:187–211

Gelfand, I.M., Kapranov, M.M., Zelevinsky, A.V. (1994): Discriminants and resultants. Birkhäuser, Boston

Havel, T.F., Hyberts, S., Najfeld, I. (1997): Recent advances in molecular distance geometry. In: Bioinformatics, Springer, Berlin (Lecture Notes in Computer Science, vol. 1278)

Nikitopoulos, T.G. (1999): Matrix perturbation theory in distance geometry. Bachelor's thesis. INRIA Sophia-Antipolis. `www.inria.fr/saga/stages/arch.html`

Sturmfels, B. (1994): On the Newton polytope of the resultant. J. Algebr. Combinat. 3:207–236

A Feasibility Result for Interval Gaussian Elimination Relying on Graph Structure

Andreas Frommer

1 Introduction

Let $\mathbf{A} = (\mathbf{a}_{ij}) \in \mathbb{IR}^{n \times n}$ be an *interval matrix*, i.e. each entry is a compact real interval. 'Usual' matrices $A \in \mathbb{R}^{n \times n}$ with real coefficients will be called *point matrices* in this paper, and a similar notation and terminology is adopted for vectors. Let $\mathbf{b} \in \mathbb{IR}^n$ be an interval vector. We are interested in computing an interval vector containing the solution set

$$S := \{x \in \mathbb{R}^n : \text{there exist a point matrix } A \in \mathbf{A}$$
$$\text{and a point vector } b \in \mathbf{b} \text{ such that } Ax = b\}.$$

In general, S is not an interval vector, so that interval arithmetic computation can only provide an interval vector enclosing S. One way to compute such an enclosure $\mathbf{x}$ (which is not necessarily the interval hull of S) is by means of interval Gaussian elimination according to the following algorithm:

ALGORITHM IGA$(\mathbf{A}, \mathbf{b})$
for $k = 1$ to $n - 1$ {*'factorization'*}
   for $i = k + 1$ to n
     $\mathbf{l}_{ik} = \mathbf{a}_{ik}^k / \mathbf{a}_{kk}^k$
     for $j = i + 1$ to n
       $\mathbf{a}_{ij}^{k+1} = \mathbf{a}_{ij}^k - \mathbf{l}_{ik}\mathbf{a}_{kj}^k$
     $\mathbf{b}_i^{k+1} = \mathbf{b}_i^k - \mathbf{l}_{ik}\mathbf{b}_k^k$
for $i = n$ downto 1   { *back substitution* }
   $\mathbf{x}_i = \left(\mathbf{b}_i^i - \sum_{j=i+1}^n \mathbf{a}_{ij}^i \mathbf{x}_j\right) / \mathbf{a}_{ii}^i$

Here, $\mathbf{A}^1 = \mathbf{A}$ and $\mathbf{b}^1 = \mathbf{b}$. Since this algorithm replaces all operations of the usual Gaussian elimination algorithm for point quantities by interval operations, the inclusion property of interval arithmetic (see Alefeld et al. (1983)) guarantees $\mathbf{x} \supseteq S$. Also note that we do not perform pivoting in algorithm IGA, i.e. we assume that the necessary permutations have been done beforehand. Even then, however, IGA may break down due to division by some $\mathbf{a}_{kk}^k$ containing zero, although point Gaussian elimination would be feasible on every point matrix from $\mathbf{A}$; see Alefeld et al. (1983) and Reichmann (1979). A standard example is the 3×3 matrix

$$\mathbf{A} = \begin{pmatrix} 1 & [0, 2/3] & [0, 2/3] \\ [0, 2/3] & 1 & [0, 2/3] \\ [0, 2/3] & [0, 2/3] & 1 \end{pmatrix}$$

of Reichmann (1979). For any point matrix in $\mathbf{A}$ Gaussian elimination is feasible without pivoting, but IGA is infeasible for any pivot choice.

One is therefore interested in conditions which guarantee IGA to be feasible. We review the most important results known so far in Section 2 (see also Mayer (1991)) and then develop a new feasibility result, relying on the graph of **A**, in Section 3. Basically, this result says that interval Gaussian elimination is feasible if the graph is a tree and if (point) Gaussian elimination can be performed on all point matrices. Special cases include tridiagonal and arrowhead matrices. We finish by discussing some extensions, examples and applications.

2 Review of known feasibility results

A nonsingular matrix $A \in \mathbb{R}^{n \times n}$ is called an M-matrix, if $a_{ij} \leq 0$ for all $i \neq j$ and $A^{-1} \geq 0$, where the last inequality is to be understood componentwise. Given an interval matrix **A**, we define its comparison matrix $\langle \mathbf{A} \rangle = (\alpha_{ij}) \in \mathbb{R}^{n \times n}$ as

$$\alpha_{ij} = \begin{cases} -\max\{|a| \mid a \in \mathbf{a}_{ij}\} & \text{for } i \neq j \\ \min\{|a| \mid a \in \mathbf{a}_{ii}\} & \text{for } i = j \end{cases} .$$

The following result was proved by Alefeld (1977).

Theorem 1 *If* $\langle \mathbf{A} \rangle$ *is an M-matrix, then* $IGA(\mathbf{A}, \mathbf{b})$ *is feasible for any right hand side* **b**.

If any $A \in \mathbf{A}$ is an M-matrix, $\langle \mathbf{A} \rangle$ is an M-matrix, too. So Theorem 1 applies in particular to this case. In addition, Barth et al. (1974) showed that then $IGA(\mathbf{A}, \mathbf{b})$ yields the best possible result, i.e. the interval hull of the solution set, if $0 \in \mathbf{b}$ or **b** lies entirely in the nonnegative or nonpositive orthant.

The next result from Frommer et al. (1993b), Theorem 5.3, relies on the notion of irreducibility, see Berman et al. (1994), e.g. We use the notation $\check{a}$ to denote the midpoint of an interval **a**.

Theorem 2 *Let* $\langle \mathbf{A} \rangle \cdot (1, \ldots, 1)^T \geq 0$, *let* **A** *be irreducible and let there exist a triple* $(i, j, k) \in \{1, \ldots, n\}^3$ *such that* $k < i, j$ *and*

$$sign(\check{a}_{ij}) \cdot sign(\check{a}_{ik}) \cdot sign(\check{a}_{kj}) = \begin{cases} sign(\check{a}_{kk}) & \text{if } i \neq j \\ -sign(\check{a}_{kk}) & \text{if } i = j \end{cases} .$$

Then $IGA(\mathbf{A}, \mathbf{b})$ *is feasible for any right hand side* **b**.

In Frommer at al. (1993a) this result was developed further using a generalization of irreducibility.

Another feasibility result goes back to Reichmann (1975) for the case of interval upper Hessenberg matrices whose entries obey certain sign restrictions. To state his result, let us define two sign-functions for an interval $\mathbf{a} = [b, c]$ as

$$\text{sign}(\mathbf{a}) = \begin{cases} +1 & \text{if } b > 0 \\ -1 & \text{if } c < 0 \\ 0 & \text{else} \end{cases} , \qquad \sigma(\mathbf{a}) = \begin{cases} +1 & \text{if } b \geq 0 \\ -1 & \text{if } c \leq 0 \\ 0 & \text{else} \end{cases} .$$

Reichmann's result then reads as follows.

Theorem 3 *Assume that* $\mathbf{A}$ *has upper Hessenberg form, i.e.* $\mathbf{a}_{ij} = 0$ *for* $i > j+1$, *and assume that* $sign(\mathbf{a}_{ii}) \neq 0$ *for* $i = 1,\ldots,n$ *as well as* $0 \neq \mathbf{a}_{i-1,i}$ *for* $i = 1,\ldots,n-1$. *Furthermore, let the following sign conditions hold for all* $i = 1,\ldots,n$ *and* $j = i+2,\ldots,n$:

(i) $\mathbf{a}_{k,i+1} \neq 0$ *and* $sign(\mathbf{a}_{ii}) \cdot sign(\mathbf{a}_{i+1,i+1}) = -\sigma(\mathbf{a}_{i+1,i}) \cdot \sigma(\mathbf{a}_{i,i+1})$

or

$\mathbf{a}_{k,i+1} = 0$ *for* $k = 1,\ldots,i,$

(ii) $\mathbf{a}_{i+1,j} \neq 0$ *and* $\mathbf{a}_{i,j} \neq 0$ *and* $sign(\mathbf{a}_{ii}) \cdot \sigma(\mathbf{a}_{i+1,j}) = -\sigma(\mathbf{a}_{i+1,i}) \cdot \sigma(\mathbf{a}_{ij})$

or

$\mathbf{a}_{i+1,j} \neq 0$ *and* $\mathbf{a}_{kj} = 0$ *for* $k = 1,\ldots,i$

or

$\mathbf{a}_{i+1,j} = 0$ *and* $\mathbf{a}_{k,i+1} = 0$ *for* $k = 1,\ldots,i.$

Then $IGA(\mathbf{A},\mathbf{b})$ *is feasible for any right hand side* $\mathbf{b}$.

3 A feasibility result relying on graph structure

With $\mathbf{A}$ we associate the following (undirected) graph $G = G(\mathbf{A}) = (V,E)$ defined through the entries of $\mathbf{A}$:

$$V := \{1,\ldots,n\} \quad \text{(vertices)},$$
$$E := \{e = \{i,j\} \mid i,j \in V, i \neq j \text{ and } (\mathbf{a}_{ij} \neq 0 \text{ or } \mathbf{a}_{ji} \neq 0)\} \quad \text{(edges)}.$$

Let us recall some basic definitions from graph theory before we proceed to define the minimum degree ordering.

The *degree* $\deg(i)$ of a vertex $i \in V$ is the number of incident edges, i.e.

$$\deg(i) = |\{e \in E \mid i \in e\}|.$$

A *path* from vertex v to vertex w is a sequence $(v_0,\ldots,v_l)$ of vertices such that $v_0 = v$, $v_l = w$ and $\{v_i,v_{i+1}\} \in E$ for $i = 0,\ldots,l-1$. The length of the path is l. A *cycle* in G is a path of length ≥ 3 with $v_0 = v_l$ and $v_i \neq v_j$ for $i \neq j$, $i,j = 0,\ldots l-1$. A graph is *connected* if for each pair of different vertices i and j there is a path from i to j. A connected graph without cycles is a *tree*.

(Interval) Gaussian elimination defines a sequence of graphs G^k if we initialize $G^1 = G(\mathbf{A})$ and take $G^{k+1} = (V^{k+1},E^{k+1})$ with

$$V^{k+1} = \{k+1,\ldots,n\},$$
$$E^{k+1} = \{e = \{i,j\} \mid i,j \in V^{k+1} \text{ and } \{i,j\} \in E^k\} \quad (1)$$
$$\cup \{e = \{i,j\} \mid i,j \in V^{k+1}, i \neq j \text{ and } \{i,k\},\{j,k\} \in E^k\}. \quad (2)$$

G^k is very closely related to the graph of the matrix $\mathbf{A}^k = (\mathbf{a}^k_{ij})_{i,j=k,\ldots,n}$, the trailing lower right $(n-k+1,n-k+1)$-submatrix in the kth step of Gaussian elimination. To state this precisely, let us say that an *accidental zero* occurs in $\mathbf{A}^{k+1}$ if $\mathbf{a}^{k+1}_{ij} = 0$ with $\mathbf{a}^k_{ij} \neq 0$. Note that this will only happen if $\mathbf{a}^k_{ij}$ is a point quantity and, in addition, $\mathbf{a}^k_{ij} = l_{ik}\mathbf{a}^k_{kj}$. The following lemma is then obvious.

Lemma 1 *For $k = 1, \ldots, n$ the graph $G(\mathbf{A}^k)$ is a subgraph of G^k. If no accidental zeros occur for all k, we even have $G(\mathbf{A}^k) = G^k$ for $k = 1, \ldots, n$.*

Definition 1 *The matrix $\mathbf{A}$ is* ordered by minimum degree, *if the following holds for all $k = 1, \ldots, n$:*

$$\deg_{G^k}(k) = \min_{i=k}^{n} \deg_{G^k}(i).$$

Here, the subscript G^k indicates that we mean the degree in the graph G^k. Any matrix can be ordered by minimum degree by an appropriate simultaneous permutation of its rows and columns, see Duff et al. (1989).

Lemma 2 *If $G(\mathbf{A})$ is a tree and $\mathbf{A}$ is ordered by minimum degree, then for $k = 1, \ldots, n$*

  (i) G^k is a tree,

  (ii) $\deg_{G^k}(k) = 1$.

Proof: This result is fairly standard, but for convenience we reproduce it here. Note that (ii) is an immediate consequence of (i), since each tree has leaves (see Cormen et al. (1989), e.g.), i.e. nodes with degree 1. This degree is minimal since a tree is connected.

To show (i) we proceed by induction. Assume then that G^k is a tree, which is correct for $k = 1$. Let us first show that G^{k+1} is connected. To this purpose, let $v, w \geq k+1$ be two different vertices in G^{k+1}. Since G^k is connected, there exists a path $(v = v_0, v_1, \ldots, v_l = w)$ in G^k from v to w. Since $\deg_{G^k}(k) = 1$, there is exactly one vertex $j > k$ for which $\{k, j\} \in E^k$. Assume that the path $(v = v_0, v_1, \ldots, v_l = w)$ contains the vertex k, i.e. $v_i = k$ for some $i \in \{1, \ldots, l-1\}$. Then $v_{i-1} = v_{i+1} = j \in \{k+1, \ldots, n\}$, and $(v_0, \ldots, v_{i-2}, j, v_{i+2}, \ldots, v_l)$ is also a path connecting v to w in G^k. Continuing in this manner we can remove all occurencies of k in the path. Then, by (1) all edges in the path are also edges from E^{k+1} so that we end up with a path connecting v to w in G^{k+1}.

To show that G^{k+1} does not have cycles, assume on the contrary that $(v = v_0, v_1, \ldots, v_l = v)$ is a cycle in G^{k+1}. But since $\deg_{G^k}(k) = 1$, the set in (2) is empty which shows that all edges from E^{k+1} are also edges from E^k. This implies that $(v_0, v_1, \ldots, v_l)$ is already a cycle in G^k which is impossible since G^k is a tree. $\quad\square$

We are now in a position to state our central result.

Theorem 4 *Assume that $\mathbf{A} \in \mathbb{IR}^{n \times n}$ is such that $G(\mathbf{A})$ is a tree and that $\mathbf{A}$ is ordered by minimum degree. Furthermore, assume that for all point matrices $A \in \mathbf{A}$, classical (point) Gaussian elimination is feasible without pivoting. Then interval Gaussian elimination is feasible, too, i.e. $IGA(\mathbf{A}, \mathbf{b})$ is feasible for all right hand sides $\mathbf{b}$.*

Proof: Let us first introduce the notation $A^k = (a_{ij}^k)_{i,j=k,\ldots,n}$ to denote the trailing $(n - k+1, n-k+1)$ submatrix in usual (point) Gaussian elimination for a matrix $A \in \mathbb{R}^{n \times n}$. Due to the inclusion property of interval arithmetic we know that $A^k \in \mathbf{A}^k$ for all k, provided $A \in \mathbf{A}$. The idea of the proof is to show that in fact

$$\mathbf{A}^k = \{A^k \mid A \in \mathbf{A}\}, \tag{3}$$

which, in particular, yields

$$\mathbf{a}_{kk}^k = \{a_{kk}^k \mid A \in \mathbf{A}\}, \tag{4}$$

and $\{a_{kk}^k \mid A \in \mathbf{A}\}$ cannot contain zero, since by assumption, Gaussian elimination is feasible for all $A \in \mathbf{A}$.

Assume that (3) is correct up to some $k < n$, which is indeed the case for $k = 1$. Since by Lemma 2 we have $\deg_{G^k}(k) = 1$, Lemma 1 shows that there is at most one $l > k$ such that $\mathbf{a}_{kl}^k \neq 0$ or $\mathbf{a}_{lk}^k \neq 0$, whereas $\mathbf{a}_{ki} = \mathbf{a}_{ik} = 0$ for all $i = k+1, \ldots, n; i \neq l$. In view of the two innermost loops of Algorithm IGA, this shows

$$\mathbf{a}_{ij}^{k+1} = \mathbf{a}_{ij}^k \text{ for all pairs } (i,j) = (k+1, k+1), \ldots, (n,n), \ (i,j) \neq (l,l). \tag{5}$$

For $(i,j) = (l,l)$ we have

$$\mathbf{a}_{ll}^{k+1} = \mathbf{a}_{ll}^k - (\mathbf{a}_{lk}^k / \mathbf{a}_{kk}^k) \cdot \mathbf{a}_{kl}^k. \tag{6}$$

Each of the four intervals which make up the term on the right hand side of (6) occurs only once. By an elementary rule of interval arithmetic (see Alefeld et al. (1983)), this implies that $\mathbf{a}_{ll}^{k+1}$ does not overestimate the corresponding range, i.e.

$$\mathbf{a}_{ll}^{k+1} = \{a_{ll}^k - (a_{lk}^k / a_{kk}^k) \cdot a_{kl}^k \mid a_{ll}^k \in \mathbf{a}_{ll}^k, a_{lk}^k \in \mathbf{a}_{lk}^k, a_{kl}^k \in \mathbf{a}_{kl}^k, a_{kk}^k \in \mathbf{a}_{kk}^k\}. \tag{7}$$

On the other hand, by definition of the (point) Gaussian elimination process, and since we assume that (3) holds for k, each element on the right hand side of (7) is nothing else but some a_{ll}^{k+1} for a certain initial point matrix $A \in \mathbf{A}$. Therefore, (7), together with (5), proves (3). $\square$

In some applications, 'symmetric' interval linear systems (where $\mathbf{a}_{ij} = \mathbf{a}_{ji}$) arise naturally and one is interested in getting an enclosure of the *symmetric* solution set

$$S_{sym} := \{x \in \mathbb{R}^n \colon \text{ there exist a symmetric point matrix } A \in \mathbf{A}$$
$$\text{and a point vector } b \in \mathbf{b} \text{ such that } Ax = b\}.$$

For this situation we get the following corollary.

Corollary 1 *Assume that $\mathbf{A} \in \mathbb{IR}^{n \times n}$ is symmetric, that $G(\mathbf{A})$ is a tree and that $\mathbf{A}$ is ordered by minimum degree. Furthermore, assume that for all symmetric point matrices $A \in \mathbf{A}$, classical (point) Gaussian elimination is feasible without pivoting. Then the 'symmetrized' interval Gaussian elimination is feasible, too, i.e. $IGA_{sym}(\mathbf{A}, \mathbf{b}) \supseteq S_{sym}$ can be computed for all right hand sides $\mathbf{b}$. Here, IGA_{sym} is a modification of Algorithm IGA, where the computation of $\mathbf{a}_{ij}^{k+1}$ in the two innermost loops is replaced by*

$$\mathbf{a}_{ll}^{k+1} = \mathbf{a}_{ll}^k - (\mathbf{a}_{lk}^k)^2 / \mathbf{a}_{kk}^k, \tag{8}$$

with $l > k$ denoting the unique vertex in G^k for which $\{l, k\} \in E^k$. The interval square function is assumed to return the range of the squares, i.e.

$$\mathbf{a}^2 = \{a^2 \mid a \in \mathbf{a}\}.$$

Proof: Observe first that now $\mathbf{a}_{ij}^k = \mathbf{a}_{ji}^k$ for $i, j = k, \ldots, n$, and similarly $a_{ij}^k = a_{ji}^k$ for symmetric point matrices $A \in \mathbf{A}$. Thus, in a manner completely analogous to the proof of Theorem 4 we now can show

$$\mathbf{A}^k = \{A^k \mid A \in \mathbf{A}, A \text{ symmetric }\}.$$

This proves the corollary. □

Note that the modified algorithm of Corollary 1 may be regarded as a variant of an interval Cholesky algorithm in the sense of Alefeld et al. (1993).

We would like to end this section by mentioning that our Theorem 4 and Corollary 1 are quite different in spirit from the results reviewed in Section 2: Our results rely on the fact that interval arithmetic does not produce overestimation during the elimination process, whereas in the other results overestimation may occur, but the pivots are shown to still not contain zero. It is also important to notice that even in our situation the back substitution process may produce overestimation (the interval quantities $\mathbf{a}_{ij}^i$ are used again there, but they already entered into $\mathbf{a}_{i+1,j}^{i+1}$ during the 'factorization'). Thus, even when the hypotheses of Theorem 4 are satisfied we cannot expect IGA$(\mathbf{A}, \mathbf{b})$ to yield the interval hull of the solution set S.

4 Discussion

For the purpose of illustration, Figure 1 gives some patterns of matrices for which the graph is a tree. The nodes are already ordered by minimum degree. We see that in particular tridiagonal (a) and arrowhead (b) matrices fall into this category. For these two cases, the feasibility of interval Gaussian elimination has been proved by Reichmann (1979) and Schäfer (2000) respectively. (The talk Schäfer (1996) on this result was the starting point for the work presented here.) The matrix (c) has a nonsymmetric pattern.

$$\begin{pmatrix} \times & \times & & & \\ \times & \times & \times & & \\ & \times & \times & \times & \\ & & \times & \times & \times \\ & & & \times & \times \end{pmatrix} \quad \begin{pmatrix} \times & & & & \times \\ & \times & & & \times \\ & & \times & & \times \\ & & & \times & \times \\ \times & \times & \times & \times & \times \end{pmatrix} \quad \begin{pmatrix} \times & & \times & & \\ & \times & \times & & \\ \times & & \times & \times & \\ & & & \times & \\ & & & \times & \times \end{pmatrix}$$

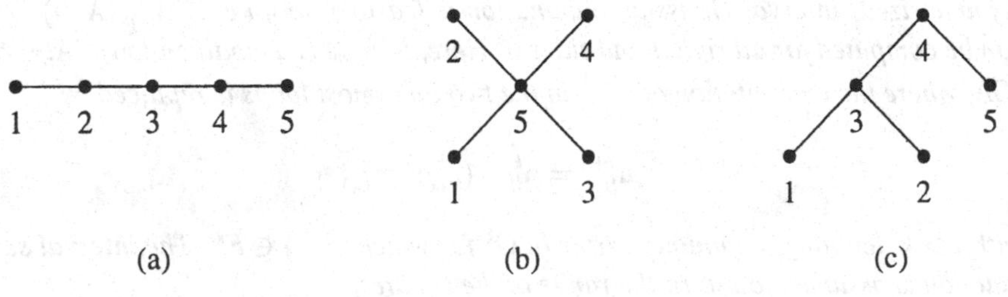

(a) (b) (c)

Figure 1: patterns for matrices and corresponding graphs

A natural question is to ask whether it is possible to extend the results of the preceding section to block Gaussian elimination with a corresponding hypothesis on the graph of the blocks of the matrix, which should again be a tree. Using capital letters to denote blocks, the crucial operation in a block version of IGA would then be the computation of A_{ij}^{k+1} satisfying

$$A_{ij}^{k+1} \supseteq \{A_{ij}^k - A_{ik}^k \left(A_{kk}^k\right)^{-1} A_{kj}^k \mid A_{ij} \in \mathbf{A}_{ij}^k, \, A_{ik}^k \in \mathbf{A}_{ik}^k, \, A_{kk}^k \in \mathbf{A}_{kk}^k, \, A_{kj}^k \in \mathbf{A}_{kj}^k\}. \quad (9)$$

A necessary condition for the argumentation of the previous section to work again is that we have in fact equality in (9). However, this seems quite difficult to satisfy, since the right hand side of (9) will in general not be an interval matrix, whereas $\mathbf{A}_{ij}^{k+1}$ is one. The only positive situation we can think of is when all diagonal blocks $\mathbf{A}_{ii}$ of $\mathbf{A}$ are themselves diagonal interval matrices and all (non-zero) off-diagonal blocks $\mathbf{A}_{ij}$ with $i \neq j$ contain at most one non-zero entry in each row and column. Then equality can be obtained in (9) by computing $\mathbf{A}_{ij}^{k+1}$ as

$$\mathbf{A}_{ij}^{k+1} = \mathbf{A}_{ij}^k - \left(\mathbf{A}_{ik}^k \mathbf{B}_{kk}^k\right) \mathbf{A}_{kj}^k,$$

where $\mathbf{B}_{kk}^k$ is the diagonal interval matrix whose diagonal elements are the interval arithmetic 'reciprocals' of the diagonal elements of $\mathbf{A}_{kk}^k$. (Note that under these assumptions each $\mathbf{A}_{kk}^k$ is diagonal.) We refrain from stating this result explicitly as a theorem.

One area of application of the results presented here is in methods for improving error bounds for the solution of (point) linear system using the Lanczos process. The method presented in Frommer et al. (1999) indeed builds up a tridiagonal interval matrix for which enclosures for the (symmetric) solution set must be obtained for certain right hand sides. Since all symmetric point matrices within the interval matrix are positive definite, so that Gaussian elimination is feasible without pivoting, this is a situation where Corollary 1 applies directly.

References

Alefeld, G. (1977): Über die Durchführbarkeit des Gausschen Algorithmus bei Gleichungen mit Intervallen als Koeffizienten. Computing Suppl. 1:15–19

Alefeld, G. and Herzberger, J. (1983): Introduction to Interval Computations. Academic Press, New York

Alefeld, G. and Mayer, G. (1993): The Cholesky method for interval data. Linear Algebra Appl. 194:161–182

Barth, W. and Nuding, E. (1974): Optimale Lösung von Intervallgleichungssystemen. Computing 12:117–125

Berman, A. and Plemmons, R. (1994): Nonnegative Matrices in the Mathematical Sciences, volume 9 of Classics in Applied Mathematics. SIAM, Philadelphia

Cormen, T., Leiserson, C. and Rivest, R. (1989): Introduction to Algorithms. The MIT Press, Cambridge Mass.

Duff, I., Erisman, A. and Reid, J. (1989): Direct Methods for Sparse Matrices. Oxford Science Publications, Oxford

Frommer, A. and Mayer, G. (1993a): Linear systems with Ω-diagonally dominant matrices and related ones. Linear Algebra Appl. 186:165–181

Frommer, A. and Mayer, G. (1993b): A new criterion to guarantee the feasibility of the interval Gaussian algorithm. SIAM J. Matrix Anal. Appl. 14:408–419

Frommer, A. and Weinberg, A. (1999): Verified error bounds for linear systems through the Lanczos process. Reliable Computing 5:255–267

Mayer, G. (1991): Old and new aspects for the interval Gaussian algorithm. In: Kaucher, E., Markov, S. and Mayer, G. (eds.): Computer Arithmetic, Scientific Computation and Mathematical Modelling, Baltzer, Basel, pp. 329–349

Reichmann, K. (1975): Ein hinreichendes Kriterium für die Durchführbarkeit des Intervall-Gauss-Algorithmus bei Intervall-Hessenberg-Matrizen ohne Pivotsuche. Z. Angew. Math. Mech. 59:373–379

Reichmann, K. (1979): Abbruch beim Intervall-Gauss-Algorithmus. Computing 22:335–361

Schäfer, U. (1996): Talk presented at the workshop on scientific computing, Riezlern, october 1996.

Schäfer, U. (2000): The feasibility of the interval Gaussian algorithm for arrowhead matrices. Reliable Computing, to appear

Solution of Systems of Polynomial Equations by Using Bernstein Expansion*

Jürgen Garloff and Andrew P. Smith

1   Introduction

Systems of polynomial equations appear in a great variety of applications, e.g.,
in geometric intersection computations (Hu et al. 1996), chemical equilibrium
problems, combustion, and kinematics, to name only a few. Examples can be
found in the monograph Morgan (1987). Following Sherbrooke and Patrikalakis
(1993), most of the methods for the solution of such a system can be classified as techniques based on elimination theory, continuation, and subdivision.
Elimination theory-based methods for constructing Gröbner bases rely on symbolic manipulations, making those methods seem somewhat unsuitable for larger
problems. This class and also the second of the methods based on continuation
frequently give us more information than we need since they determine all complex solutions of the system, whereas in applications often only the solutions
in a given area of interest - typically a box - are sought. In the last category we collect all methods which apply a domain-splitting approach: Starting
with the box of interest, such an algorithm sequentially splits it into subboxes,
eliminating infeasible boxes by using bounds for the range of the polynomials
under consideration over each of them, and ending up with a union of boxes
that contains all solutions to the system which lie within the given box. Methods utilising this approach include interval computation techniques as well as
methods which apply the expansion of a multivariate polynomial into Bernstein
polynomials. In principle, each interval computation method for solving a system of nonlinear equations, cf. the monographs Kearfott (1996) and Neumaier
(1990), can be applied to a polynomial system. Not surprisingly, techniques
specially designed for polynomial systems are often more efficient in computing
time. So we concentrate here on these methods. Jäger and Ratz (1995) combine
the method of Gröbner bases with interval computations. Van Hentenryck et
al. (1997) present a branch and prune approach which can be characterised as a
global search method using intervals for numerical correctness and for pruning

---
*Support from the Ministry of Education, Science, Research, and Technology of the Federal
Republic of Germany under contract no.1706998 is gratefully acknowledged.

the search space early.

As in the method to be presented in this paper, Sherbrooke and Patrikalakis (1993) use Bernstein expansion. Sequences of bounding boxes for the solutions to the polynomial system are generated by two different approaches: the first method projects control polyhedra onto a set of coordinate planes and the second exploits linear programming. But no use of the relationship between the Bernstein coefficients on neighbouring subboxes, cf. Subsection 2.2 below, is made and no existence test for a box to contain a solution, cf. Section 3, is provided.

Other applications of Bernstein expansion include applications to Computer Aided Geometric Design (e.g., Hu et al. 1996), to robust stability problems, cf. the survey article Garloff (2000), and to the solution of systems of polynomial inequalities (Garloff and Graf 1999).

The organisation of this paper is as follows: In the next section we briefly recall the Bernstein expansion, cf. Garloff (1993) and Zettler and Garloff (1998) and the references therein. The method is presented in Section 3. Examples are given in Section 4.

We concentrate here on real solutions, but we note that complex solutions can be found simply by separating each variable and each polynomial into their real and imaginary parts, doubling the order of the system.

2   Bernstein expansion

For compactness, we will use multi-indices $I = (i_1, \ldots, i_l)$ and multi-powers $\mathbf{x}^I = x_1^{i_1} x_2^{i_2} \cdot \ldots \cdot x_l^{i_l}$ for $\mathbf{x} \in \mathbf{R}^l$. Inequalities $I \leq N$ for multi-indices are meant componentwise, where $0 \leq i_k, k = 1, \cdots, l$, is implicitly understood. With $I = (i_1, \ldots, i_{r-1}, i_r, i_{r+1}, \ldots, i_l)$ we associate the index $I_{r,k}$ given by $I_{r,k} = (i_1, \ldots, i_{r-1}, i_r + k, i_{r+1}, \ldots, i_l)$, where $0 \leq i_r + k \leq n_r$. We can then write an l-variate polynomial p in the form

$$p(\mathbf{x}) = \sum_{I \leq N} a_I \mathbf{x}^I, \quad \mathbf{x} \in \mathbf{R}^l, \tag{1}$$

and refer to N as the *degree* of p. Also, we write $\binom{N}{I}$ for $\binom{n_1}{i_1} \cdot \ldots \cdot \binom{n_l}{i_l}$.

2.1   Bernstein transformation of a polynomial

In this subsection we expand a given l-variate polynomial (1) into Bernstein polynomials to obtain bounds for its range over an l-dimensional box. Without loss of generality we consider the unit box $\mathbf{U} = [0, 1]^l$ since any nonempty box of $\mathbf{R}^l$ can be mapped affinely onto this box.

For $\mathbf{x} = (x_1, \ldots, x_l) \in \mathbf{R}^l$, the Ith *Bernstein polynomial* of degree N is defined as

$$B_{N,I}(\mathbf{x}) = b_{n_1,i_1}(x_1) b_{n_2,i_2}(x_2) \cdot \ldots \cdot b_{n_l,i_l}(x_l),$$

where for $i_j = 0, \ldots, n_j, \; j = 1, \ldots, l$

$$b_{n_j,i_j}(x_j) = \binom{n_j}{i_j} x_j^{i_j}(1 - x_j)^{n_j - i_j}.$$

The transformation of a polynomial from its power form (1) into its *Bernstein form* results in

$$p(\mathbf{x}) = \sum_{I \leq N} b_I(\mathbf{U}) B_{N,I}(\mathbf{x}),$$

where the *Bernstein coefficients* $b_I(\mathbf{U})$ of p over $\mathbf{U}$ are given by

$$b_I(\mathbf{U}) = \sum_{J \leq I} \frac{\binom{I}{J}}{\binom{N}{J}} a_J, \quad I \leq N. \tag{2}$$

We collect the Bernstein coefficients in an array $B(\mathbf{U})$, i.e., $B(\mathbf{U}) = (b_I(\mathbf{U}))_{I \leq N}$. In analogy to Computer Aided Geometric Design we call $B(\mathbf{U})$ a *patch*. For an efficient calculation of the Bernstein coefficients, which does not use (2), see Garloff (1986). All rounding errors appearing in the computation of the Bernstein coefficients can be taken into account similarly as in Fischer (1990).

In the following lemma, we list some useful properties of the Bernstein coefficients. Property (i) was given by Cargo and Shisha (1966) and property (ii) by Farouki and Rajan (1988).

Lemma 1 *Let p be a polynomial (1) of degree N. Then the following properties hold for its Bernstein coefficients $b_I(\mathbf{U})$ (2):*

i) Range enclosing property:

$$\forall \mathbf{x} \in \mathbf{U} : \; \min_{I \leq N} b_I(\mathbf{U}) \leq p(\mathbf{x}) \leq \max_{I \leq N} b_I(\mathbf{U}) \tag{3}$$

with equality in the left (resp., right) inequality if and only if $\min_{I \leq N} b_I(\mathbf{U})$ (resp., $\max_{I \leq N} b_I(\mathbf{U})$) is attained at a Bernstein coefficient $b_I(\mathbf{U})$ with $i_k \in \{0, n_k\}, \; k = 1, \ldots, l$.

ii) Partial derivative:

$$\frac{\partial p}{\partial x_r}(\mathbf{x}) = n_r \sum_{I \leq N_{r,-1}} [b_{I_{r,1}}(\mathbf{U}) - b_I(\mathbf{U})] B_{N_{r,-1},I}(\mathbf{x}). \tag{4}$$

Lemma 2 *Let p be an l-variate polynomial and let $B(\mathbf{U})$ be the patch of its Bernstein coefficients on $\mathbf{U}$. Then the Bernstein coefficients of p on the m-dimensional faces of $\mathbf{U}$ are just the coefficients on the respective m-dimensional faces of the patch $B(\mathbf{U}), 0 \leq m \leq l - 1$.*

Proof: It is sufficient to prove the statement only for $m = l - 1$. Let $k \in \{1, \ldots, l\}$. In the sequel we indicate by (k) that the quantity under consideration is taken without the contribution of the kth component, e.g., $I_{(k)} = (i_1, \ldots, i_{k-1}, i_{k+1}, \ldots, i_l)$. The Ith Bernstein coefficient of

$$p(x_1, \ldots, x_{k-1}, 0, x_{k+1}, \ldots, x_l) = \sum_{I \leq N, i_k = 0} a_I(\mathbf{x}^I)_{(k)}$$

considered as a polynomial in $l - 1$ variables is given by

$$\sum_{J \leq I, j_k = 0} \frac{\binom{I}{J}_{(k)}}{\binom{N}{J}_{(k)}} a_J$$

which coincides with $b_{i_1 \ldots i_{k-1} 0 \, i_{k+1} \ldots i_l}$. Similarly, we obtain for $x_k = 1$

$$p(x_1, \ldots, x_{k-1}, 1, x_{k+1}, \ldots, x_l) = \sum_{I_{(k)} \leq N_{(k)}} c_{I_{(k)}}(\mathbf{x}^I)_{(k)},$$

where $c_{I_{(k)}} = \sum_{i_k=0}^{n_k} a_I$. The Ith Bernstein coefficient of this polynomial is given by

$$\sum_{J_{(k)} \leq I_{(k)}} \frac{\binom{I}{J}_{(k)}}{\binom{N}{J}_{(k)}} c_{J_{(k)}}. \tag{5}$$

On the other hand, we have

$$b_{i_1 \ldots i_{k-1} n_k i_{k+1} \ldots i_l} = \sum_{J \leq I, i_k = n_k} \frac{\binom{I}{J}}{\binom{N}{J}} a_J$$

which coincides with (5).

Remark: Application of Lemma 2 for bounding the range of p over an edge of $\mathbf{U}$ was given in the Edge Lemma in Zettler and Garloff (1998).

2.2 Sweep procedure

The bounds obtained by the inequalities (3) can be tightened if the unit box $\mathbf{U}$ is bisected into subboxes and Bernstein expansion is applied to the polynomial p on these subboxes, i.e., to the polynomial shifted from each subbox back to $\mathbf{U}$. A sweep in the rth direction ($1 \leq r \leq l$) is a bisection perpendicular to this direction and is performed by recursively applying a linear interpolation. Let

$$\mathbf{D} = [\underline{d}_1, \overline{d}_1] \times \ldots \times [\underline{d}_l, \overline{d}_l]$$

be any subbox of $\mathbf{U}$ generated by sweep operations (at the beginning, we have $\mathbf{D} = \mathbf{U}$). Starting with $B^{(0)}(\mathbf{D}) = B(\mathbf{D})$ we set for $k = 1, \ldots, n_r$

$$b_I^{(k)}(\mathbf{D}) = \begin{cases} b_I^{(k-1)}(\mathbf{D}) : i_r < k \\ (b_{I_{r,-1}}^{(k-1)}(\mathbf{D}) + b_I^{(k-1)}(\mathbf{D}))/2 : k \leq i_r. \end{cases}$$

To obtain the new coefficients, this is applied for $i_j = 0, \ldots, n_j$, $j = 1, \ldots, r - 1, r + 1, \ldots, l$. Then the Bernstein coefficients on $\mathbf{D}_0$, where the subbox $\mathbf{D}_0$ is given by

$$\mathbf{D}_0 = [\underline{d}_1, \overline{d}_1] \times \ldots \times [\underline{d}_r, \hat{d}_r] \times \ldots \times [\underline{d}_l, \overline{d}_l],$$

with $\hat{d}_r$ denoting the midpoint of $[\underline{d}_r, \overline{d}_r]$, are obtained as $B(\mathbf{D}_0) = B^{(n_r)}(\mathbf{D})$. The Bernstein coefficients $B(\mathbf{D}_1)$ on the neighbouring subbox $\mathbf{D}_1$

$$\mathbf{D}_1 = [\underline{d}_1, \overline{d}_1] \times \cdots \times [\hat{d}_r, \overline{d}_r] \times \cdots \times [\underline{d}_l, \overline{d}_r]$$

are obtained as intermediate values in this computation, since for $k = 0, \ldots, n_r$ the following relation holds (Garloff 1993):

$$b_{i_1, \ldots, n_r - k, \ldots, i_l}(\mathbf{D}_1) = b_{i_1, \ldots, n_r, \ldots, i_l}^{(k)}(\mathbf{D}).$$

A sweep needs $O(\hat{n}^{l+1})$ additions and multiplications, where $\hat{n} = \max\{n_i : i = 1, \ldots, l\}$, cf. Zettler and Garloff (1998). Note that by the sweep procedure the explicit transformation of the subboxes generated by the sweeps back to $\mathbf{U}$ is avoided. Fig. 1 illustrates the sweeping process for $l = 2$ and $r = 1$.

3 The method

Let n polynomials p_i, $i = 1, \ldots, n$, in the real variables $x_1, \ldots, x_n$ and a box $\mathbf{Q}$ in the $\mathbf{R}^n$ be given. We want to know the set of all solutions to the equations

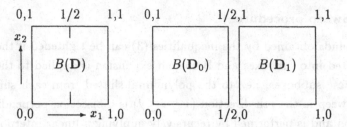

Fig. 1. Two new patches are obtained by a sweep in the first direction

$p_i(x) = 0, i = 1, \ldots, n$, within $\mathbf{Q}^1$. Without loss of generality we can assume that $\mathbf{Q}$ is the unit box.

Our procedure is very simple: We take away from $\mathbf{Q}$ all subboxes generated by sweeps for which there is a polynomial p_i being (strictly) positive or negative over the subbox. We check the sign of the polynomials by their Bernstein coefficients according to Lemma 1: If all Bernstein coefficients of a polynomial p_i are either positive or negative over a box, this box can not contain a solution.

After this pruning step we end up with a set of boxes of sufficiently small volume. All these boxes now undergo an existence test. In a first attempt we exploit the existence test given by Miranda (1941) which provides a generalisation of the fact that if a univariate continuous function f has a sign change at the endpoints of an interval then this interval contains a zero of f:

Theorem 1 (Miranda) *Let* $\mathbf{X} = [\underline{x}_1, \overline{x}_1] \times \ldots \times [\underline{x}_n, \overline{x}_n]$. *Denote by* $\mathbf{X}_i^- :=$ $\{\mathbf{x} \in \mathbf{X} \mid x_i = \underline{x}_i\}$ *and* $\mathbf{X}_i^+ := \{\mathbf{x} \in \mathbf{X} \mid x_i = \overline{x}_i\}$ *the pair of opposite parallel faces of* $\mathbf{X}$ *perpendicular to the ith coordinate direction.*

Let $\mathbf{F} = (f_1, \ldots, f_n)^T$ *be a continuous function defined on* $\mathbf{X}$. *If there is a permutation* $(v_1, \ldots, v_n)$ *of* $(1, \ldots, n)$ *such that*

$$f_i(\mathbf{x}) f_i(\mathbf{y}) \leq 0 \text{ for all } \mathbf{x} \in \mathbf{X}_{v_i}^-, \mathbf{y} \in \mathbf{X}_{v_i}^+, i = 1, \ldots, n, \tag{6}$$

then the equation $\mathbf{F}(\mathbf{x}) = 0$ *has a solution in* $\mathbf{X}$.

A short proof of Miranda's Theorem was given by Vrahatis (1989). An efficient method for checking all permutations is to be presented in Garloff and Smith (2001). Kioustelidis (1978), cf. Moore and Kioustelidis (1980) and Zuhe and Neumaier (1988), argued that the system $\mathbf{F}(\mathbf{x}) = 0$ should be preconditioned, i.e., it should be replaced by $A\mathbf{F}(\mathbf{x}) = 0$ with a suitably chosen matrix A. If $\mathbf{F}$ is differentiable on $\mathbf{X}$, a reasonable choice for A is to take an approximation to the inverse of the Jacobian of $\mathbf{F}$ at the midpoint of $\mathbf{X}$. If we apply

---

[1] If the number of the equations does not coincide with the number of the variables we could indeed find an enclosure for the set of the solutions. This enclosure would consist of a union of boxes. But we would not be able to check easily whether such a box contains a solution.

Miranda's Theorem to the given polynomial system and use Bernstein expansion we can then make use of the easy calculation of the Bernstein form of the partial derivatives of a polynomial from its Bernstein form , cf. (4). Furthermore, the test required in (6) costs nearly nothing since the Bernstein coefficients of p on the faces of $\mathbf{X}$ are known once the Bernstein coefficients of p on $\mathbf{X}$ are computed, cf. Lemma 2.

We employ a heuristic sweep direction selection rule in an attempt to mini-mise the total number of subboxes which need to be processed. Such a rule may favour directions in which polynomials have large partial derivatives and in which the box edge lengths are larger, to avoid repetitive sweeps in a single direction. The method is tested with the following direction selection rule variants:

- **C**: The direction is set equal to the subdivision depth modulo the number of variables, plus one, viz. each direction is afforded an equal bias and is chosen in sequence. This is used as a control rule.

- **D1**: We compute an upper bound for the absolute value of the partial derivative (from its Bernstein form, cf. Zettler and Garloff (1998)) for each direction on each polynomial (patch). In each direction, we sum these values over all polynomials, and select the direction for which the product of box edge length and partial derivative sum is maximal.

- **D2**: As **D1**, except that we take the maximum of the upper bounds for the absolute value of the partial derivatives over all polynomials for each direction, and then multiply by the box edge length, as before.

4 Examples

The method was tested for some of the sample problems from Sherbrooke and Patrikalakis (1993) (**SP**) and Jäger and Ratz (1995) (**JR**), see Table 1.

The maximum subdivision depth is chosen to achieve the same tolerance as used in **SP** and **JR**, respectively. In each case, we record in Table 2 the total number of boxes processed (which is equal to twice the number of sweep operations, plus one), the number of Miranda tests performed, and the execution time (averaged over 5 repeat runs). All examples were run on a PC equipped with a 450MHz Pentium III processor.

Some categories of problems seem to require subdivision in all directions equally; for these cases we observe no appreciable difference in the output data between the control and the derivative methods. In other cases we notice that

Table 1. Example problems

Name	Q	Tolerance	Max subdivision depth	#Solutions
SP1	$[0, 1]^2$	10^{-8}	53	1
SP2	$[0, 21]$	10^{-7}	28	20
SP3	$[0, 1]^2$	10^{-8}	53	9
SP4	$[0, 1]^6$	10^{-8}	159	4
SP5	$[0, 1]^2$	10^{-14}	93	1
JR2	$[-1, 1]^3$	10^{-12}	123	2
JR4	$[-1, 1]^3$	10^{-12}	123	8

Table 2. Results for example problems

Example	Method	C	D1	D2
SP1	Number of boxes	205	183	183
	Miranda tests	2	1	1
	Time	0.02	0.02	0.02
SP2	Number of boxes	983	identical results	
	Miranda tests	20	(sweep in one direction only)	
	Time	0.12		
SP3	Number of boxes	2493	2245	2245
	Miranda tests	28	20	20
	Time	0.29	0.27	0.29
SP4	Number of boxes	6901	6315	6789
	Miranda tests	15	12	9
	Time	11.99	10.15	8.69
SP5	Number of boxes	5141	5131	5131
	Miranda tests	31	32	32
	Time	0.80	0.80	0.85
JR2	Number of boxes	3705	3571	3655
	Miranda tests	16	16	16
	Time	0.77	0.77	0.82
JR4	Number of boxes	8759	5895	6173
	Miranda tests	27	22	22
	Time	3.05	2.23	2.49

the choice of the sweep direction based on the absolute value of the partial derivative is effective in reducing the overall number of boxes that are processed and the number of Miranda tests required. There is very little difference between the sum and maximum variants. The methods were also tested for a range of subdivision depths, and it is worth noting that by making a small change, a greater variance in the number of Miranda tests (and the time taken) between them may be observed. We do not present the results here, since they would require tolerances which would not coincide with those used in **SP** and **JR**. The timings compare mostly favourably to those reported by **SP** and **JR**, but we should note that the processor capability available to us is approximately an order of magnitude faster.

5 Conclusions

In this paper, we have presented a further application of Bernstein expansion. It is an advantage of this approach that continua of solutions can also be enclosed. With its range enclosing property, the Bernstein form provides an alternative to the narrowing operators used by Van Hentenryck et al. (1997), cf. Granvilliers (2000), most likely speeding up the algorithm presented therein. On the other hand, a preprocessing step as used by Van Hentenryck et al. (1997) seems to be required in order to avoid unnecessarily many bisections and in order to approach the vicinity of the solutions earlier.

References

Cargo, G.T., Shisha, O. (1966): The Bernstein form of a polynomial. J. Res. Nat. Bur. Standards Sect. B, vol. 70B (Math. Sci.), 1: 79-81

Farouki, R.T., Rajan, V.T. (1988): Algorithms for polynomials in Bernstein form. Computer Aided Geometric Design 5: 1-26

Fischer, H.C. (1990): Range computations and applications. In: Ullrich, C. (ed.): Contributions to computer arithmetic and self-validating numerical methods. J.C.Baltzer, Amsterdam, pp. 197-211

Garloff, J. (1986): Convergent bounds for the range of multivariate polynomials. In: Nickel, K. (ed.): Interval mathematics 1985. Springer, Berlin Heidelberg New York, pp. 37-56 (Lecture notes in computer science, vol. 212)

Garloff, J. (1993): The Bernstein algorithm. Interval Comp. 2: 154-168

Garloff, J. (2000): Applications of Bernstein expansion to the solution of control problems. Reliable Comp. 6: 303-320

Garloff, J., Graf, B. (1999): Solving strict polynomial inequalities by Bernstein expansion. In: Munro, N. (ed.): The use of symbolic methods in control system analysis and design. IEE, London, pp. 329-352

Garloff, J., Smith, A.P. (2001): Improvements of a subdivision-based algorithm for solving systems of polynomial equations, to appear in the Proceedings of the 3rd World Congress of Nonlinear Analysts, July 19-26 2000, Catania, Italy, special series of the Journal of Nonlinear Analysis, Elsevier Sci. Publ.

Granvilliers, L. (2000): Towards cooperative interval narrowing. In: Proceedings 3rd Intern. Workshop on Frontiers of Combining Systems, FroCoS'2000, Nancy, France, Springer, Berlin Heidelberg New York (Lecture notes in artificial intelligence, vol. 1794)

Hu, Chun-Yi, Maekawa, T., Sherbrooke, E.C., Patrikalakis, N.M. (1996): Robust interval algorithm for curve intersections. Computer-Aided Design 28:495-506

Jäger, C., Ratz, D. (1995): A combined method for enclosing all solutions of nonlinear systems of polynomial equations, Reliable Comp. 1: 41-64

Kearfott, R.B. (1996): Rigorous global search: continuous problems, Kluwer Acad. Publ., Dordrecht Boston London

Kioustelidis, J.B. (1978): Algorithmic error estimation for approximate solutions of nonlinear systems of equations. Computing 19: 313-320

Miranda, C. (1941): Un' osservazione su un teorema di Brouwer. Boll. Un. Mat. Ital. Ser.2, 3: 5-7

Moore, R.E., Kioustelidis, J.B. (1980): A simple test for accuracy of approximate solutions to nonlinear (or linear) systems. SIAM J. Numer. Anal. 17:521-529

Morgan, A.P.(1987): Solving polynomial systems using continuation for engineering and scientific problems. Prentice Hall, Englewood-Cliffs, N.J.

Neumaier, A. (1990): Interval methods for systems of equations. Cambridge University Press, Cambridge (Encyclopedia of mathematics and its applications)

Sherbrooke, E.C., Patrikalakis, N.M. (1993): Computation of the solutions of nonlinear polynomial systems. Computer Aided Geometric Design 10: 379-405

Van Hentenryck, P., McAllester, D., Kapur, D. (1997): Solving polynomial systems using a branch and prune approach. SIAM J. Numer. Anal. 34:797-827

Vrahatis, M.N. (1989): A short proof and a generalization of Miranda's existence theorem. Proc. Amer. Math. Soc. 107: 701-703

Zettler, M., Garloff, J. (1998): Robustness analysis of polynomials with polynomial parameter dependency using Bernstein expansion. IEEE Trans. Automat. Contr. 43:425-431

Zuhe, S., Neumaier, A. (1988): A note on Moore's interval test for zeros of nonlinear systems. Computing 40: 85-90

Symbolic-Algebraic Computations in a Modeling Language for Mathematical Programming

1 Introduction

AMPL is a language and environment for expressing and manipulating *mathematical programming* problems, i.e., minimizing or maximizing an algebraic objective function subject to algebraic constraints. The AMPL processor simplifies problems, as discussed in more detail below, but calls on separate *solvers* to actually solve problems. Solvers obtain information about the problems they solve, including first and, for some solvers, second derivatives, from the AMPL/solver interface library.

This paper gives an overview of symbolic and algebraic computations within the AMPL processor and its associated solver interface library. The next section gives a more detailed overview of AMPL. Section 3 discusses communications with solvers. Correctly rounded decimal-to-binary and binary-to-decimal conversions reduce one possible source of confusion and are discussed in section 4. An overview of AMPL's problem simplifications appears in section 5. Directed roundings help these simplifications, as described in section 6; a short discussion of the inconvenience this currently entails appears in section 7. Finally, section 8 provides concluding remarks and section 9 gives references.

2 AMPL overview

AMPL is a language and computing environment designed to simplify the tasks of stating, solving, and generally manipulating mathematical programming problems, such as the problem of finding $x \in \mathbb{R}^n$ to

$$\text{minimize } f(x)$$
$$\text{subject to } \ell \leq x \leq u \tag{1}$$

where $f: \mathbb{R}^n \to \mathbb{R}$ and $c: \mathbb{R}^n \to \mathbb{R}^m$ are algebraically defined functions.

AMPL began when Bob Fourer spent a sabbatical at Bell Labs. He had written in Fourer (1983) about the need for modeling languages in the context of mathematical programming, and his sabbatical came at a time of interest in "little languages" at Bell Labs — see Bentley (1986). AMPL permits stating problem (1) in a notation close to conventional mathematical notation that can by typed on an ordinary keyboard. In the spirit of a little language, AMPL does not solve problems itself, but rather translates them to a form that is easily manipulated with the help of the AMPL/solver interface library: see Gay (1997). AMPL can invoke various *solvers* that use the interface library to obtain information about (1), such as objective function values $f(x)$, constraint bodies $c(x)$, the constraint bounds ℓ and u, gradients $\nabla f(x)$, Lagrangian Hessians

$$W(x, \lambda) = \nabla^2 f(x) - \sum_{i=1}^{m} \lambda_i \nabla^2 c_i(x),$$

etc. After computing a solution, a solver uses the interface library to return solution information, such as "optimal" primal (x) and dual (λ) variable values, to the AMPL processor. A growing AMPL command language facilitates inspecting solutions and other problem data and modifying problems, perhaps to solve a sequence of problems.

One powerful feature of the AMPL modeling language is its ability to state problems involving sets of entities without knowledge of the specific values of the sets. This permits separating a *model*, i.e., a class of optimization problems parameterized by some fundamental sets and "parameters", from the data needed to specify a particular problem instance. Models usually involve both fundamental and derived sets and parameters; the derived entities are computed from fundamental entities or previously computed derived ones. The AMPL command language permits changing the values of fundamental entities, in which case derived entities are recomputed as necessary.

Though it started as a little language, by the early 1990s AMPL had grown sufficiently that we felt it reasonable to write the book of Fourer et al. (1993) about it. AMPL continues to grow. Extensions since the AMPL book appeared are described in the AMPL web site

http://www.ampl.com/ampl/

which contains much more information about AMPL, including pointers to various papers about it.

3 Communication with solvers

Solvers run as separate processes, possibly on remote machines. AMPL communicates with solvers via files; it encodes problem descriptions in ".nl" files, which include sparse-matrix representations of the nonzero structure and linear parts of constraints and objectives and expression graphs for nonlinear expressions. The AMPL/solver interface library offers several .nl file readers that read .nl files and may prepare for nonlinear function and derivative evaluations. Expression graphs are encoded in a Polish prefix form, but this detail is invisible outside of the .nl file readers.

The interface library's computations of first derivatives (gradients) proceed by backwards automatic differentiation, as Gay (1991) describes. Preparations for Hessian computations are more elaborate; they involve several "tree walks", i.e., passes over the expression graphs, in part to identify and exploit partially separable structure: objective functions (and constraint bodies) often have the form

$$f(x) = \sum_i f_i(A_i x), \tag{2}$$

in which each A_i matrix has just a few rows and represents a linear change of variables. As Griewank and Toint (1981, 1984) have pointed out, partially separable structure is well worth exploiting when one computes or approximates the Hessian matrix (of second partial derivatives) $\nabla^2 f$, which has the form

$$\nabla^2 f(x) = \sum_i A_i^T \nabla^2 f_i A_i. \tag{3}$$

Indeed, computing or approximating each term in (3) sometimes leads to substantially faster Hessian computations than would otherwise occur. Gay (1996) gives many more details about how the AMPL/solver interface library finds and exploits partially separable structure, including a more elaborate form, "group partial separability", that Conn et al. (1992) exploit in their solver LANCELOT:

$$f(x) = \sum_i f_i \left(\sum_{j \in S_i} \phi_{i,j}(A_{i,j}x) \right),$$

in which $\phi_{i,j}$ is a function of one variable.

Some solvers deal only with linear and quadratic objectives. The interface library provides a special reader for such solvers. It does a tree walk that extracts the (constant) Hessian of a quadratic function and complains if the function is nonlinear but not quadratic.

4 Binary ↔ decimal conversions

By default, AMPL writes binary .nl files, but it can also write equivalent ASCII (text) files. Binary files are faster to read and write, but ASCII files are more portable: they can be written by one kind of computer and read by another. Both kinds begin with 10 lines of text that provide problem statistics and a code that indicates the format in which the rest of the file is written.

Most current computers use binary "IEEE arithmetic", or at least the representation for floating-point numbers described in the IEEE (1985) arithmetic standard. The AMPL/solver interface library will automatically swap bytes if necessary so that a binary .nl file written on a machine that uses big-endian IEEE arithmetic can be read on a machine that uses little-endian IEEE arithmetic and vice versa.

To remove one source of confusion and inaccuracy and make binary and ASCII .nl files completely interchangeable, AMPL and its solver interface library use correctly rounded binary ↔ decimal conversions, which is now possible on all machines where AMPL has run other than old Cray machines. Details are described in Gay (1990).

Part of the reason for mentioning binary ↔ decimal conversions here is to point out a recent extension to Gay (1990) that carries out correctly rounded conversions for other arithmetics with properties similar to binary IEEE arithmetic. This includes correct directed roundings and rounding of a decimal string to a floating-point interval of width at most one unit in the last place, both of which are obviously useful for rigorous interval computations. There is no paper yet about this work, but the source files are available as

```
ftp://netlib.bell-labs.com/netlib/fp/gdtoa.tgz
```

which includes a README file for documentation.

5 Presolve

The AMPL processor simplifies problems in some ways before sending them to solvers, a process called "presolving". The original motivation was just to permit flexibility in stating bounds on variables: a solver should see the same problem

101

independently of whether bounds on a variable are stated in the variable's declaration or in explicit constraints. But we have incorporated all of the "primal" simplifications described by Brearley et al. (1975), since this sometimes permits diagnosing infeasibility without invoking a solver and it sometimes permits transmitting a significantly smaller problem to the solver. Moreover, while some solvers have their own presolver (generally also based on Brearley et al. (1975)), others do not, and these other solvers sometimes solve problems significantly faster after AMPL's presolve phase has acted.

Fourer and Gay (1994) describe much of AMPL's presolve algorithm, and Ferris et al. (1999) describe extensions to it for complementarity constraints. For this paper, a short overview will suffice. The simplification method of Brearley et al. (1975) applies to linear constraints and objectives; it proceeds to recursively

1. fold singleton rows into variable bounds;
2. omit slack inequalities;
3. deduce bounds from other rows;
4. deduce bounds on dual variables.

AMPL currently omits step 4, since in general there can be several objectives (e.g., for a multi-objective solver) and some solvers can be asked to reverse the sense of optimization, maximizing an objective declared as intended to be minimized and vice versa.

AMPL's presolver currently treats nonlinearities quite primitively, assuming simply that a nonlinear expression can produce values in all of $(-\infty, +\infty)$; this is clearly an area in which there is plenty of opportunity to do better. But even now it is possible for a variable that appears in a nonlinear expression to be "fixed", i.e., have its value determined, by other constraints, in which case AMPL replaces the nonlinear variable by a constant, which may turn previously nonlinear constraints into linear ones that can now participate in presolve simplifications.

In slightly more detail, AMPL maintains a stack of constraints to process, which permits folding simple bound constraints into variable bounds in linear time. For linear constraints involving more than one one variable, AMPL makes several passes, which amount to Gauß-Seidel iterations. For example, consider the constraints

$$x + y \geq 2$$
$$x - y \leq 0$$
$$0.1 \cdot x + y \leq 1.1 \tag{4}$$
$$x \geq 0$$

System (4) has a single feasible point, $(1,1)$, which the Gauß-Seidel iterations only approximate; AMPL's default 9 passes deduce the bounds

$$0.99999 \leq x \leq 1.00001,$$

$$0.99999 \leq y \leq 1.00001.$$

Such bounds, while an overestimate, sometimes permit deducing that other inequalities can never be tight — or are inconsistent.

Some solvers use an active-set strategy: they maintain a "basis" that involves inequality constraints currently holding as equalities. Degeneracy is said to occur when the choice of basis is not unique; solvers sometimes must carry out extra iterations when degeneracy occurs. Conveying the tightest deduced bounds to solvers can introduce extra degeneracy, so by default AMPL does not convey the tightest variable bounds it has deduced when those bounds are implied by other constraints that the solver sees. Solvers that do not use a basis, such as interior-point solvers, may not be bothered by degeneracy, and it is possible to tell AMPL to send the tightest deduced bounds to such solvers.

Table 1 illustrates several of the above points. It shows results for two variants of a small shipping problem, one (git2) with variable bounds stated in separate constraints, the other (git3) with variable bounds stated in variable declarations, and with various settings of AMPL's presolve and var_bounds options in the columns labeled ps and vb, respectively: $ps = 0$ omits all presolve simplifications; this is the only ps setting under which git2 and git3 differ in the solver's eyes. For $ps = 1$, presolve deductions involving two or more nonzeros per constraint are omitted, whereas for $ps = 10$ (the default, permitting 9 Gauß-Seidel iterations), they are allowed; only one iteration is required for this particular problem, but the stronger deductions do permit reducing the problem size. Under the (default) setting of $vb = 1$, variable bounds implied by constraints that the solver sees are not transmitted to the solver, whereas they are transmitted for $vb = 2$. The columns labeled m, n, and nz give the number of constraints, variables, and constraint nonzeros in the problem seen by the solver; in this particular example, the number of variables does not change, but it does in other examples. The time column shows execution times for MINOS 5.5, a nonlinear solver by Murtagh and Saunders (1982) that uses an active-set strategy and solves linear problems (such as the git problems) as a special case. MINOS benefits from AMPL's presolve phase, as it does not have its own presolver, and it is affected by the var_bounds setting. The times are CPU seconds on a machine with a 466 MHz DEC Alpha 21164A processor, running Red Hat Linux 6.0 (with compilation by the egcs-2.91.66 C compiler after conversion of the MINOS source from Fortran to C by the Fortran 77 to C converter f2c of Feldman et al. (1990)).

Table 1. Illustration of presolve settings (see text).

Problem	ps	vb	m	n	nz	iters	time
git2	0	1	1299	1089	4645	341	1.03
git3	0	1	410	1089	3756	383	0.39
git3	1	1	376	1089	3746	359	0.36
git3	10	1	286	1089	3051	324	0.30
git3	10	2	286	1089	3051	402	0.32

6 Directed rounding benefits

Rounding errors could lead to incorrect deductions in AMPL's presolve algorithm. Assume the i-th constraint has the form

$$b_i \leq f_i(x) \leq d_i. \tag{5}$$

AMPL's presolve algorithm deduces bounds $\tilde{b}_i \leq f_i(x) \leq \tilde{d}_i$ on $f_i(x)$ from bounds

103

on x; if the arithmetic were exact, then $\tilde{b}_i \geq b_i$ would imply that the lower inequality in (5) could be discarded without changing the set of feasible x values, and similarly $\tilde{d}_i \leq d_i$ would imply that the upper inequality in (5) could be discarded. Initially we attempted to cope with rounding errors by introducing a tolerance τ (option constraint_drop_tol, which is 0 by default) and requiring

$$\tilde{b}_i - b_i \geq \tau$$

or

$$d_i - \tilde{d}_i \geq \tau$$

before discarding the lower or upper inequality in (5). For example, on the *maros* test problem of *netlib*'s lp/data collection of Gay (1985), under binary IEEE arithmetic, $\tau = 10^{-13}$ suffices, whereas the default $\tau = 0$ leads to incorrect deductions. (Problems in *netlib*'s lp/data collection can be fed to AMPL with the help of a model, mps.mod, and *awk* script, m2a, that are available from *netlib*.)

Rather than requiring users to guess suitable values for τ, it is safer to use directed roundings in computing deduced bounds $\tilde{b}_i$ and $\tilde{d}_i$. On machines with IEEE arithmetic, AMPL has been using directed roundings in this context since the early 1990s. These roundings sometimes eliminate incorrect warnings about infeasibility.

Directed roundings can also affect the number of Gauß-Seidel iterations in AMPL's presolve algorithm. Table 2 shows the numbers of such iterations required without (column "near") and with directed roundings on the problems in *netlib*'s lp/data directory where directed roundings matter here.

Table 2. Presolve iterations with
IEEE nearest and directed rounding.

problem	Rounding	
	near	directed
80bau3b	10	9
blend	6	5
czprob	10	5
israel	6	5
kb2	5	4

Table 3 compares the performance of MINOS, running on the previously described machine, without and with directed roundings in AMPL's presolve algorithm on the problems in *netlib*'s lp/data directory where this rounding makes a difference. It is disappointing that the directed roundings often lead MINOS to take more time (CPU seconds) and iterations.

7 Directed rounding frustrations

The directed roundings discussed in the previous section are merely a special case of interval computations; directed roundings are obviously important for interval computations in general. Although machines that with IEEE arithmetic must provide directed roundings, the means of accessing them remain system dependent and sometimes even require use of assembly code. The recently updated C standard does

Table 3. MINOS times and iterations affected by presolve rounding on *netlib*'s lp/data problems.

problem	rounding	rows	cols	nonzeros	iters	time
greenbea	near	1962	4153	24480	15072	103
	dir.	2014	4270	24145	15342	116
greenbeb	near	1966	4151	24474	7942	54
	dir.	2033	4311	25393	8862	62
maros	near	691	1112	7554	1235	2.6*
	dir.	697	1127	7717	1504	3.3
perold	near	600	1276	5678	3395	11.1
	dir.	597	1269	5637	3388	8.5

* Unbounded

finally provide facilities for controlling rounding direction, but (as of this writing) these facilities are not yet widely available. Similarly, a forthcoming update to the Fortran standard will probably include control of the rounding direction (where possible), but portable specification of the rounding direction so far remains elusive.

The situation is even worse in the currently popular Java world. Although Java makes a big fuss about using part of the IEEE arithmetic standard, it makes no provision for directed roundings. To get them, one must resort to using the Java Native Interface to call functions written in another language.

8 Conclusion

AMPL permits separating a model, i.e., a symbolic representation of a class of problems, from the data required to specify a particular problem instance. Once AMPL has a problem instance, it can make simplifications before transmitting the problem to a solver; these simplifications sometimes benefit from use of directed roundings. Expression graphs sent to the solver can be manipulated by the AMPL/solver interface library to arrange for efficient gradient and Hessian computations. The net effect is that hidden symbolic and algebraic manipulations play a significant role in making life easier for AMPL users, who usually want to concentrate on formulating and using their intended mathematical programming problems, rather than worrying about arcane technical details.

Acknowledgement. I thank Bob Fourer for helpful comments on the manuscript.

9 References

IEEE Standard for Binary Floating-Point Arithmetic, Institute of Electrical and Electronics Engineers, New York, NY, 1985. ANSI/IEEE Std 754-1985.

Bentley, J. L. (Aug. 1986), ''Little Languages,'' *Communications of the ACM* **29** #8: 711–721.

Conn, A. R.; Gould, N. I. M.; and Toint, Ph. L., *LANCELOT, a Fortran Package for Large-Scale Nonlinear Optimization (Release A),* Springer-Verlag, 1992. Springer Series in Computational Mathematics 17.

Feldman, S. I.; Gay, D. M.; Maimone, M. W.; and Schryer, N. L., "A Fortran-to-C Converter," Computing Science Technical Report No. 149 (1990), Bell Laboratories, Murray Hill, NJ.

Ferris, Michael C.; Fourer, Robert; and Gay, David M. (1999), "Expressing Complementarity Problems in an Algebraic Modeling Language and Communicating Them to Solvers," *SIAM Journal on Optimization* **9** #4: 991–1009.

Fourer, R. (1983), "Modeling Languages Versus Matrix Generators for Linear Programming," *ACM Trans. Math. Software* **9** #2: 143–183.

Fourer, Robert; Gay, David M.; and Kernighan, Brian W., *AMPL: A Modeling Language for Mathematical Programming,* Duxbury Press/Wadsworth, 1993. ISBN: 0-89426-232-7.

Gay, D. M. (1985), "Electronic Mail Distribution of Linear Programming Test Problems," *COAL Newsletter* #13: 10–12.

Gay, D. M., "Correctly Rounded Binary-Decimal and Decimal-Binary Conversions," Numerical Analysis Manuscript 90-10 (11274-901130-10TMS) (1990), Bell Laboratories, Murray Hill, NJ.

Gay, David M., "Automatic Differentiation of Nonlinear AMPL Models," pp. 61–73 in *Automatic Differentiation of Algorithms: Theory, Implementation, and Application,* ed. A. Griewank and G. F. Corliss, SIAM (1991).

Gay, D. M., "More AD of Nonlinear AMPL Models: Computing Hessian Information and Exploiting Partial Separability," in *Computational Differentiation: Applications, Techniques, and Tools*, ed. George F. Corliss, SIAM (1996).

Gay, David M., "Hooking Your Solver to AMPL," Technical Report 97-4-06 (April, 1997), Computing Sciences Research Center, Bell Laboratories. See http://www.ampl.com/ampl/REFS/hooking2.ps.gz.

Griewank, A. and Toint, Ph. L., "On the Unconstrained Optimization of Partially Separable Functions," pp. 301–312 in *Nonlinear Optimization 1981*, ed. M. J. D. Powell, Academic Press (1982).

Griewank, A. and Toint, Ph. L. (1984), "On the Existence of Convex Decompositions of Partially Separable Functions," *Math. Programming* **28**: 25–49.

Murtagh, B. A. and Saunders, M. A. (1982), "A Projected Lagrangian Algorithm and its Implementation for Sparse Nonlinear Constraints," *Math. Programming Study* **16**: 84–117.

Translation of Taylor series into LFT expansions

Reinhold Heckmann

1 Introduction

In Exact Real Arithmetic, real numbers are represented as potentially infinite streams of information units, called *digits*. In this paper, we work in the framework of *Linear Fractional Transformations* (LFT's, also known as Möbius transformations) that provide an elegant approach to real number arithmetic (Gosper 1972, Vuillemin 1990, Nielsen and Kornerup 1995, Potts and Edalat 1996, Edalat and Potts 1997, Potts 1998b). One-dimensional LFT's are used as digits and to implement basic unary functions, while two-dimensional LFT's provide binary operations such as addition and multiplication, and can be combined to obtain infinite expression trees denoting transcendental functions. Peter Potts (1998a, 1998b) derived these expression trees from *continued fraction expansions* of the transcendental functions. In contrast, we show how to derive LFT expression trees from *power series*, which are available for a greater range of functions.

In Section 2, we present the LFT approach in some detail. Section 3 contains the main results of the paper. We first derive an LFT expansion from a power series using Horner's scheme (Section 3.1). The results are not very satisfactory. Thus, we show how LFT expansions may be modified using algebraic transformations (Section 3.2). A particular such transformation, presented in Section 3.3, yields satisfactory results for standard functions, as shown in the final examples section 4.

2 Exact real arithmetic by linear fractional transformations

In this section, we recall the framework of exact real arithmetic via LFT's (Gosper 1972, Vuillemin 1990, Nielsen and Kornerup 1995). We do not follow exactly the version used by the group of Edalat and Potts at Imperial College (Potts and Edalat 1996, Edalat and Potts 1997, Potts 1998a, Potts 1998b), but change the *base interval* from $[0, \infty]$ to $[-1, 1]$. The reasons for this change and its pros and cons were discussed in (Heckmann 2000).

2.1 LFT's and matrices

Linear Fractional Transformations (LFT's) are functions $x \mapsto \frac{ax+c}{bx+d}$ from reals to reals, parametrised by numbers a, b, c, and d. In general, these numbers may be reals, but in this paper, we shall only consider LFT's with integer parameters, as it is usually done in practical implementations of exact real arithmetic.

More precisely, LFT's are functions from $\mathbb{R}_\perp^*$ to $\mathbb{R}_\perp^*$ where $\mathbb{R}^* = \mathbb{R} \cup \{\infty\}$ is the one-point compactification of the real line, and $\mathbb{R}_\perp^* = \mathbb{R}^* \cup \{\perp\}$ is $\mathbb{R}^*$ plus an additional 'undefined' value $\perp$. The value ∞ arises as $r/0$ with $r \neq 0$, and on the other hand, the value of the LFT at ∞ is defined as a/b. The undefined value $\perp$ arises as $0/0$, and the value of any LFT at $\perp$ is $\perp$.

The four parameters of an LFT may be presented as a 2-2-matrix $M = \begin{pmatrix} a & c \\ b & d \end{pmatrix}$ of integers, hereafter *matrix*. Thus, any matrix $M = \begin{pmatrix} a & c \\ b & d \end{pmatrix}$ denotes an LFT $\langle M \rangle$ with $\langle M \rangle (x) = \frac{ax+c}{bx+d}$. Clearly, common factors of the four entries of the matrix do not matter; the matrices M and kM denote the same LFT if k is a non-zero integer. Thus, we get an equivalence $M \cong kM$ for $k \neq 0$. In particular, $M \cong -M$ holds. As a slight normalisation, we usually present matrices in a way such that the lower right entry is non-negative ($d \geq 0$). In the sequel, we shall identify matrices and LFT's, writing $M(x)$ instead of $\langle M \rangle (x)$.

Multiplication of matrices preserves equivalence and corresponds to composition of LFT's. LFT's with non-zero determinant (non-singular LFT's) map $\mathbb{R}^*$ to $\mathbb{R}^*$ and are invertible. An integer representation of the inverse is given by $\begin{pmatrix} a & c \\ b & d \end{pmatrix}^* = \begin{pmatrix} d & -c \\ -b & a \end{pmatrix}$.

In the LFT approach to Exact Real Arithmetic, LFT's (matrices) are used for two purposes: first, they implement some basic arithmetic functions, e.g., $M(x) = -x$ for $M = \begin{pmatrix} -1 & 0 \\ 0 & 1 \end{pmatrix}$ and $M(x) = 1/x$ for $M = \begin{pmatrix} 0 & 1 \\ 1 & 0 \end{pmatrix}$, and second, they are used to represent real numbers.

2.2 Representing reals by LFT's

We like to approximate real numbers by nested sequences of intervals. A sequence A_0, $A_1, \ldots$ of LFT's induces a sequence $I_1, I_2, \ldots$ of intervals which appear as the images of a fixed *base interval* B:

$$I_n = A_0 \cdots A_{n-1}(B). \tag{1}$$

For instance, with the standard base interval $B = [-1, 1]$, the matrices $A_0 = \begin{pmatrix} 1 & 2 \\ 0 & 1 \end{pmatrix}$ and $A_n = \begin{pmatrix} 1 & 1 \\ 0 & n+1 \end{pmatrix}$ for $n \geq 1$ induce the intervals

$I_1 = A_0(B) = \begin{pmatrix} 1 & 2 \\ 0 & 1 \end{pmatrix} ([-1, 1]) = [1, 3],$

$I_2 = A_0 A_1(B) = \begin{pmatrix} 1 & 2 \\ 0 & 1 \end{pmatrix} \begin{pmatrix} 1 & 1 \\ 0 & 2 \end{pmatrix} (B) = \begin{pmatrix} 1 & 5 \\ 0 & 2 \end{pmatrix} ([-1, 1]) = [2, 3],$

$I_3 = A_0 A_1 A_2(B) = \begin{pmatrix} 1 & 5 \\ 0 & 2 \end{pmatrix} \begin{pmatrix} 1 & 1 \\ 0 & 3 \end{pmatrix} (B) = \begin{pmatrix} 1 & 16 \\ 0 & 6 \end{pmatrix} ([-1, 1]) = [\frac{15}{6}, \frac{17}{6}] \subseteq [2.5, 2.834],$

$I_4 = A_0 A_1 A_2 A_3(B) = \begin{pmatrix} 1 & 16 \\ 0 & 6 \end{pmatrix} \begin{pmatrix} 1 & 1 \\ 0 & 4 \end{pmatrix} (B) = \begin{pmatrix} 1 & 65 \\ 0 & 24 \end{pmatrix} ([-1, 1]) = [\frac{64}{24}, \frac{66}{24}] \subseteq [2.666, 2.75],$

...

The sequence of matrices used in this example will be obtained in Section 4.1 as a representation of the number e. This means that the induced sequence of intervals $(I_n)_{n \geq 1}$ is nested with intersection $\{e\}$. From this representation, a representation of, say, $1/e$ can be easily obtained by putting the matrix $R = \begin{pmatrix} 0 & 1 \\ 1 & 0 \end{pmatrix}$ with $R(x) = 1/x$ at the front. For instance, the second interval of the resulting interval sequence will then be

$RA_0(B) = \begin{pmatrix} 0 & 1 \\ 1 & 0 \end{pmatrix} \begin{pmatrix} 1 & 2 \\ 0 & 1 \end{pmatrix} (B) = \begin{pmatrix} 0 & 1 \\ 1 & 2 \end{pmatrix} ([-1, 1]) = [1/3, 1].$

We now return to the general case. For the purposes of the arithmetic, it is important that the sequence of intervals $(I_n)_{n \geq 1}$ is *nested*, i.e., $I_n \supseteq I_{n+1}$ holds for all $n \geq 1$. This inclusion means $A_0 \cdots A_{n-1}(B) \supseteq A_0 \cdots A_n(B)$. If the matrices $A_0, \ldots, A_{n-1}$ are non-singular, this is equivalent to $A_n(B) \subseteq B$. Matrices with this property are called *refining* (w.r.t. B). (This notion was studied in (Edalat and Potts 1997) for arbitrary intervals B.) Note that A_0 need not be refining to obtain a nested sequence of intervals.

Later, it will be useful to represent intervals by center c and radius r, written as $[c \pm r]$:

$$[c \pm r] = \{x \in \mathbb{R} \mid |x - c| \le |r|\} . \tag{2}$$

Note that we allow negative radii with the convention $[c \pm r] = [c \pm (-r)] = [c \pm |r|]$. In the formulae above, this was realised by using $|r|$ on the right hand side. Negative radii will prove convenient when dealing with power series, where the individual summands may be negative as well as positive.

A nested sequence of intervals $I_n = [c_n \pm r_n]$ converges to a single point x iff $r_n \to 0$ as $n \to \infty$. Even if the sequence of intervals $[c_n \pm r_n]$ is not nested, it may converge to a single point x in the sense that $c_n \to x$ and $r_n \to 0$. But this is not sufficient to say that the intervals approximate x (consider for instance the case $I_n = [2/n \pm 1/n] = [1/n, 3/n]$). Therefore, we require the interval sequence to be nested. In theory, it is sufficient that the interval sequence is *eventually nested*, i.e., $I_n \supseteq I_{n+1}$ for almost all n, because we may omit the first few intervals and obtain a nested sequence. Yet this procedure is not always practical as we shall see later.

In the sequel, we shall write $I_n \to y$ if the two end points converge to y, $I_n \downarrow y$ if $I_n \to y$ and the intervals are eventually nested, and $I_n \Downarrow y$ if $I_n \to y$ and the intervals are nested from the beginning.

To prove convergence of the sequence of intervals induced by a sequence of matrices, the notion of contractivity is helpful. The *contractivity* of a refining matrix M is defined as $\sup_{x \ne y \in B} \frac{|M(x) - M(y)|}{|x - y|}$. If the contractivity of M is c, then the radius r' of the image $[x', r'] = M([x, r])$ of an interval $[x, r] \subseteq B$ is bounded by $c \cdot r$. Hence, the interval sequence $I_n = A_0 \cdots A_{n-1}(B)$ converges if the infinite product of the contractivities of the matrices $(A_n)_{n \ge 0}$ is 0. The contractivities of the matrices provide also information on how many matrices must be considered to obtain an interval I_n whose radius is below some given threshold.

A sequence of matrices that generates a nested sequence of intervals converging to x may be considered as a representation of x. Because of the usage of matrix multiplication in (1), we write $x = \prod_{n=0}^{\infty} A_n$, using a (formal) infinite product notation. Many real numbers can be elegantly represented by such infinite products.

To control the information flow in computations with reals, it turned out to be useful to convert these representations into a kind of standard form, using a leading *exponent matrix* (Heckmann 2000) or *sign matrix* followed by a sequence of *digit matrices* (Potts and Edalat 1997, Edalat and Potts 1997). For the purposes of the present paper, it is not necessary to delve into the details of these special representations. Instead, we may assume that real numbers are represented by arbitrary infinite products where all matrices but the first one are refining (to obtain a nested sequence of intervals) with contractivities bounded by some $c < 1$ (to ensure the interval sequence shrinks to a single point).

2.3 A note on affine LFT's

Affine LFT's form an important special case. An LFT $M = \begin{pmatrix} a & c \\ b & d \end{pmatrix}$ is called *affine* if $b = 0$ and $d \neq 0$. By normalisation, any affine LFT can be written as $\begin{pmatrix} a & c \\ 0 & 1 \end{pmatrix}$ which we'll sometimes abbreviate as $\langle c; a \rangle$; the corresponding function maps x to $c + ax$. Composition (multiplication) and inversion of affine LFT's are obtained as follows:

$$\langle c; a \rangle \cdot \langle c'; a' \rangle = \langle c + ac'; aa' \rangle \tag{3}$$

$$\langle c; a \rangle^* = \langle -c/a; 1/a \rangle \quad \text{if } a \neq 0 \tag{4}$$

Note that the image of the base interval $B = [-1, 1] = [0 \pm 1]$ under $\langle c; a \rangle$ is $[c \pm a]$. Recall that an LFT M is *refining* if $M(B) \subseteq B$. Thus,

$$\langle c; a \rangle \text{ is refining} \iff [c \pm a] \subseteq B \iff c - |a| \geq -1 \ \& \ c + |a| \leq 1$$

$$\iff -c \leq 1 - |a| \ \& \ c \leq 1 - |a| \iff |c| + |a| \leq 1 \tag{5}$$

Consequently, $\begin{pmatrix} a & c \\ 0 & d \end{pmatrix}$ is refining iff $|a| + |c| \leq |d|$.

The contractivity of $\begin{pmatrix} a & c \\ 0 & d \end{pmatrix}$ is $|a/d|$, and hence the contractivity of $\langle c; a \rangle$ is $|a|$.

2.4 Tensors

To represent sums, products, and quotients, *two-dimensional LFT's* are employed. They are characterised by 8 integer parameters, and thus can be represented by 2-4-matrices of integers, called *tensors*. A tensor $T = \begin{pmatrix} a & c & e & g \\ b & d & f & h \end{pmatrix}$ denotes the bistrict function $T :$ $\mathbb{R}_1^* \times \mathbb{R}_1^* \to \mathbb{R}_1^*$ given by $T(x, y) = \frac{axy + cx + ey + g}{bxy + dx + fy + h}$. Again, common factors do not matter. For $T = \begin{pmatrix} 0 & 1 & 1 & 0 \\ 0 & 0 & 0 & 1 \end{pmatrix}$, we get $T(x, y) = x + y$. Similarly, $\begin{pmatrix} 1 & 0 & 0 & 0 \\ 0 & 0 & 0 & 1 \end{pmatrix}$ realises multiplication and $\begin{pmatrix} 0 & 1 & 0 & 0 \\ 0 & 0 & 1 & 0 \end{pmatrix}$ division. Given arguments x and y represented by infinite products $\prod_{n=0}^{\infty} A_n$ and $\prod_{n=0}^{\infty} B_n$, there are methods to absorb the matrices A_n and B_n gradually into a tensor T and to emit matrices C_n from the tensor, thereby producing a representation $\prod_{n=0}^{\infty} C_n$ for the result $T(x, y)$.

For fixed x, the function $T|_x := (y \mapsto T(x, y))$ is an LFT with $T|_x = \begin{pmatrix} ax + e & cx + g \\ bx + f & dx + h \end{pmatrix}$. Conversely, any matrix M_x whose entries linearly depend on a parameter x can be turned into a tensor T such that $M_x(y) = T(x, y)$. The representations of the values of transcendental functions will generally involve such parametrised matrices. For instance, we shall later derive the representation

$$e^x = \begin{pmatrix} x & x+1 \\ 0 & 1 \end{pmatrix} \prod_{n=1}^{\infty} \begin{pmatrix} x & x \\ 0 & n+1 \end{pmatrix}$$

from the Taylor series $e^x = \sum_{n=0}^{\infty} \frac{x^n}{n!}$. Such a product of parametrised matrices is open to two interpretations: For rational arguments x, it is (equivalent to) an infinite product of integer matrices, e.g., $e = \begin{pmatrix} 1 & 2 \\ 0 & 1 \end{pmatrix} \prod_{n=1}^{\infty} \begin{pmatrix} 1 & 1 \\ 0 & n+1 \end{pmatrix}$. For general (real) arguments however, it should be turned into the infinite tensor expression $e^x = T_0(x, T_1(x, T_2(x, \ldots)))$ with $T_0 = \begin{pmatrix} 1 & 1 & 0 & 1 \\ 0 & 0 & 0 & 1 \end{pmatrix}$ and $T_n = \begin{pmatrix} 1 & 1 & 0 & 0 \\ 0 & 0 & 0 & n+1 \end{pmatrix}$.

3 From power series to LFT expansions

We first recall the basic properties of power series. Let $a_0, a_1, \ldots$ be a sequence of real numbers which is used to form the power series $\sum_{n=0}^{\infty} a_n x^n$ for $x \in \mathbb{R}$. The number $R = 1/\limsup_{n\to\infty} \sqrt[n]{|a_n|}$ is called *convergence radius*. If $|x| < R$, then $\sum_{n=0}^{\infty} a_n x^n$ converges, for $|x| > R$, it diverges, while for $|x| = R$, anything can happen. If almost all coefficients a_n are non-zero and $|a_n/a_{n+1}|$ converges, its limit is R.

3.1 The Horner evaluation of power series

Horner's rule to evaluate polynomials can easily be adapted to obtain an LFT expansion from a power series $f(x) = \sum_{n=0}^{\infty} a_n x^n$ with convergence radius R:

$$\sum_{n=0}^{\infty} a_n x^n = a_0 + x \cdot \sum_{n=1}^{\infty} a_n x^{n-1} = \langle a_0; x \rangle \left(\sum_{n=0}^{\infty} a_{n+1} x^n \right) \tag{6}$$

with the affine LFT $\langle a_0; x \rangle = (y \mapsto a_0 + xy)$. This suggests that $f(x)$ is given by

$$f(x) = \prod_{n=0}^{\infty} \langle a_n; x \rangle = \prod_{n=0}^{\infty} \begin{pmatrix} x & a_n \\ 0 & 1 \end{pmatrix} \tag{7}$$

To verify this guess, we first determine $A_0 \cdots A_{n-1}$ for $A_i = \langle a_i; x \rangle$:

$$\langle a_0; x \rangle \cdots \langle a_{n-1}; x \rangle = \left\langle \sum_{i=0}^{n-1} a_i x^i; x^n \right\rangle \tag{8}$$

This is obvious for $n = 1$. The inductive step follows from $\langle \sum_{i=0}^{n-1} a_i x^i; x^n \rangle \langle a_n; x \rangle = \langle \sum_{i=1}^{n-1} a_i x^i + x^n \cdot a_n; x^n \cdot x \rangle$ by (3).

From (8), we immediately obtain $I_n = A_0 \cdots A_{n-1}(B) = [s_n \pm x^n]$ where $s_n = \sum_{i=0}^{n-1} a_i x^i$. For $|x| < R$, we know $s_n \to f(x)$ as $n \to \infty$, and $x^n \to 0$ holds if and only if $|x| < 1$. Thus, $I_n \to f(x)$ holds for $|x| < \min(R, 1)$. For $|x| \geq 1$, the radii x^n do not tend to 0. Thus, we have already lost part of the convergence area if $R > 1$.

The LFT's $\langle a_n; x \rangle$ are non-singular iff $x \neq 0$. For $x = 0$, we have $f(x) = a_0$ and $I_n = I_1 = [a_0 \pm 0]$. Thus, we always have a nested sequence of intervals converging to the right value in this singular situation.

The LFT's $\langle a_n; x \rangle$ are refining iff $|a_n| + |x| \leq 1$. Thus in case $x \neq 0$, the generated sequence of intervals is nested iff $|x| \leq 1 - |a_n|$ for all $n \geq 1$, iff $|x| \leq 1 - \sup_{n\geq 1} |a_n|$, and it is eventually nested if $|x| < 1 - \limsup_{n\to\infty} |a_n|$.

Summary of the properties of the Horner evaluation: For $|x| < R$, we have

- $I_n \to f(x)$ if $|x| < 1$;

- $I_n \downarrow f(x)$ if $|x| < 1 - \limsup_{n\to\infty} |a_n|$;

- $I_n \Downarrow f(x)$ if $|x| \leq 1 - \sup_{n\geq 1} |a_n|$.

Let us consider some examples:

111

- For $f(x) = \frac{1}{1-x} = \sum_{n=0}^{\infty} 1 \cdot x^n$ with convergence radius $R = 1$, the LFT expansion is $\prod_{n=0}^{\infty} \langle 1; x \rangle = \prod_{n=0}^{\infty} \left(\begin{smallmatrix} x & 1 \\ 0 & 1 \end{smallmatrix} \right)$. The generated intervals are $I_n = [\sum_{i=0}^{n-1} x^i \pm x^n] = [\frac{1-x^n}{1-x} \pm x^n]$. For $|x| < 1$, the center points converge to $\frac{1}{1-x}$ and the radii to 0; thus both end points converge to the correct value $\frac{1}{1-x}$. Yet A_n is refining only if $x = 0$. Thus, for $x \neq 0$, the sequence of intervals is not nested—not even eventually nested. In fact, for $0 < x < 1$,

$$\left| \frac{1}{1-x} - \frac{1-x^n}{1-x} \right| = \frac{|x^n|}{|1-x|} > |x^n|$$

which means that $f(x) = \frac{1}{1-x}$ is not even contained in any of the intervals I_n!

- For $f(x) = e^x = \sum_{n=0}^{\infty} \frac{1}{n!} \cdot x^n$ with convergence radius $R = \infty$, the LFT expansion is $\prod_{n=0}^{\infty} \langle \frac{1}{n!}; x \rangle = \prod_{n=0}^{\infty} \left(\begin{smallmatrix} n!x & 1 \\ 0 & n! \end{smallmatrix} \right)$ with uncomfortably big entries. The general theory tells us that the end points of I_n tend to e^x iff $|x| < 1$ (since for $|x| \geq 1$, the radii do not converge to 0). Matrix A_n is refining iff $|x| \leq 1 - \frac{1}{n!}$. On the negative side, this means that the intervals are only nested if $x = 0$, since A_1 is only refining if $x = 0$. On the positive side, it means that for $|x| < 1$, the interval sequence is eventually nested since $|x| \leq 1 - \frac{1}{n!}$ will hold for sufficiently large n. Hence, in contrast to the first example, the function value e^x is contained in almost all intervals I_n if $|x| < 1$.

3.2 Transformation of the LFT expansion

The LFT expansions derived from Horner's rule in the previous section are not very good: they are rarely nested, loose out part of the convergence circle, and contain uncomfortably large entries for important functions like e^x. There is a general way to transform LFT expansions which may lead to improvements.

Given a sequence $A_0, A_1, A_2, \ldots$ of matrices, the corresponding sequence of intervals $I_1, I_2, \ldots$ is given by $I_n = P_n(B)$ ($n \geq 1$) where $P_n = A_0 \cdots A_{n-1}$ is the product of the first n matrices. Now, we may choose any sequence $U_1, U_2, \ldots$ of non-singular matrices and throw in products $U_i U_i^*$ of U_i and its inverse U_i^* into the original product:

$$P_n = A_0 \cdots A_{n-1} = A_0 U_1 U_1^* A_1 U_2 U_2^* \cdots U_{n-1} U_{n-1}^* A_{n-1}$$

With the new names $Q_n = P_n U_n$ for $n \geq 1$, $B_0 = A_0 U_1$, and $B_i = U_i^* A_i U_{i+1}$ for $i \geq 1$, we obtain

$$Q_n = P_n U_n = B_0 B_1 \cdots B_{n-1}.$$

This means we have defined a new sequence $B_0, B_1, \ldots$ of matrices which denotes the sequence of intervals $J_n = Q_n(B)$.

We want to apply this technique to the affine matrices $A_n = \langle a_n; x \rangle$ coming from Horner's rule. Since we want to preserve affinity here, we choose affine matrices $U_n = \langle c_n; r_n \rangle$ (with $r_n \neq 0$ for non-singularity). Then

$$A_n U_{n+1} = \langle a_n; x \rangle \langle c_{n+1}; r_{n+1} \rangle = \langle a_n + c_{n+1} x; r_{n+1} x \rangle$$

Therefore
$$B_0 = A_0 U_1 = \langle a_0 + c_1 x; r_1 x \rangle$$
while multiplication with U_n^* from the left yields
$$B_n = \left\langle -\frac{c_n}{r_n}; \frac{1}{r_n} \right\rangle \langle a_n + c_{n+1}x; r_{n+1}x \rangle = \left\langle \frac{a_n - c_n + c_{n+1}x}{r_n}; \frac{r_{n+1}}{r_n} x \right\rangle$$
The matrices Q_n are given by
$$Q_n = P_n U_n = \langle s_n; x^n \rangle \langle c_n; r_n \rangle = \langle s_n + c_n x^n; r_n x^n \rangle$$
so that the denoted intervals are $J_n = [(s_n + c_n x^n) \pm r_n x^n]$. Note that for $x = 0$, we have $J_n = [a_0 \pm 0] \Downarrow a_0 = f(0)$.

3.3 The standard choice for c_n and r_n

There are two possibilities to reduce the term $a_n - c_n + c_{n-1}x$ in B_n to just one summand, namely $c_n = 0$ and $c_n = a_n$. In the first case, J_n becomes $[s_n \pm r_n x^n]$, while in the second, we have $J_n = [(s_n + a_n x^n) \pm r_n x^n] = [s_{n+1} \pm r_n x^n]$. This shows that the two possibilities are actually not that different. We prefer the second possibility because usually, its center s_{n+1} provides a better approximation to $f(x)$ than s_n.

The choice $c_n = a_n$ yields
$$B_0 = \begin{pmatrix} r_1 x & a_0 + a_1 x \\ 0 & 1 \end{pmatrix} \quad \text{and} \quad B_n = \begin{pmatrix} r_{n+1}x & a_{n+1}x \\ 0 & r_n \end{pmatrix} \quad \text{for } n \geq 1.$$
The resulting intervals were already determined to be $J_n = [s_{n+1} \pm r_n x^n]$. Thus, $J_n \to f(x)$ if $|x| < R$ and $r_n x^n \to 0$. The matrices B_n are refining for $|x| \leq \frac{|r_n|}{|r_{n+1}| + |a_{n+1}|}$.
We now come to a choice for r_n. If all a_n for $n \geq 1$ are non-zero (a_0 may be 0), we may choose $r_n = k a_n$ with some constant $k > 0$. This leads to
$$B_0 = \begin{pmatrix} k a_1 x & a_0 + a_1 x \\ 0 & 1 \end{pmatrix} \quad \text{and} \quad B_n = \begin{pmatrix} k a_{n+1}x & a_{n+1}x \\ 0 & k a_n \end{pmatrix} = \begin{pmatrix} kx & x \\ 0 & k q_n \end{pmatrix} \quad \text{for } n \geq 1$$
where $q_n = \frac{a_n}{a_{n+1}}$. The resulting intervals are $J_n = [s_{n+1} \pm k a_n x^n]$. If $|x| < R$, then $\sum_{n=0}^{\infty} a_n x^n$ converges, which implies $a_n x^n \to 0$. Thus $J_n \to f(x)$ holds for all $|x| < R$.

The matrices B_n are refining for $|x| \leq \frac{|k a_n|}{|k a_{n+1}| + |a_{n+1}|} = \frac{k}{k+1} |q_n|$. Thus the interval sequence is eventually nested if $|x| < \frac{k}{k+1} \liminf_{n \to \infty} |q_n|$. Now if $|q_n|$ converges, its limit is R, so that the condition for eventually nesting becomes $|x| < \frac{k}{k+1} R$. The condition for a nested sequence is similar, with lim inf replaced by inf.

- $J_n \to f(x)$ for $|x| < R$;
- $J_n \downarrow f(x)$ for $|x| < \frac{k}{k+1} \liminf_{n \to \infty} |q_n|$ (or $|x| < \frac{k}{k+1} R$ if $|q_n|$ converges);
- $J_n \Downarrow f(x)$ for $|x| \leq \frac{k}{k+1} \inf_{n \geq 1} |q_n|$.

The factor $\frac{k}{k+1}$ increases with k, yet the simplest possibility is $k = 1$ with $\frac{k}{k+1} = \frac{1}{2}$. Values of $|x|$ near R should be avoided anyway since the contractivity of B_n is $|x|/|q_n|$ which converges to $|x|/R$ if $|q_n|$ is convergent; in this case, we have poor contractivity for $|x| \approx R$, and the restriction to $|x| \leq \frac{1}{2} R$ is reasonable.

113

4 Case studies

Throughout this section, we use the name q for $\inf_{n \geq 1} |q_n|$ which is important for determining the region where a nested sequence of intervals is obtained.

4.1 Exponential function

Recall $e^x = \sum_{n=0}^{\infty} \frac{1}{n!} x^n$. Thus $q_n = \frac{a_n}{a_{n+1}} = \frac{1/n!}{1/(n+1)!} = n+1$, whence $R = \infty$, while $q = \inf_{n \geq 1} |q_n| = 2$. In the standard approach with $k = 1$, we obtain

$$
e^x = \begin{pmatrix} x & x+1 \\ 0 & 1 \end{pmatrix} \prod_{n=1}^{\infty} \begin{pmatrix} x & x \\ 0 & n+1 \end{pmatrix} \qquad \text{nested for } |x| \leq 1. \tag{9}
$$

All B_n with $n \geq 1$ have contractivity $\frac{|x|}{n+1} \leq \frac{1}{n+1}$, which seems reasonable. The function can be extended from $[-1, 1]$ to the whole real line using the equation $e^{2x} = (e^x)^2$.

With a higher value of k, the area where the intervals are nested can be increased to $\frac{k}{k+1} q = \frac{2k}{k+1}$, but this is still less than $q = 2$. With $k = 3$ for instance, we obtain

$$
e^x = \begin{pmatrix} 3x & x+1 \\ 0 & 1 \end{pmatrix} \prod_{n=1}^{\infty} \begin{pmatrix} 3x & x \\ 0 & 3(n+1) \end{pmatrix} \qquad \text{nested for } |x| \leq \frac{3}{2}.
$$

The contractivity is not affected by varying k. Even with $k = 1$, representation (9) yields an eventually nested convergent sequence of intervals for all x; the condition $|x| \leq 1$ is only needed to ensure that the sequence is nested from the beginning.

As already mentioned, a representation such as (9) is open to two different interpretations. For rational arguments x, it is (equivalent to) an infinite product of integer matrices, e.g., $e = \begin{pmatrix} 1 & 2 \\ 0 & 1 \end{pmatrix} \prod_{n=1}^{\infty} \begin{pmatrix} 1 & 1 \\ 0 & n+1 \end{pmatrix}$. In the rational case, it is not absolutely necessary that all matrices but the first one are refining. Consider for instance $x = 2$, where the first three matrices $\begin{pmatrix} 2 & 3 \\ 0 & 1 \end{pmatrix}$, $\begin{pmatrix} 2 & 2 \\ 0 & 2 \end{pmatrix}$, and $\begin{pmatrix} 2 & 2 \\ 0 & 3 \end{pmatrix}$ are not refining, while the next matrix $\begin{pmatrix} 2 & 2 \\ 0 & 4 \end{pmatrix}$ and all subsequent ones are refining. We may multiply the first three matrices into one which gives $e^2 = \begin{pmatrix} 4 & 19 \\ 0 & 3 \end{pmatrix} \prod_{n=1}^{\infty} \begin{pmatrix} 2 & 2 \\ 0 & n+3 \end{pmatrix}$ which generates a nested sequence of intervals.

For general (real) arguments however, representation (9) should be turned into the infinite tensor expression $e^x = T_0(x, T_1(x, T_2(x, \ldots)))$ with

$$
T_0 = \begin{pmatrix} 1 & 1 & 0 & 1 \\ 0 & 0 & 0 & 1 \end{pmatrix} \quad \text{and} \quad T_n = \begin{pmatrix} 1 & 1 & 0 & 0 \\ 0 & 0 & 0 & n+1 \end{pmatrix}
$$

In this case, the above trick to obtain a nested sequence of intervals even for $|x| > 1$ cannot be applied because there is no way to multiply the first few tensors into one (at least in the standard LFT approach). If you try to multiply the corresponding matrices, whose entries depend linearly on x, the entries of the product matrix will be non-linear polynomials in x.

4.2 Cosine and hyperbolic cosine

The power series of the cosine function is $\cos x = \sum_{n=0}^{\infty} \frac{(-1)^n}{(2n)!}(x^2)^n$. By writing this in terms of x^2 instead of x, zero coefficients are avoided. We get $a_0 = 1$, $a_1 = -\frac{1}{2}$, and $q_n = -(2n+1)(2n+2)$, whence $q = 12$, which for the standard case gives a bound of 6 for a nested sequence. Hence with a bit of scaling

$$\cos x = \begin{pmatrix} -u & 2-u \\ 0 & 2 \end{pmatrix} \prod_{n=1}^{\infty} \begin{pmatrix} -u & -u \\ 0 & 2(n+1)(2n+1) \end{pmatrix} \quad \text{nested for } u = x^2 \le 6.$$

The function can be extended to the whole real line by $\cos(2x) = 2\cos^2 x - 1$, or more efficiently, $\cos(x \pm 2\pi) = \cos x$.

The power series for cosh is similar to that for cos: $\cosh x = \sum_{n=0}^{\infty} \frac{1}{(2n)!}(x^2)^n$. Hence, the results look similar; only replace $-u$ by u throughout. The extension formula is again $\cosh(2x) = 2\cosh^2 x - 1$.

4.3 Sine and hyperbolic sine

The power series for sine is $\sin x = \sum_{n=0}^{\infty} \frac{(-1)^n x^{2n+1}}{(2n+1)!}$. Here, the exponents of x are odd. They can be expressed in terms of x^2 if one factor of x is taken out: $\frac{\sin x}{x} = \sum_{n=0}^{\infty} \frac{(-1)^n}{(2n+1)!}(x^2)^n$. We get $a_0 = 1$, $a_1 = -\frac{1}{3}$, and $q_n = -(2n+2)(2n+3)$, whence $q = 20$, and thus

$$\frac{\sin x}{x} = \begin{pmatrix} -u & 3-u \\ 0 & 3 \end{pmatrix} \prod_{n=1}^{\infty} \begin{pmatrix} -u & -u \\ 0 & 2(n+1)(2n+3) \end{pmatrix} \quad \text{nested for } u = x^2 \le 10.$$

For extension to the whole real line, the formula $\sin(x \pm 2\pi) = \sin x$ may be used. Again, sinh is analogous; just replace $-u$ by u.

4.4 Inverse of tangent

The power series for $\arctan x$ is $\sum_{n=0}^{\infty} \frac{(-1)^n x^{2n+1}}{2n+1}$. As with $\sin x$, we consider $\frac{\arctan x}{x} = \sum_{n=0}^{\infty} \frac{(-1)^n}{2n+1}(x^2)^n$ instead. Here, $a_0 = 1$, $a_1 = -\frac{1}{3}$, $q_n = -\frac{2n+3}{2n+1}$, whence $q = 1$, and thus

$$\frac{\arctan x}{x} = \begin{pmatrix} -u & 3-u \\ 0 & 3 \end{pmatrix} \prod_{n=1}^{\infty} \begin{pmatrix} -(2n+1)u & -(2n+1)u \\ 0 & 2n+3 \end{pmatrix} \quad \text{for } u = x^2 \le \frac{1}{2}.$$

Again, the hyperbolic variant artanh can be obtained by changing $-u$ into u throughout.

4.5 Natural logarithm

Recall $\ln(1+x) = \sum_{n=1}^{\infty} \frac{(-1)^{n+1}}{n} x^n$. Thus $q_n = -\frac{n+1}{n}$ with $q = 1$, and we obtain after some scaling

$$\ln(1+x) = \begin{pmatrix} x & x \\ 0 & 1 \end{pmatrix} \prod_{n=1}^{\infty} \begin{pmatrix} -nx & -nx \\ 0 & n+1 \end{pmatrix} \quad \text{nested for } |x| \le 1/2.$$

An arbitrary positive argument of ln can be moved towards 1 by the equations $\ln(2x) = \ln x + \ln 2$ and $\ln(\frac{1}{2}x) = \ln x - \ln 2$. While $\ln 2 = \ln(1+1)$ is outside the scope of the above formula, it can be obtained as $\ln 2 = -\ln\frac{1}{2} = -\ln(1-\frac{1}{2})$ which gives

$$\ln 2 = \begin{pmatrix} 1 & 1 \\ 0 & 2 \end{pmatrix} \prod_{n=1}^{\infty} \begin{pmatrix} n & n \\ 0 & 2(n+1) \end{pmatrix}$$

where the matrices have contractivities of about $\frac{1}{2}$.

A better way to obtain $\ln 2$ is at follows: Recall $\operatorname{artanh} x = \frac{1}{2}\ln\left(\frac{1+x}{1-x}\right)$, and therefore, $\ln 2 = \ln\left(\frac{1+1/3}{1-1/3}\right) = 2\operatorname{artanh}(1/3) = \frac{2}{3}\frac{\operatorname{artanh}(1/3)}{1/3}$. Using the results of the previous subsection (with $u = 1/9$) yields

$$\ln 2 = \begin{pmatrix} 2 & 56 \\ 0 & 81 \end{pmatrix} \prod_{n=1}^{\infty} \begin{pmatrix} 2n+1 & 2n+1 \\ 0 & 9(2n+3) \end{pmatrix}$$

where the matrices have contractivities of about $\frac{1}{9}$.

References

Edalat, A., Potts, P. (1997): A new representation for exact real numbers. In: Thirteenth Annual Conference on Mathematical Foundations of Programming Semantics, MFPS XIII, Pittsburgh, Pennsylvania. Elsevier Science B.V., Amsterdam, URL: http://www.elsevier.nl/locate/entcs/volume6.html (Electronic Notes in Theoretical Computer Science, vol. 6)

Gosper, W. (1972): Continued fraction arithmetic. In: Technical Report HAKMEM Item 101B, MIT Artificial Intelligence Memo 239, MIT.

Heckmann, R. (2000): How many argument digits are needed to produce n result digits? In: Real Number Computation, RealComp '98, Indianapolis, Indiana. Elsevier Science B.V., Amsterdam, URL: http://www.elsevier.nl/locate/entcs/volume24.html (Electronic Notes in Theoretical Computer Science, vol. 24)

Nielsen, A., Kornerup, P. (1995): MSB-first digit serial arithmetic. J. of Univ. Comp. Scien. 1(7): 523–543

Potts, P.J. (1998a): Efficient on-line computation of real functions using exact floating point. Draft report, Imperial College, London.
URL: http://www.purplefinder.com/~potts/pub/phd/efficient.ps.gz

Potts, P.J. (1998b): Exact Real Arithmetic using Möbius Transformations. PhD thesis, Imperial College, London. URL: http://www.purplefinder.com/~potts/thesis.ps.gz

Potts, P.J., Edalat, A. (1996): Exact real arithmetic based on linear fractional transformations. Draft report, Imperial College, London.
URL: http://www.purplefinder.com/~potts/pub/phd/stanford.ps.gz

Potts, P.J., Edalat, A. (1997): Exact real computer arithmetic. Draft report, Imperial College, London. URL: http://www.purplefinder.com/~potts/pub/phd/normal.ps.gz

Vuillemin, J.E. (1990): Exact real computer arithmetic with continued fractions. IEEE Transactions on Computers 39(8): 1087–1105

Quasi Convex-Concave Extensions

Christian Jansson

1 Introduction

Convexity and its generalizations have been considered in many publications during the last decades. In this paper we discuss the problem of bounding functions from below by quasiconvex functions and from above by quasiconcave functions. Moreover, applications for nonlinear systems and constrained global optimization problems are considered briefly.

For a given real-valued function $f(x)$, which is defined on some subset of $\mathbf{R}^n$ (the set of real vectors with n components), the gradient and the Hessian is denoted by $f'(x)$ and $f''(x)$, respectively. Sometimes, a function depends on parameters. In such situations the argument is seperated from the parameters by a semicolon. For example, the function $f(x; X)$ has the argument x and depends on the parameter X.

A function $f : S \to \mathbf{R}$ which is defined on a convex set $S \subseteq \mathbf{R}^n$ is called *quasiconvex*, if for all $x, y \in S, 0 \le \lambda \le 1$ the inequality

$$f(\lambda x + (1 - \lambda)y) \le \max\{f(x), f(y)\} \tag{1}$$

holds true.

Obviously, each convex function is quasiconvex. The function f is said to be *quasiconcave* if $-f$ is quasiconvex. A function which is both quasiconvex and quasiconcave is called quasilinear. A differentiable function $f : S \to \mathbf{R}$ is said to be *pseudoconvex*, if for all $x, y \in S$ with $f'(x)^T (x - y) \ge 0$ it is $f(y) \ge f(x)$. It can be shown that pseudoconvex functions are quasiconvex, and differentiable convex functions are pseudoconvex (see Avriel et al. (1988)).

Many quasiconvex functions are known which are not convex; for example $f(x) = -x_1 x_2$ is quasiconvex on the positive orthant, and the ratio of a non-negative convex and a positive concave function is quasiconvex. Quasiconvex functions have among others the following useful properties: (i) each strict local minimum is a global minimum, (ii) the level sets are convex, and (iii) the necessary Kuhn-Tucker conditions for a nonlinear optimization problem are sufficient for global optimality, provided that the objective and the constraints are defined by certain quasiconvex and quasiconcave functions. Quasiconvexity (see Avriel (1976), Avriel et al. (1988), and Schaible (1972), (1981)) is of great importance in optimization theory, engineering and management science. Many applications can be found in Cambini et al.(1989).

Following, we introduce a new notion which can be viewed as a generalization of interval extensions. We assume that the reader is familiar with the basic concepts of interval arithmetic (see Alefeld and Herzberger (1983), Moore (1979), Neumaier (1990), and Ratschek and Rokne (1984)). A *quasi convex-concave extension* of a function $f : S \to \mathbf{R}$ is a mapping $[\underline{f}, \overline{f}]$ which delivers for each interval vector $X := [\underline{x}, \overline{x}] := \{x \in \mathbf{R}^n : \underline{x} \le x \le \overline{x}\}$ with $X \subseteq S$ a quasiconvex function $\underline{f}(x; X)$ and a quasiconcave function $\overline{f}(x; X)$, which are both defined on X such that

$$\underline{f}(x; X) \leq f(x) \leq \overline{f}(x; X) \text{ for all } x \in X. \tag{2}$$

In other words, the pair $\underline{f}(x; X)$ and $\overline{f}(x; X)$ contain the range of f over X. The functions $\underline{f}(x; X), \overline{f}(x; X)$ are called *lower bound function* and *upper bound function* on X, respectively.

By definition, interval extensions of f deliver real lower and upper bounds $\underline{f}(X), \overline{f}(X)$ for the range of f over X. These bounds can be interpreted as constant functions, and constant functions are both quasiconvex and quasiconcave. Hence, it follows that an interval extension is a special case of a quasi convex-concave extension.

The major goal of this paper is to give a short survey about the construction of quasi convex-concave extensions with nonconstant bound functions by using the tools of interval arithmetic. These extensions can be applied in a flexible manner such that the overestimation due to interval arithmetic can be reduced in many cases.

2 Relaxations

We will motivate quasi convex-concave extensions in this section by some applications. Consider the constrained global optimization problem

$$\min_{x \in F} f(x) \text{ where } F := \{x \in X : g_i(x) \leq 0 \text{ for } i = 1, \dots, m\}, \tag{3}$$

and assume that f, g_i are differentiable real-valued functions defined on $X \subseteq \mathbf{R}^n$.

Constrained global optimization problems are NP-hard, even for quadratic functions. One approach for solving NP-hard optimization problems is to use branch and bound methods. During the last decade several methods and programs for constrained global optimization problems became available which use branch and bound techniques and are mainly based on interval arithmetic (see e.g. Hansen (1992), Kearfott (1996), Neumaier (1997), Dallwig, Neumaier and Schichl (1997), Ratschek and Rokne (1988), Schnepper and Stadtherr (1993), and Van Hentenryck, Michel and Deville (1997)).

Our approach (see Jansson (1999a),(1999b), and (2000)) for solving constrained global optimization problems uses quasiconvex relaxations, which permit to compute lower bounds for subproblems generated during a branch and bound process.

A *quasiconvex relaxation* of problem (3) is an optimization problem

$$\min_{x \in R} \underline{f}(x) \text{ where } R := \{x \in X : \underline{g}_i(x) \leq 0 \text{ for } i = 1, \dots, m\}, \tag{4}$$

with the properties that $\underline{f}$ is a pseudoconvex lower bound function of f on X, and $\underline{g}_i$ are quasiconvex lower bound functions of g_i for $i = 1, \dots, m$ on X. These lower bound functions can be obtained by using the results about quasi convex-concave extensions which are presented in the following sections. By definition

(4) it follows immediately that the quasiconvex relaxation has the properties (i) $F \subseteq R$, and (ii) the global minimum value of the relaxation (4) provides a lower bound for the global minimum value of the original problem (3).

The Kuhn-Tucker points x^* of the relaxation (4) are characterized as the solutions of the nonlinear system

$$\underline{f}'(x^*) + \sum_{i=1}^{m} \lambda_i \underline{g}_i'(x^*) = 0, \; \lambda_i \underline{g}_i(x^*) = 0, \; \text{for } i = 1, \ldots, m \qquad (5)$$

with $\lambda_i \geq 0$ and $\underline{g}_i(x^*) \leq 0$ for $i = 1, \ldots, m$. It follows (cf. Avriel (1976)) that each Kuhn-Tucker point of a quasiconvex optimization problem (4) is a global minimum point. There are very efficient methods (for example interior-point methods or SQP-methods) for calculating Kuhn-Tucker points. Hence, an approximate lower bound of the global minimum value for problem (3) can be computed efficiently. Using such an approximation, rigorous bounds for the global minimum value of (4) can be obtained by applying some verified nonlinear system solver to the equations (5).

Roughly speaking, our method for solving problem (3) consists of a branch and bound framework using quasiconvex relaxations. The lower bounds of the subproblems, which are required by this branch and bound scheme, are calculated by computing the Kuhn-Tucker points of the corresponding quasiconvex relaxations. For a detailed treatment of this approach the reader is referred to Jansson (1999b) and (2000).

Methods using convex relaxations for special structured continuous global optimization problems were first introduced by Falk and Soland (1969). Their approach concerns separable nonconvex programming problems. Later, in the case of concave minimization, convex relaxations have been used by Bulatov (1977), Bulatov and Kasinkaya (1982), Emelichev and Kovalev (1970), Falk and Hoffmann (1976), and Horst (1976). For recent developements and improvements see also Zamora and Grossmann (1998) and the references cited over there.

If additionally nonlinear equations $h_i(x) = 0$ are added to (3), then these can be represented as two nonlinear inequalities $h_i(x) \leq 0, h_i(x) \geq 0$. The latter two inequalities are replaced by $\underline{h}_i(x) \leq 0$ and $\overline{h}_i(x) \geq 0$ where $\underline{h}_i(x), \overline{h}_i(x)$ are a quasiconvex lower and a quasiconcave upper bound function, respectively. Then, it can be shown that for the resulting relaxation the necessary Kuhn-Tucker conditions are sufficient for global optimality. Hence, the aforementioned method can be applied for solving problems involving also nonlinear equations.

The zeros of nonlinear systems can be viewed as the set of feasible solutions of a constrained global optimization, where the constraints consist only of nonlinear equations and the objective is the zero function. Therefore, our branch and bound method can be applied for solving nonlinear systems.

3 Bound functions of first and second order

This section treats the construction of affine and quadratical convex lower bound functions. For arithmetical expressions, these bound functions can be automatically generated on a computer by using interval arithmetic.

For an interval vector $X = [\underline{x}, \overline{x}] \in \mathbf{IR}^n$ the 2^n vertices $x(\sigma)$ can be described by

$$x(\sigma) = \underline{x} + \sum_{i=1}^{n} \sigma_i(\overline{x}_i - \underline{x}_i)e_i, \tag{6}$$

where $\sigma \in \{0,1\}^n$ is an n-dimensional vector with components σ_i equal to 0 or 1, and e_i denotes the i-th unit vector. We simply denote the vector with all components equal to 0 by $0 \in \{0,1\}^n$, and the vector with all components equal to 1 by $1 \in \{0,1\}^n$; this will cause no confusion. It follows that $x(0) = \underline{x}$, $x(1) = \overline{x}$.

Theorem 1 *Given a continuously differentiable function $f : S \to \mathbf{R}$ with $S \subseteq \mathbf{R}^n$, and given an interval vector $X \subseteq S$. Suppose further that there exist two vectors $\underline{d}, \overline{d} \in \mathbf{R}^n$ such that the inequalities*

$$\underline{d} \leq f'(x) \leq \overline{d} \tag{7}$$

are valid for all $x \in X$. For a fixed vector $\sigma \in \{0,1\}^n$ let $x(\sigma) \in X$, $d(\sigma) \in D := [\underline{d}, \overline{d}]$ be the vertices of the interval vectors X, D, respectively. Then the affine function

$$\underline{f}(x; X, \sigma) := d(\sigma)^T \cdot x + \{f(x(\sigma)) - d(\sigma)^T \cdot x(\sigma)\} \tag{8}$$

satisfies for all $x \in X$ the inequality

$$\underline{f}(x; X, \sigma) \leq f(x), \tag{9}$$

and moreover

$$\underline{f}(x(\sigma); X, \sigma) = f(x(\sigma)). \tag{10}$$

This theorem is a special case of Theorem 1 in Jansson (2000), and it allows to construct several affine lower bound functions. Bounds $\underline{d}, \overline{d}$ for the gradient $f'(x)$ over X can be calculated by using some interval extension of $f'(x)$. Moreover, the calculation of these bounds can be fully automatized using automatic differentiation (see for example Griewank and Corliss (1991). Then, a fixed vector $\sigma \in \{0,1\}^n$ is chosen. Formula (8) yields an affine function $\underline{f}(x; X, \sigma)$. The inequality (9) implies that $\underline{f}(x; X, \sigma)$ is a lower bound function of f over X. Equation (10) shows that this lower bound function coincides with the original function f in the vertex $x(\sigma)$.

A similar formula can be derived for affine upper bound functions. One way is to construct a corresponding lower bound function of $-f$, and then to take the negative bound function. Therefore, in this paper we will present only formulae for lower bound functions.

Summarizing, we obtain a convex-concave extension of $f : S \to \mathbf{R}$ in the following way: fix a vector $\sigma \in \{0,1\}^n$, and for each interval vector $X \subseteq S$ calculate, by using some interval extension, an interval vector $D := [\underline{d}, \overline{d}]$ such that (7) is satisfied. Then the lower bound function (cf. (8)) and the upper bound function provide a corresponding quasi convex-concave extension where the bounds are affine functions.

In a large variety of engineering applications nonlinear systems and constrained global optimization problems occur where many of the constraints are defined by bilinear functions. In Jansson (2000) it is proved that for the bilinear functions $f(x_1, x_2) := x_1 \cdot x_2$ Theorem 1 generates automatically the convex envelope, that is the uniformly best possible underestimating function: The function

$$\underline{f}(x) := \max\{\underline{f}(x; X, 0), \underline{f}(x, X, 1)\}, \tag{11}$$

where

$$\begin{aligned} \underline{f}(x; X, 0) &= \underline{x}_2 x_1 + \underline{x}_1 x_2 - \underline{x}_1 \underline{x}_2 \\ \underline{f}(x; X, 1) &= \overline{x}_2 x_1 + \overline{x}_1 x_2 - \overline{x}_1 \overline{x}_2 \end{aligned} \tag{12}$$

is the convex envelope of f on $X = [\underline{x}_1, \overline{x}_1] \times [\underline{x}_2, \overline{x}_2]$. A similar formula is valid for the concave envelope. Originally, the convex envelope of a bilinear function is given in Al-Kayal and Falk (1983). However, it is interesting that this convex envelope is generated by applying Theorem 1.

Last, we mention that Theorem 1 remains also valid if in (7) the bounds $[\underline{d}, \overline{d}]$ are replaced by corresponding bounds for slopes (cf. for example Hansen (1992), Krawczyk and Neumaier (1985), and Rump (1996)). Therefore, also nondifferentiable functions may be bounded by affine functions.

The following theorem, which is proved in Jansson (1999b), is concerned with quadratic lower bound functions.

Theorem 2 *Let $f : S \to \mathbf{R}$ be a twice continuously differentiable function where $S \subseteq \mathbf{R}^n$. Let X be an interval vector with $X \subseteq S$, and let $\sigma \in \{0,1\}^n$. Suppose further that*

1. *two real $n \times n$ matrices $\underline{H}, \overline{H}$ satisfy the inequalities*

$$\underline{H} \leq f''(x) \leq \overline{H} \quad \text{for all } x \in X; \tag{13}$$

2. *a real $n \times n$ matrix $\underline{H}(\sigma)$ is componentwise defined by*

$$\underline{H}_{ij}(\sigma) := \begin{cases} \underline{H}_{ij} & \text{if } \sigma_i = \sigma_j \\ \overline{H}_{ij} & \text{otherwise} \end{cases} \tag{14}$$

for $0 \leq i, j \leq n$;

3. an interval vector $Y := [\underline{y}, \overline{y}]$ is defined by

$$Y := \frac{1}{2}\underline{H}(\sigma) \cdot (X - x(\sigma)), \tag{15}$$

where all operations are interval operations, and $x(\sigma)$ is the vertex of X corresponding to σ.

Then the following results are valid:

(a) The quadratic function

$$
\begin{aligned}
\underline{f}(x; X, \sigma) \; &:= \; f(x(\sigma)) + f'(x(\sigma))^T(x - x(\sigma)) \\
&+ \; \tfrac{1}{2}(x - x(\sigma))^T \underline{H}(\sigma)(x - x(\sigma))
\end{aligned} \tag{16}
$$

is a lower bound function of f with $\underline{f}(x(\sigma); X, \sigma) = f(x(\sigma))$.

(b) If $\underline{H}(\sigma)$ is positive semidefinite, then $\underline{f}(x; X, \sigma)$ is a convex lower bound function of f on X.

(c) The function

$$g(x; X, \sigma) := f(x(\sigma)) + f'(x(\sigma))^T(x - x(\sigma)) + \sum_{i=1}^{m} \max\{q_i^1(x), q_i^2(x)\} \tag{17}$$

where

$$
\begin{aligned}
q_i^1(x) &:= \underline{y}_i(x_i - x_i(\sigma)) + \tfrac{1}{2}(\underline{x}_i - x_i(\sigma))\underline{H}_{i\cdot}(\sigma)(x - x(\sigma)) - (\underline{x}_i - x_i(\sigma))\underline{y}_i, \\
q_i^2(x) &:= \overline{y}_i(x_i - x_i(\sigma)) + \tfrac{1}{2}(\overline{x}_i - x_i(\sigma))\underline{H}_{i\cdot}(\sigma)(x - x(\sigma)) - (\overline{x}_i - x_i(\sigma))\overline{y}_i
\end{aligned}
$$

is a convex lower bound function of f with $g(x(\sigma); X, \sigma) = f(x(\sigma))$.

As in the case of the previous affine bound functions, this theorem gives the possibility to generate convex quadratical bound functions. The only difference is that now bounds $\underline{H}, \overline{H}$ for the Hessian of f with respect to X have to be calculated by using interval arithmetic. As before, also these bound functions define a quasi convex-concave extension.

4 Splittings

Interval arithmetic is capable to compute lower and upper bounds for the range of arithmetical expressions over a box. These bounds can be obtained by replacing each real operation and real variable in an arithmetical expression by a corresponding interval operation and interval variable, respectively. This process yields bounds which can be computed very fast, but which may overestimate the range because of the dependence of variables (see e.g. Neumaier (1990)).

In order to reduce the overestimation due to the problem of dependence, the question arises whether it is possible to bound and to work appropriately with arithmetical expressions itself rather than with interval variables and interval operations. In other words: Is it possible to bound an arithmetical expression by quasiconvex lower and quasiconcave upper bound functions, provided the arithmetical expression is splitted into two parts with respect to one of the real operations $+, -, *, /$.

In general this seems to be not possible for nonconstant bound functions: Indeed, the sum of two convex arithmetical expressions is convex, but this property is not valid for the product and the ratio.

However, the following theorem shows that at least in several cases quasiconvexity of lower bound functions for the product and the ratio of arithmetical expressions can be proved, provided these expressions satisfy certain sign and convexity properties.

Theorem 3 *Let $X \subseteq \mathbf{R}^n$ be convex, $g, h : X \to \mathbf{R}$, and assume that $\underline{g}, \underline{h}, \overline{g}, \overline{h}$ are convex lower and concave upper bound functions of g and h on X, respectively. Then:*

1. *$\underline{g} + \underline{h}$ is a convex lower bound function of the sum $g + h$ on X.*

2. *$\underline{g} - \overline{h}$ is a convex lower bound function of the difference $g - h$ on X.*

3. *If g is nonpositive and h is nonnegative on X, then $\underline{g} \cdot \overline{h}$ is a quasiconvex lower bound function of the product $g \cdot h$ on X.*

4. *If $\underline{g}$ and $\underline{h}$ are positive, and $1/\underline{g}$ or $1/\underline{h}$ is concave on X, then $\underline{g} \cdot \underline{h}$ is a quasiconvex lower bound function of the product $g \cdot h$ on X.*

5. *If $\overline{g}$ and $\overline{h}$ are negative, and $1/\overline{g}$ or $1/\overline{h}$ is convex on X, then $\overline{g} \cdot \overline{h}$ is a quasiconvex lower bound function of the product $g \cdot h$ on X.*

6. *If h is positive on X, then $1/\overline{h}$ is a convex lower bound function of the reciprocal $1/h$ on X.*

7. *If g is nonnegative and h is positive on X, then $\underline{g}/\overline{h}$ is a quasiconvex lower bound function of the ratio g/h on X.*

PROOF. 1. and 2. follow by observing that the sum of two convex functions is convex.

3. Using the assumptions $\underline{g}(x) \leq g(x) \leq 0, 0 \leq h(x) \leq \overline{h}(x)$ it follows that $\underline{g}(x) \cdot \overline{h}(x) \leq \underline{g}(x) \cdot h(x) \leq g(x) \cdot h(x)$ for every $x \in X$.

Hence, $\underline{g} \cdot \overline{h}$ is a lower bound function of $g \cdot h$ on X. Since $-\underline{g}$ is nonnegative and concave, and $\overline{h}$ is nonnegative and concave, Table 5.1 in Avriel et al. (1988) implies the quasiconcavity of the product $-\underline{g} \cdot \overline{h}$. Therefore, $\underline{g} \cdot \overline{h}$ is quasiconvex.

4. The positivity of $\underline{g}$ and $\underline{h}$ implies that $\underline{g} \cdot \underline{h}$ is a lower bound function of $g \cdot h$. Since $1/\underline{g}$ or $1/\underline{h}$ is concave, Table 5.1 in Avriel et al. (1988) yields the quasiconvexity of the product $\underline{g} \cdot \underline{h}$.

5. The assumption is equivalent to $-\overline{g}, -\overline{h}$ are positive and $-1/\overline{g}$ or $-1/\overline{h}$ is concave. Table 5.1 in Avriel et al. (1988) yields the quasiconvexity of the

product $-\overline{g} \cdot (-\overline{h})$. Observing that $-\overline{g}(x) \leq -g(x), -\overline{h}(x) \leq -h(x)$ implies that $-\overline{g} \cdot (-\overline{h})$ is a quasiconvex lower bound function of $-g \cdot (-h)$.

Property 6. follows from the fact that the reciprocal function of a positive concave function is convex on X.

To prove 7., we observe that $g/\overline{h}$ is a lower bound function of g/h because of the sign restrictions. The quasiconvexity of $g/\overline{h}$ follows from Table 5.4 in Avriel et al. (1988) by using the convexity of g and the concavity of $\overline{h}$. □

The following theorem is concerned with the composition of functions.

Theorem 4 *Let $X \subseteq \mathbf{R}^n$ be convex, and let $f_i : X \to \mathbf{R}$ with convex lower bound functions $\underline{f}_i$ for $i = 1, \ldots, m$ be given. Let g be a nondecreasing quasiconvex function on the convex hull of the range of $(f_1, \ldots, f_m)$ over X. Then $g(\underline{f}_1, \ldots, \underline{f}_m)$ is a quasiconvex lower bound function of the composition $g(f_1, \ldots, f_m)$ on X.*

PROOF. Since g is nondecreasing, the inequalities $\underline{f}_i(x) \leq f_i(x)$ for $i = 1, \ldots, m$ imply $g(\underline{f}_1(x), \ldots, \underline{f}_m(x)) \leq g(f_1(x), \ldots, f_m(x))$ for all $x \in X$. Hence, $g(\underline{f}_1(x), \ldots, \underline{f}_m(x))$ is a lower bound function of the composition. The remaining quasiconvexity follows from Theorem 5.3 in Avriel et al. (1988). □

Several other special rules can be derived in a similar manner. But this is out of the scope of this paper. However, there are situations where a function can be decomposed into two parts, but only one part can be well bounded by a convex or quasiconvex function. In this case the other part should be bounded by using an interval extension yielding a constant bound function. Then, because one of the parts is a constant, the sum, difference, product or ratio of these two parts can be bounded as usual.

5 Example

In order to illustrate how the previous results may be applied, we consider the function

$$f(x) = (5 + 2x(1 - x) + 0.5x \exp(1.5x) + \frac{1}{4} \cos(6\pi x))/(x + 0.5)$$

on the interval $X = [0, 1]$. The function f (see Figure 1, solid) is a nonconvex function. The nonconvex numerator of f can be split into two parts: $g_1(x) := 5 + 2x + 0.5x \exp(1.5x)$ and $g_2(x) := -2x^2 + \frac{1}{4} \cos(6\pi x)$. The first part is convex, and we set $\underline{g}_1(x) := g_1(x)$ for all $x \in X$. Using Theorem 1, we bound g_2 from below by the affine functions

$$\underline{g}_2(x; X, \sigma) = d(\sigma)^T x + \{g_2(x(\sigma)) - d(\sigma)^T \cdot x(\sigma)\},$$

where $\sigma \in \{0, 1\}$, and the interval $D = [\underline{d}, \overline{d}]$ is computed by using the slope arithmetic which is contained in INTLAB (Rump (1999)). By Theorem 3, assertion 1 it follows that the two functions

$$\underline{g}(x; X, \sigma) := \underline{g}_1(x) + \underline{g}_2(x; X, \sigma), \quad \sigma \in \{0, 1\}$$

are convex lower bound functions of the numerator of f. A short calculation using interval arithmetic shows that both functions $\underline{g}(x; X, \sigma)$ are positive. The denominator $h(x) := x + 0.5$ is also positive on X. Setting $\overline{h}(x) = h(x)$ for all $x \in X$, Theorem 3 assertion 7 implies that the functions $\underline{g}(x; X, \sigma)/\overline{h}(x)$ are quasiconvex lower bound functions of the ratio g/h on X. These lower bound functions are displayed for $\sigma = 0$ (dotted) and for $\sigma = 1$ (dash-dotted) in Figure 1. In the case $\sigma = 1$ the lower bound function is not convex but quasiconvex.

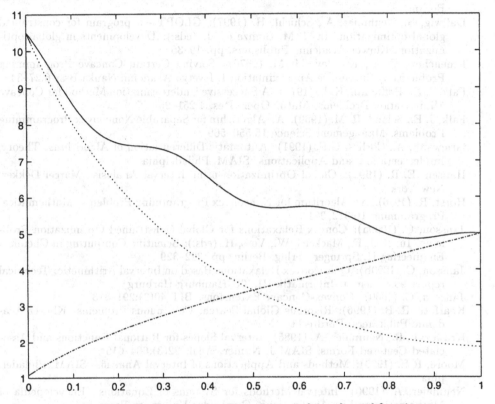

Fig. 1. Function f and two quasiconvex lower bound functions

Several other numerical examples including constrained global optimization problems of large scale are presented in Jansson (1999b), (2000).

References

Al-Khayyal, F. A., Falk, J. E. (1983): Jointly Constrained Biconvex Programming. Math. Oper. Res. 8(2):273–286

Alefeld, G., Herzberger, J. (1983): Introduction to Interval Computations. Academic Press, New York

Avriel, M. (1976): Nonlinear Programming: Analysis and Methods. Prentice-Hall, Inc., New Jersey

Avriel, M., Diewert, W. W., Schaible, S., Zang, I. (1988): Generalized Concavity. Plenum Press, New York

Bulatov, V. P. (1977): Embedding Methods in Optimization Problems. Nauka, Novosibirsk (in Russian)

Bulatov, V. P., Kasinkaya, L. I. (1982): Some Methods of Concave Minimization and Their Applications. In: Methods of Optimization and Their Applications. Nauka, Novosibirsk, pp. 71–80

Cambini, A., Castagnoli, E., Martein, L., Mazzoleni, P., Schaible, S. (1989): Generalized Convexity and Fractional Programming with Economic Applications. Lecture Notes in Economics and Mathematical Systems, No. 345. Springer-Verlag, Berlin

Dallwig, S., Neumaier, A., Schichl, H. (1997): GLOPT - a program for constrained global optimization. In: I. M. Bomze et al. (eds.): Developments in global optimization. Kluwer Academic Publishers, pp. 19–36

Emelichev, V. A., Kovalev, M. M. (1970): Solving Certain Concave Programming Problems by Successive Approximation I. Izvetya Akademii Nauka Bssr 6:27–34

Falk, J. E., Hoffmann, K. L. (1976): A Successive Underestimation Method for Concave Minimization Problems. Math. Oper. Res. 1:251–259

Falk, J. E., Soland, R. M. (1969): An Algorithm for Separable Nonconvex Programming Problems. Management Science 15:550–569

Griewank, A., Corliss, G.F. (1991): Automatic Differentiation of Algorithms. Theory, Implementation, and Applications. SIAM, Philadelphia

Hansen, E. R. (1992): Global Optimization using Interval Analysis. Marcel Dekker, New York

Horst, R. (1976): An Algorithm for Nonconvex Programming Problems. Mathematical Programming 10:312–321

Jansson, C. (1999a): Convex Relaxations for Global Constrained Optimization Problems. In: Keil, F., Mackens, W., Voss, H. (eds.): Scientific Computing in Chemical Engineering II. Springer Verlag, Berlin, pp. 322–329

Jansson, C. (1999b): Quasiconvex Relaxations Based on Interval Arithmetic. Technical report 99.4, Inst. f. Informatik III, TU Hamburg-Harburg

Jansson, C. (2000): Convex-Concave Extensions. BIT 40(2):291–313

Kearfott, R. B. (1996): Rigorous Global Search: Continuous Problems. Kluwer Academic Publishers, Dordrecht

Krawczyk, R., Neumaier, A. (1985): Interval Slopes for Rational Functions and Associated Centered Forms. SIAM J. Numer. Anal. 22(3):604–616

Moore, R.E. (1979): Methods and Applications of Interval Analysis. SIAM, Philadelphia

Neumaier, A. (1990): Interval Methods for Systems of Equations. Encyclopedia of Mathematics and its Applications. Cambridge University Press

Neumaier, A. (1997): NOP - a compact input format for nonlinear optimization problems. In: I. M. Bomze et al. (eds.): Developments in global optimization. Kluwer Academic Publishers, pp. 1–18

Ratschek, H., Rokne, J. (1984): Computer Methods for the Range of Functions. Halsted Press (Ellis Horwood Limited), New York (Chichester)

Ratschek, H., Rokne, J. (1988): New Computer Methods for Global Optimization. John Wiley & Sons (Ellis Horwood Limited), New York (Chichester)

Rump, S. M. (1996): Expansion and Estimation of the Range of Nonlinear Functions. Mathematics of Computation 65(216):1503–1512

Rump, S.M. (1999): INTLAB - INTerval LABoratory. In: Tibor Csendes (ed.): Developements in Reliable Computing. Kluwer Academic Publishers, pp. 77–104

Schaible, S. (1972): Quasi-convex Optimization in General Real Linear Spaces. Zeitschrift für Operations Research 16:205–213

Schaible, S. (1981): Quasiconvex, Pseudoconvex, and Strictly Pseudoconvex Quadratic Functions. Journal of Optimization Theory and Applications 35(3): 303–338

Schnepper, C. A., Stadtherr, M. A. (1993): Application of a Parallel Interval Newton/Generalized Bisection Algorithm to Equation-Based Chemical Process Flowsheeting. Interval Computations 4:40–64

Van Hentenryck, P., Michel, P., Deville, Y. (1997): Numerica: A Modelling Language for Global Optimization. MIT Press Cambridge

Zamora, J. M., Grossmann, I. E. (1998): Continuous Global Optimization of Structured Process Systems Models. Computers Chem. Engineering 22(12):1749–1770

Rewriting, Induction and Decision Procedures: A Case Study of Presburger Arithmetic

Deepak Kapur*

1 Introduction

Theorem provers and automated reasoning tools are not as widely used in specification analysis, debugging and verification of hardware and software as one would hope. A major reason is perhaps that these tools are often found difficult to use. The learning curve for a typical theorem prover is quite high; it takes a great deal of effort by typical users to effectively use such a tool for their application domain. Even then, considerable resources must be expended by building a large knowledge base and library of useful properties in a representation suitable for the prover, to bring it to a level so that it can start playing a useful role.

For theorem provers to be acceptable to application experts, the key requirement is that a theorem prover should be able to easily perform reasoning steps considered routine in an application domain; performing such reasoning should definitely not become a burden on the expert. Kapur (1997b) called this *reasoning in the large* and the inference steps used for such reasoning as *large* inference steps, to distinguish them from easy and application-independent inference steps typically supported in most proof checkers as well as many theorem provers, which we have called *reasoning in the small*.

In the computer-aided verification literature, there has recently been a surge of interest in decision procedures in order to enhance the capabilities of reasoning tools which operate in push-button mode. The main focus of these works has been mostly to fine-tune methods and frameworks for combining decision procedures, thus reviving earlier work by Shostak (1984) and Nelson and Oppen (1979), or integrating decision procedures with model-checkers and OBDD based tools (Bryant, 1986). Little attention has been paid to integrate decision procedures with methods for mechanizing induction, and/or with the use of the definitions of function symbols and their known properties in the form of lemmas.

We consider the following features crucial in building *large* inference steps:

1. Integrating decision procedures for equality, numbers, bits, bit vectors, finite sets, finite lists, and other frequently used data structures in applications,
2. performing case analyses automatically,
3. finding instances of definitions and lemmas using decision procedures for simplification, and
4. discharging conditions in definitions and lemmas.

Another approach for building large inference steps is by providing mechanisms for writing tactics which can be used to combine small inference steps to have the effect of a large inference step. Many proof checkers including PVS, ISABELLE, HOL, do provide linguistic constructs for writing tactics. Every user however must define such tactics, or use tactics from the library developed

* Partially supported by the National Science Foundation Grant nos. CCR-9712396, CCR-9712366, CCR-9996150, and CDA-9503064.

by other users. Further, it is unclear how large inference steps discussed below that can require intricate integration between decision procedures, simplification (rewriting) and induction, can be written as efficient tactics.

In this paper, we use the case of Presburger arithmetic for illustrating interaction between decision procedures, rewriting (simplification) and induction. The discussion is in the context of our rewrite-based theorem prover *Rewrite Rule Laboratory*, (RRL) (Kapur and Zhang, 1995), which has been used to automatically verify properties of arithmetic hardware circuits including a family of adders, multipliers as well as the SRT division circuit (Kapur and Subramaniam, 1996b; Kapur and Subramaniam, 1997; Kapur and Subramaniam, 1998a; Kapur and Subramaniam, 2000b). After briefly reviewing conditional rewriting, the cover set induction method, and Presburger arithmetic, we discuss how a decision procedure can be used to automatically find appropriate instantiations of rewrite rules serving as function definitions and lemmas, thus building large inference steps. Intertwining of simplification by rewriting and decision procedures is illustrated, thus exhibiting tight integration, something we conjecture, is nontrivial to achieve using tactics in tactic-based provers.

The next key issue is the interaction between decision procedures and induction, a topic that has not gotten much attention. We show how methods based on extracting induction schemes from function definitions can be generalized to exploit semantic information expressed as decision procedures. It is also shown how decision procedures can be augmented with induction schemes extracted from a family of function definitions so that a certain subclass of conjectures can be automatically decided, thus enlarging the scope of formulas that can be decided using traditional decision procedures. The proposed approach has the ability of making decision procedures extensible by enriching them with different induction schemes suggested by different terminating function definitions.

RRL supports decision procedures for (i) equality on ground terms (congruence closure), (ii) propositional reasoning, (iii) bits, (iv) freely generated data structures, as well as (v) a procedure for quantifier-free theory of Presburger arithmetic over integers and naturals (Kapur and Nie, 1994; Kapur and Zhang, 1995). A distinguishing feature of these decision procedures is that if a formula cannot be found unsatisfiable, then equalities are generated as consequences to be used as rewrite rules for simplification. We believe that it is the integration of decision procedures with rewriting and induction which makes *RRL* effective for the application of reasoning about arithmetic circuits.

2 Background: Rewriting

Below, we give an overview of the rewrite-based approach to theorem proving and simplification.

Let $T(F, X)$ be the set of terms constructed using a finite set F of function symbols and a set X of variables. A term is either a variable $x \in X$ or $f(t_1, \ldots, t_n)$, where $f \in F$, and each $t_i, 1 \leq i \leq n$, is a term. Let $Vars(t)$ denote the variables appearing in a term t. The *subterms* of a term are the term itself and the subterms of its arguments, and they can be uniquely identified using *positions,* finite sequences of positive integer separated by ".".'s. The subterm of t at the empty sequence ϵ is t itself. If $f(t_1, \cdots, t_n)$ is a subterm at position p, then t_j is the subterm at position $p.j$. A term $f(t_1, \cdots, t_n)$ is called *basic* if each t_i is a distinct variable.

A substitution θ is a mapping from a finite set of variables to terms, denoted as $\{x_1 \leftarrow t_1, \cdots, x_n \leftarrow t_n\}$, $n \geq 0$, and x_i's are distinct. θ applied on $s =$

$f(s_1, \cdots, s_m)$, written as $\theta(s)$, is $f(\theta(s_1), \cdots, \theta(s_m))$. Term s *matches* t under θ if $\theta(s) = t$. Terms s and t *unify* under θ if $\theta(s) = \theta(t)$.

An (unconditional) rewrite rule is $s \to t$ with $Vars(t) \subseteq Vars(s)$ and $s \notin X$. A rule $s \to t$ is *applicable* to a term u iff for some substitution θ and position p in u, $\theta(s) = u|_p$, the subterm at position p in u.

The application of the rule *rewrites* u to $u[p \leftarrow \theta(t)]$, the term obtained after replacing the subterm at position p in u by $\theta(t)$. A rewrite system R is a finite set of rewrite rules. R induces a relation among terms denoted $\to_R$. $s \to_R t$ denotes rewriting of s to t by a single application of a rule in R. $\to_R^+$ and $\to_R^*$ denote the transitive and the reflexive, transitive closure of $\to_R$, respectively.

Conditional rewrite rules are interesting because of ways their conditions can be discharged. A conditional rewrite rule

$$s \to t \ if \ cond$$

is *applicable* to a term u iff for some substitution θ and position p in u, $\theta(s) = u|_p$ and $\theta(cond)$ can be reduced to *true*. Different definitions of conditional rewriting rely on different ways to ensure the truth of $\theta(cond)$ to check for the applicability of the conditional rule. In *RRL*, the congruence closure over ground terms, propositional calculus and Presburger arithmetic (see below) are used for discharging conditions. When u has a *context* in the form of a conjunction of literals, then the context can also be used to establish conditions; this is how contextual rewriting is defined; for details, see (Zhang, 1992; Kapur and Subramaniam, 2000b).

3   Cover Set Induction: Cover Sets and Induction Schemes

The *cover set* method is used to implement well-founded induction in *RRL*, and has been used to successfully perform proofs by induction in a variety of nontrivial application domains including verification of arithmetic circuit descriptions (Kapur and Subramaniam, 1996b; Kapur and Subramaniam, 1997; Kapur and Subramaniam, 1998a). Below, we give an overview of the cover set induction method. For details, the reader can refer to (Zhang et al, 1988; Kapur and Subramaniam, 1996a).

Function symbols in F are partitioned into *defined* and *interpreted* functions. An interpreted function symbol comes from a decidable theory $\mathcal{T}$ whose decision procedure is being integrated into rewriting and induction. Constructor symbols are viewed as interpreted symbols; the quantifier-free theory for free constructor symbols is decidable.

A defined function symbol f is specified by a finite set of terminating rewrite rules, and it is assumed that some rule in the definition covers every tuple of arguments to f (see below for a discussion on checking completeness of a definition). The termination of rewrite rules defining functions is established using well-founded rewrite orderings. *RRL*, for instance, uses the lexicographic recursive path ordering on terms induced by a well-founded precedence ordering on function symbols (Dershowitz, 1986).

Given a complete function definition as a finite set of terminating rules

$$\{l_i \to r_i \ if \ cond_i \mid l_i = f(s_1, \cdots, s_n), \ 1 \leq i \leq k\},$$

the main steps of the cover set induction method are

1. *Generate a Cover Set from a Function Definition:* A cover set associated with f is a finite set of triples. For a rule $l \to r$ *if cond*, where $l = f(s_1, \cdots, s_n)$, and $f(t_1^i, \cdots, t_n^i)$ is the ith recursive call to f in r, the corresponding triple is $\langle \langle s_1, \cdots, s_n \rangle, \{\cdots, \langle t_1^i, \cdots, t_n^i \rangle, \cdots\}, \{cond\}\rangle$. The second component of a triple is the empty set if there is no recursive call to f in r. The third component is the empty set for unconditional rules. The positions in the recursive calls to f where the arguments change are called *inductive* (or *changeable*) positions. Other positions are called *unchangeable* or *noninductive* positions.

2. *Generate an Induction Scheme using a Cover Set:* Given a conjecture C with a basic subterm $u = f(x_1, \cdots, x_n)$, the cover set of f can be chosen for generating an induction scheme as follows.
   An induction scheme is a finite set of induction cases, each of the form $\langle \sigma_c, \{\theta_i\}, \{cond\} \rangle$ generated from a cover set triple $\langle \langle s_1, \cdots, s_n \rangle, \{\cdots, \langle t_1^i, \cdots, t_n^i \rangle, \cdots\}, \{cond\} \rangle$ as follows[1]:
   $\sigma_c = \{x_1 \leftarrow s_1, \cdots, x_n \leftarrow s_n\}$ and $\theta_i = \{x_1 \leftarrow t_1^i, \cdots, x_n \leftarrow t_n^i\}$.[2] The variables in the basic subterm u in inductive argument positions are then the *induction variables* of the conjecture C.

3. *Generate Induction Subgoals using an Induction Scheme:* Each induction case generates an induction subgoal: σ_c is applied to the conjecture to generate the induction conclusion, whereas each substitution θ_i applied to the conjecture generates an induction hypothesis. Basis subgoals come from induction cases whose second component in the corresponding triple of the induction scheme is the empty set.

Heuristics have been developed and implemented in *RRL*, which in conjunction with failure analysis of induction schemes and backtracking in case of failure, have been found appropriate for prioritizing induction schemes, automatically selecting the "most appropriate" induction scheme (thus selecting induction variables), and generating proofs of conjectures.

4 Presburger Arithmetic

Throughout this paper, Presburger arithmetic stands for the quantifier-free theory of numbers (rationals, integers or natural numbers) with numeric variables, the arithmetic operations (successor(s), predecessor(p) and addition($+$)) and the arithmetic relations ($>, <, \neq, =, \geq, \leq$).[3] It also goes under the name of *linear arithmetic* (Boyer and Moore, 1988).

The pure subtheory includes the above symbols, but for verification purposes, we need to include uninterpreted symbols as well. By a *linear* term, we mean either a number, a variable, or a term of the form $f(t_1, \cdots, t_n)$ where f is an arithmetic operation and each t_i, $1 \leq i \leq n$, is a (not necessarily linear) term. If

---

[1] The variables in a cover set triple are suitably renamed if necessary.

[2] To generate an induction scheme, the subterm u in a conjecture C is unified with the left side of each rule in the definition of f as well as with the recursive calls to f in the right side of the rule. This is always possible if u is a basic term $f(x_1, \cdots, x_n)$. It also works if only variables in the inductive positions of u are distinct. The method can be used to generate induction schemes from other nonbasic subterms in a conjecture as well.

[3] Presburger proved in 1929 that the full first-order theory of numbers with the above stated symbols is decidable (Enderton, 1972). In our work, we have used only a quantifier-free subtheory of it.

p is an arithmetic relation, $p(t_1, \cdots, t_n)$ denotes a linear atom. Linear terms and atoms are called *pure* if they do not involve any uninterpreted symbols. It is easy to see that every atom can be translated to $s_i = t_i$ or $s_i \leq t_i$.

In this paper, by a Presburger arithmetic decision procedure PA, we assume an algorithm specified in terms of input and output as follows.

- *Input:* A set of equations and inequalities $LC : \{s_i \; op \; t_i | 1 \leq i \leq n\}$, where $op \in \{=, \leq\}$, s_i's and t_i's are pure linear terms.
- *Output:* No, if LC cannot be solved. Otherwise, yes with $\langle \sigma, NLI \rangle$, a solved (simplified) form of LC (i.e., $\sigma \cup NLI \leq LC$ in a well-founded ordering $<$ on terms and literals), where $\sigma = \{x_j \leftarrow u_j \mid 1 \leq i \leq k\}$ satisfying the occur-check (i.e., x_j does not appear in any u_i's), and (possibly, a disjunction of conjunction of) a finite set of normalized inequalities NLI obtained after applying σ (again implying that no x_j appears in NLI). Further, any implicit equality derivable from LC is made explicit, i.e., any equality $s = t$ derivable from LC should be such that $\sigma(s) - \sigma(t)$ simplifies to 0. For instance, if LC includes $E \leq F - G$ and $F \leq G + E$, then $E - F + G = 0$ must be derivable from σ.
- *Completeness and Soundness:* The solution sets $Soln(LC) = Soln(\langle \sigma, NLI \rangle)$, i.e., any m-tuple $e = \langle e_1, \cdots, e_m \rangle$ of numbers is a solution to LC (for every $s_i = t_i \in LC$, $\theta(s_i) - \theta(t_i) = 0$ and for every $s_i \leq t_i \in LC$, $\theta(s_i) - \theta(t_i) \leq 0$, where $\theta = \{x_1 \leftarrow e_1, \ldots, x_m \leftarrow e_m\}$) iff e is a solution of σ and NLI.

For equations, a solved form can be obtained using an algorithm for solving linear Diophantine equations and conditions on variables. For inequalities, Fourier's algorithm can be used to check for unsatisfiability; a byproduct of this algorithm is the detection of implicit equalities. Details can be found in (Kapur and Nie, 1994; Kapur and Subramaniam, 1996a).

5 Rewriting and Decision Procedures

Apart from deciding formulas in its theory, another important role a decision procedure can play is to determine whether a rewrite rule in a function definition or proved as a lemma can be applied to simplify a given conjecture with the ultimate goal of checking the validity of the conjecture. In order to do this, it is imperative that the decision procedure be tightly integrated with rewriting. Most theorem provers based on decision procedures including PVS, STEP, SVC, do not seem to perform such tight integration between rewriting/simplification and decision procedures.

In RRL, decision procedures are not only used for checking the validity of a goal and simplifying it to a normal form, but also for selecting which rewrite rule is applicable on the goal. Decision procedures for congruence closure and Presburger arithmetic interact with the rewriting/matching algorithm. A rewrite rule may often appear not to be applicable for simplifying a goal; but with the aid of decision procedures, it can indeed be used to show the validity of the goal. In contrast, this has to be done by the user manually in other decision procedures based theorem provers; see an example below. Capabilities such as these are perhaps the main reason why RRL has been successful in automatically proving properties of arithmetic circuits, especially the SRT division circuit (Kapur and Subramaniam, 2000b) as well as in generating readable and compact proofs.

Below, we illustrate two aspects of cooperation of decision procedures. As the reader would notice, the decision procedures for equality on ground terms, propositional calculus, and Presburger arithmetic interact with each other for simplification and deducing additional equalities. Further, appropriate instantiations for the rules are found by establishing the conditions of the rules. What would take many human-guided steps in many provers and proof checkers can be done in a single step in RRL.[4]

Assume the following rewrite rules are already known (either as a part of definitions and/or lemmas known about function symbols appearing in the conjecture):

$$1.\ min(x, y) \to y \quad if \quad max(x, y) = x,$$
$$2.\ f(x) \le g(x) \to true \quad if \quad p(x).$$

Suppose the following conjecture over the integers is attempted:

$$(p(x) \wedge (x \le max(x, y)) \wedge (z \le f(max(x, y))) \wedge (0 < min(x, y)) \wedge (max(x, y) \le x))$$

$$\supset (z < g(x) + y).$$

A proof by contradiction is attempted. First, the conjecture is negated and Skolemized; so variables are replaced by Skolem constants. The following goal must be shown to be unsatisfiable.

$$p(A) \wedge (A \le max(A, B)) \wedge (L \le f(max(A, B))) \wedge (0 < min(A, B))$$

$$\wedge (max(A, B) \le A)) \wedge \neg(L < g(A) + B).$$

Linear inequalities are first transformed by PA to the form $s_i \le c_i$, where s_i is a linear term and c_i is a number.

$$p(A) \wedge (A - max(A, B) \le 0) \wedge (L - f(max(A, B)) \le 0) \wedge (-min(A, B) \le -1)$$

$$\wedge (max(A, B) - A \le 0)) \wedge (g(A) + B - L \le 0).$$

Below we give a brief overview of the interaction among the three decision procedures (propositional satisfiability, ground congruence closure and Presburger arithmetic PA) for simplification, along with a sketchy detail of steps performed by each decision procedure. For more details, the reader should consult (Kapur and Nie, 1994).

1. Simplify using the propositional calculus decision procedure and check for propositional unsatisfiability of literals.
2. Process equalities on ground terms using the congruence closure procedure. In RRL, this is done using a ground completion procedure by first turning equalities into terminating rewrite rules, and then, normalizing them (Kapur and Nie, 1994; Kapur, 1997a). Trivial equalities such as $t = t$ are reduced to $true$.
3. Equalities are propagated to the rest of the formula. This is done by normalizing the rest of the literals in the goal using the rewrite rules in the ground canonical system obtained from the equalities.
   Repeat steps 1-3 until there is no change.
4. Process arithmetic constraints using PA by eliminating variables. If not found unsatisfiable, deduce all implicit equalities by obtaining a solved form for equalities. Go to step 2.
   Repeat all of the above steps until nothing changes.

---

[4] Boyer and Moore's prover (Boyer and Moore, 1988) may also be able to prove this conjecture without any user guidance.

The above procedure terminates since an invocation of a decision procedure either has no effect, or produces an output smaller than its input in a well-founded order on literals. We illustrate the procedure on the example.

Propositional satisfiability and congruence closure procedures have no effect in the first iteration on the above example. From the linear inequalities, an implicit equality $max(A, B) = A$ is derived by PA. This equality is used by the congruence closure procedure as a rewrite rule to simplify the inequality set to:

$$L - f(A) \leq 0, -min(A, B) \leq -1, g(A) + B - L \leq 0.$$

The key inference step in PA is to pair up linear inequalities in which a variable appears positively as well as negatively, and eliminate the variable from the pair of inequalities. Given $a_1 X + q_1 \leq e_1$ and $-b_1 X + q_2 \leq e_2$, where $a_1, b_1 > 0$, deduce a new inequality $b_1 q_1 + a_1 q_2 \leq b_1 e_1 + a_1 e_2$. In the above example, the linear term L is eliminated:

$$-min(A, B) \leq -1, -f(A) + g(A) + B \leq 0.$$

If during elimination, an inequality of the form $0 \leq 0$ is deduced, all inequalities participating in the derivation are then equalities, and they are made explicit. If an inequality of the form $0 \leq -c$, where $c > 0$, is deduced, then unsatisfiability is detected. If on the other hand, a variable appears only positively (or negatively) in inequalities, then the formula is likely to be satisfiable except for unsatisfiability arising due to solutions of equations over integers (or natural numbers) being sought. More details can be found in (Kapur and Nie, 1994).

For the above example, no new equality can be deduced by the decision procedures. Further, the formula cannot be shown to be unsatisfiable. The result of the interaction among the three decision procedures is:

$$max(A, B) = A, P(A), -min(A, B) \leq -1, -f(A) + g(A) + B \leq 0.$$

When a goal cannot be shown unsatisfiable and no new equalities can be deduced from the interaction of the decision procedures, it is then checked whether a rewrite rule from the data base of definitions and lemmas applies. Literals in the goal are analyzed for matching the rewrite rules, starting with the maximal literals (in the same well-founded ordering used to ensure termination of the rewrite rules in the data base) in the goal. If a match with the left side of some rule is found, then a proof of the condition of the rule, if any, under the match is attempted. If successful, the instance of the rewrite rule is added as a new constraint, and simplification by decision procedures is repeated. Since every literal in the condition of a rewrite rule is smaller than its left side, and the result of rewriting by a rewrite rule is smaller in the well-founded ordering, the termination of simplification is ensured.

Continuing with the above example, rule 1 is applicable on the literal $min(A, B)$ in the goal, using $\{x \leftarrow A, y \leftarrow B\}$ since the condition of rule 1 under this instantiation, $max(A, B) = A$, follows from the equalities in the rest of the goal. Literal $min(A, B)$ is reduced to B; $min(A, B) = B$ is added to the equality set. Simplification using the three decision procedures (after replacing $min(A, B) = A$ and eliminating B) then produces:

$$max(A, B) = A, P(A), -f(A) + g(A) \leq -1.$$

Assuming the ordering $f > g$, the next literal to consider in the subgoal is: $-f(A) + g(A) \leq -1$; the maximal subterm $f(A)$ in it matches a maximal term

135

in the second linear rule since $P(A)$ is true. The instance of rule 2 is added to the inequality set, giving:

$$-f(A) + g(A) \leq -1,\, f(A) - g(A) \leq 0.$$

The decision procedure PA detects a contradiction from these inequalities, thus asserting the validity of the original conjecture.

This example is a good illustration of the tight integration of contextual rewriting with decision procedures. As the reader would notice, the original goal can be automatically established. The verification of the properties of the SRT division circuit includes many such proofs of formulas; details can be found in (Kapur and Subramaniam, 1997;Kapur and Subramaniam, 1998b; Kapur and Subramaniam, 2000b). More details of the integration of contextual rewriting with decision procedures are given in (Kapur and Nie, 1994).

6　Induction and Decision Procedures

We show how a decision procedure such as PA can be used to generalize the cover set method, to make it more effective for generating and analyzing induction schemes using unification modulo a decidable theory. Below, we provide a brief overview; more details can be found in (Kapur and Subramaniam, 1996a).

Many data structures can be represented in many different ways; for example, numbers can be represented using 0 and s (to stand for the *successor* function) as constructors as well as in terms of $0, 1$ and $+$. Some functions, such as $gcd, divides, rem$, etc., are more convenient, and easier to define using $+$, instead of s. Finite sets can be represented using the *null* set, a constructor that *inserts* an element, or alternatively, using the null set, *singleton* sets and the *union* operation. Similarly, lists can be represented using the *empty* list and the *cons* constructor, or alternatively, the empty list, *singleton* lists and the *append* operator on lists. While reasoning about functions defined on such data structures, there is a need to convert from one representation to the other. This could be because (i) a function definition is given using a mixture of operators defined on different representations, (ii) a conjecture is expressed using functions which are defined on two different representations, or (iii) a conjecture involves a function that is defined using one representation but its arguments in the conjecture use a different representation.

Using PA, we illustrate how semantic information about a data structure can reconcile its different representations while attempting a proof of a conjecture. It thus becomes critical to generate an induction scheme using semantic information, that enables the application of induction hypotheses which otherwise cannot be applied because of different uses of a function.

We can generalize the algorithm for generating an induction scheme for a conjecture C by performing semantic unification, instead of syntactic unification, on a selected subterm $t = f(t_1, \cdots, t_n)$ in C, and the cover set associated with f. Given a cover set triple,

$$c = \langle \langle s_1, \cdots, s_n \rangle, \{ \cdots, \langle s_1^j, \cdots, s_n^j \rangle, \cdots \}, cond \rangle,$$

constraint equations $\{ t_i = s_i \mid 1 \leq i \leq n \}$ with context $cond$ are solved; PA generates a substitution σ_c of $Vars(t) \cup Vars(c)$ (if the third component of c is empty then the context is assumed to be $true$). Similarly, substitutions θ_j

of $Vars(t) \cup Vars(c)$ for induction hypotheses are generated using PA on the constraint equations $\{t_i = \sigma_c(s_i^j) \mid 1 \le i \le n\}$ with context $\sigma_c(cond)$.

As an example, consider the definition of gcd (greatest common divisor):

```
1. gcd(0, z)    --> z,                  2. gcd(z, 0)    --> z,
3. gcd(w, z+w) --> gcd(w, z) if w>0,   4. gcd(z+w, z) --> gcd(w,z) if z>0.
```

The cover set generated from gcd is:

```
Cover(gcd):{<<0,z>,{},{}>, <<z,0>,{},{}>, <<w, z+w>, {<w, z>}, {w>0}>,
           <<z+w, z>,{<w, z>},{z>0}>}}.
```

As the reader would have noticed, the induction scheme generated from this cover set is very different from the principle of mathematical induction. While proving a simple conjecture such as gcd(u, v) = gcd(v, u), one would notice that the principle of mathematical induction is often not helpful, whereas the induction scheme proposed by the above cover set method immediately gives a proof, which shows the power of the cover set method.

For attempting a proof of the conjecture gcd(u + u, s(s(0))) = s(s(0)), however, the principle of mathematical induction suffices. This scheme can also be generated from the above cover set and the conjecture using the above algorithm:

```
{<{ u <- 0}, {}, {}>, <{ u <- 1}, {}, {}>, <{ u <- v + 1}, {{ u <- v}}, {}>}.
```

For instance, the first component of the first triple is the most general unifier (mgu) of $u + u = 0, s(s(0)) = z$, obtained from the first triple of the cover set. The second triple of the cover set does not yield any substitution since $0 = s(s(0))$ cannot be solved. The third triple of the cover set leads to $u + u = w, s(s(0)) = z + w, w > 0$ which gives the substitution for the conclusion in the second triple in the induction scheme; there is no induction hypothesis available for use since $u + u = s(s(0)), s(s(0)) = 0$ has no solution. The fourth triple in the cover set gives $u + u = z + w, s(s(0)) = z, z > 0$, giving the substitution $u \leftarrow v + 1$ for the conclusion; the induction hypothesis is generated from the substitution solving $u + u = v + v, s(s(0)) = s(s(0)), s(s(0)) > 0$.

6.1 Checking Completeness using a Decision Procedure

Before a function definition can be used to generate a sound and complete induction scheme using the cover set method, it must be proved terminating as well as complete. It must be the case that for any ground substitution σ_g of variables in $f(x_1, \ldots, x_n)$, some rule $f(s_1^i, \ldots, s_n^i) \to r_i$ if $cond_i$ from the definition of f must be applicable on $\sigma_g(f(x_1, \ldots, x_n))$. Below we review a generalization of the methods discussed by Kapur et al. (1991) so that definitions given using linear terms can also be checked for completeness using PA.

Checking for completeness amounts to constructing a formula that is a disjunction of conditions under which each rule covers the arguments of f. Assuming that $x_1, \ldots, x_n$ are new variables not appearing in $s_1^i, \ldots, s_n^i$, a function f is *completely defined* using rules

$$f(s_1^i, \ldots, s_n^i) \to r_i \text{ if } cond_i, \ 1 \le i \le k$$

if and only if the formula

$$\bigvee_i \exists y_1, \ldots y_m, (cond_i \wedge (x_1 = s_1^i \wedge \ldots x_n = s_n^i)),$$

137

is valid, where $y_1, \ldots, y_m$ are free variables in $s_1^i, \ldots, s_n^i$. If this formula is in a decidable theory, then its validity can be decided.

For instance, gcd(u, v) above is completely defined on natural numbers as shown below: We get

$$[\exists z(u = 0 \wedge v = z) \vee \exists z(u = z \wedge v = 0) \vee \exists z, w(w > 0 \wedge (u = w \wedge v = z + w))$$
$$\vee \exists z, w(z > 0 \wedge (u = z + w \wedge v = z))].$$

Using the decision procedure for Presburger arithmetic, the existential quantifiers can be eliminated, and the above formula simplifies to:

$$((u = 0) \vee (v = 0) \vee (v \geq u \wedge u > 0) \vee (u \geq v \wedge v > 0)),$$

a valid formula in Presburger arithmetic. The cover set generated from the definition of *gcd* is, thus, complete.

Methods discussed by Kapur et al. (1991) cannot be used to prove the completeness of *gcd* since they depend upon the use of completion procedures; the completion procedure would not terminate on the definition of *gcd*.

If a formula generated for checking completeness is not valid, then the definition of f may or may not be complete depending upon whether the rewrite rules defining f are confluent or not. However, the associated cover set is incomplete. For more details, an interested reader may consult (Kapur, 1994; Kapur and Subramaniam, 1996a).

If the arguments to f in the left and right sides of every rewrite rule in the definition of f are linear terms, and $cond_i$ is a conjunction of linear literals, then the formula corresponding to the completeness of the definition of f can be decided using a decision procedure for Presburger arithmetic.

The above method has the advantage that in case of an incomplete definition, it can list a finite representation of argument tuples on which the function f needs to be defined.

Incomplete Cover Sets In order to draw sound conclusions using the cover set method, it is important to use a complete cover set. Otherwise, unsound conclusions can be made. For example, if the cover set based on the following definition of *divides*:

```
1. divides(u, 0)       ->  true,
2. divides(u, v)       ->  false if (v < u) and not (v = 0),
3. divides(u, u + v)   ->  divides(u, v) if 0 < u.
```

is used without caring about its completeness, it is possible to wrongly conclude: $divides(x, y + y)$ *if* $divides(x, y)$, even though the conjecture does not hold if $x = 0 \wedge y > 0$. The formula is true under the condition that $x > 0 \vee y = 0$, the values on which *divides* is defined. For $x = 0 \wedge y \neq 0$, $divides(x, y)$ is not defined; unless the above formula is viewed in a logic in which all undefined values are treated the same, the validity of the formula cannot be determined. The modified conjecture $divides(x, y + y)$ *if* $divides(x, y) \wedge (x > 0 \vee y = 0)$ is a theorem provable by induction.

Using the induction scheme generated from the incomplete cover set of *divides*, $divides(2, x) = not(divides(2, s(x)))$ can be proved since *divides* is completely defined if its first argument is 2 (this also follows from the fact that the instantiation of the above condition, $(2 > 0 \vee x = 0)$, is true). This induction scheme can also be used to prove $divides(1, y)$. These examples suggest that an incomplete cover set can be used for proofs insofar as conjectures are relativized with respect to the condition characterizing the input subdomain on which the function definition is given; see (Kapur and Subramaniam, 1996a) for a discussion.

6.2 Generalization Heuristic and Lemma Speculation

One of the key challenges in mechanizing proofs by induction is how to discover intermediate lemmas needed for a proof. Intermediate conjectures are generated during proof attempts. In some cases, however, it is possible to prove a generalization of a conjecture (from which the conjecture follows). Most induction theorem provers support heuristics for generalizing conjectures. Semantic analysis using a decision procedure can be used to improve upon the generalization heuristics.

An easy but useful heuristic is to abstract common subterms in a conjecture by a variable (Boyer and Moore, 1988; Kapur and Zhang, 1995). Generalizations so obtained can be first quickly tried on a few test cases to make sure that they are at least true for some instantiations before a proof is attempted. If a particular generalization cannot be proved (or turns out to be not valid), another one can be tried after backtracking.

Given a conjecture C of the form, $l = r$ if $cond$, RRL looks for a maximal nonvariable subterm s occurring in at least two of l, r and $cond$. In case of many such maximal subterms, many generalizations are possible. A generalized version C_g of C can be obtained by simultaneously abstracting (a subset of) distinct common subterms in C by distinct new variables. If C_g can be proved, then C is valid, whereas the converse is not necessarily true.

A weakness of the above heuristic is the requirement that term being generalized to a variable, have multiple syntactic occurrences. A subterm may not have multiple occurrences syntactically in a conjecture, even though semantically equivalent subterms may occur in it. Consider the following conjecture about gcd:

$$gcd(x + y + 1, 2 * x + 2 * y + 2) = x + y + 1.$$

If no semantic analysis is performed, a possible generalization by abstracting occurrences of $x + y + 1$ by a variable u is $gcd(u, 2 * x + 2 * y + 2) = u$. This generalization can be easily shown to be false.

From a semantic standpoint, however, the second argument of gcd is closely related to its first argument. Using PA for identifying equivalent subterms, this relationship can be identified, and the first argument of gcd can be found to appear twice in its second argument. The above conjecture can thus be generalized to: $gcd(u, u + u) = u$, which can be proved.[5]

The generalization heuristic can be modified to exploit semantic analysis as follows. Given a conjecture $l = r$ if $cond$, find a maximal nonvariable subterm s of l that also appears in r and/or $cond$. This can be done by checking whether s appears syntactically, or whether for some other linear subterm t, $t \geq s$. If the answer is no, then s does not occur in t; otherwise, we find the number of times s appears in t (this can be done by repeated queries and subtraction from t until the result becomes smaller than s). Let $t = k * s + tr$, where k is a positive integer and tr is a linear term $< s$. Subterm s can be abstracted to be a new variable u, and t can be replaced by $k * u + tr$. The extended generalization procedure using PA is described in detail by Kapur and Subramaniam (1996a).

7 Extending Induction Schemes with Decision Procedures

Most fully-automatic reasoning tools implementing OBDDs and model checkers have a major weakness, namely they are unable to perform any proof by

---

[5] Under the condition $u > 0$, the conjecture rewrites using the fourth rule of gcd to give $gcd(u, u) = u$, which can be proved. For the condition $u = 0$, the conjecture is again established by any of the rules 1 and 2 after simplification.

induction; see (Gupta, 1994). As anyone attempting proofs by hand must have experienced, there are many challenges in discovering a proof by induction—choosing variables to perform induction, induction schemes (e.g., whether to use the principle of mathematical induction or the principle of complete (Noetherian) induction in the case of natural numbers), and speculating intermediate lemmas needed in proofs.

Syntactic conditions can be developed which guarantee that certain equational conjectures whose proofs need the use of induction, can be decided automatically. By putting restrictions on conjectures and the definitions of function symbols appearing in conjectures, it can be ensured that using the cover set method, every subgoal in a proof attempt can be simplified to a formula in a decidable theory. We review this below. A detailed discussion can be found in (Kapur and Subramaniam, 2000a).

Definition 7.1 *A definition of a function symbol f is $\mathcal{T}$-based in a decidable theory $\mathcal{T}$ if for each rule $f(t_1, \cdots, t_m) \to r$ in the definition, every t_i, $1 \le i \le m$, is an interpreted term in $\mathcal{T}$, any recursive call to f in r only has interpreted terms as arguments, and the abstraction of r defined as replacing recursive calls to f in r by variables is an interpreted term in $\mathcal{T}$.*

For examples, the definitions of `double`, `log`, `exp2`, and `*` below are $\mathcal{T}$-based over Presburger arithmetic; `double` is also $\mathcal{T}$-based over the theory of free constructors `0`, `s`.

```
1. double(0)           --> 0,      2. double(s(x)) --> s(s(double(x))).
3. log(0)              --> 0,      4. log(s(0))    --> 0,
5. log(s(x)+s(x))      --> s(log(s(x))),
6. log(s(s(x)+s(x)))   --> s(log(s(x))).
7. exp2(0)             --> s(0),   8. exp2(s(x))   --> exp2(x)+exp2(x).
9. x * 0               --> 0,      10. x * s(y)    --> x + (x * y).
```

Definition 7.2 *A conjecture $f(x_1, \cdots, x_n) = r$ is called **simple** over $\mathcal{T}$ if f has a $\mathcal{T}$-based definition, and r is interpreted in $\mathcal{T}$.*

The conjecture (C1) below about `double` is simple over Presburger arithmetic, but (C2) about `log` is not.

(C1): `double(m) = m + m`,   (C2): `log(exp2(m)) = m`.

For a simple conjecture, the cover set method proposes only one induction scheme, which is generated from the cover set derived from the definition of f. Further, formulas arising from induction subgoals can be simplified, and after the application of the induction hypothesis(es), the resulting formulas are in a decidable theory. Thus,

Theorem 1. *A simple conjecture C over $\mathcal{T}$ is decidable by the cover set method.*

7.1 Complex $\mathcal{T}$-based Conjectures

To decide complex conjectures, the choice of induction schemes must be limited, and the interaction among the definitions of function symbols in the conjectures has to be analyzed. Such an analysis is undertaken in (Subramaniam, 1997) for predicting a priori failure of proof attempts when certain induction schemes are used. The notion of **compatibility** of function definitions is introduced for identifying intermediate steps in a proof which get blocked in the absence of additional lemmas. The same idea is used below to identify conditions under which the decidability of conjectures can be predicted a priori.

We first consider compatibility among the definitions of two function symbols. Later this is generalized to a sequence of function definitions.

Definition 7.3 *A $\mathcal{T}$-based term t is composed if*

1. *t is a basic term $f(x_1, \cdots, x_k)$, where f is $\mathcal{T}$-based, or*
2. *(a) $t = f(s_1, \cdots, t', \cdots, s_k)$, where t' is in an inductive position of a $\mathcal{T}$-based f and is composed, and each s_i is an interpreted term, and*
   (b) variables in the inductive positions of the basic subterm (in the innermost position) of t do not appear elsewhere in t. Other variables in unchangeable positions of the basic subterm may appear elsewhere in t.

The left side of (C2), $\log(\exp2(m))$, is a composed term, for instance, in which $\exp2(m)$ is a basic subterm. The definition of a composed term guarantees that there is exactly one basic subterm, and that too in its innermost position.

Given a conjecture of the form $l = r$, where l is composed and r is interpreted, there is only one basic term in it whose outermost symbol is a defined symbol and $\mathcal{T}$-based. For a conjecture $f(t_1, \cdots, g(x_1, \cdots, x_k), \cdots, t_m) = r$, interaction between the right sides of rules defining g and the left side of rules defining f must be considered, as captured below in the property of *compatibility*.

Definition 7.4 *A definition of f is compatible with a definition of g in its i-th argument in $\mathcal{T}$ if for each right side r_g of a rule defining g:*

1. *if r_g is interpreted in $\mathcal{T}$, then $f(x_1, \cdots, r_g, \cdots, x_m)$ rewrites to an interpreted term in $\mathcal{T}$, and*
2. *if $r_g = h(s_1, \cdots, g(t_1, \cdots, t_k), \cdots, s_n)$ with a single recursive call to g, the definition of f rewrites $f(x_1, \cdots, h(s_1, \cdots, y, \cdots, s_n), \cdots, x_m)$ to $h'(u_1, \cdots, f(x_1, \cdots, y, \cdots, x_m), \cdots, u_n)$, where x_i's are distinct variables, h, h' are interpreted symbols in $\mathcal{T}$, and s_i, u_j's are interpreted terms of $\mathcal{T}$.[6]*

These requirements ensure that induction hypothesis(es) are applicable after function definitions have been applied. The above definition also applies to an interpreted function symbol (i.e., one in $\mathcal{T}$) and a $\mathcal{T}$-based function symbol. As an example, + in Presburger arithmetic is compatible with * with the usual recursive definition because of the associativity and commutativity properties of +, which are valid formulas of PA.

A conjecture in which a composed term is equated to an interpreted term, can be decided if all the symbols from the root to the position p of the basic subterm can be pushed in so that the induction hypothesis is applicable. The notion of compatibility of a function definition with another function definition is extended to a *compatible sequence* $\langle f_1, \cdots, f_d \rangle$, of definitions of function symbols, where each f_i is compatible with f_{i+1} at j_i-th argument, $1 \le i \le d-1$. The above ideas are best illustrated using the following example:

(C3): $\texttt{bton}(\texttt{pad0}(\texttt{ntob}(\texttt{m}))) = \texttt{m}$.

---

[6] A more general requirement can be given in case of multiple recursive calls.

Functions bton and ntob convert binary representations to numbers, and vice versa, respectively. The function pad0 adds a leading binary zero to a bit vector. These functions are commonly used to reason about number-theoretic properties of parameterized arithmetic circuits (Kapur and Subramaniam, 1996b; Kapur and Subramaniam, 1998a). Padding of output bit vectors of one stage with leading zeros before using them as input to the next stage is common in multiplier circuits realized using a tree of carry-save adders. The above equation captures an important property that the padding does not affect the number output by a circuit. The underlying theory is the combination of the quantifier-free theory of Presburger arithmetic, the quantifier-free theory of bit vectors with the quantifier-free theory of free constructors nil, cons, and b0, b1, to stand for binary 0 and 1. (In the definitions below, bits increase in significance in a bit vector (list) with its first element being the least significant.) Function symbols bton, ntob, and pad0 are $\mathcal{T}$-based

```
1.  bton(nil)                    --> 0,
2.  bton(cons(b0, y1))           --> bton(y1) + bton(y1),
3.  bton(cons(b1, y1))           --> s(bton(y1) + bton(y1)),
4.  ntob(0)                      --> cons(b0, nil),
5.  ntob(s(0))                   --> cons(b1, nil),
6.  ntob(s(s(x2+x2)))            --> cons(b0,ntob(s(x2))),
7.  ntob(s(s(s((x2+x2)))))       --> cons(b1,ntob(s(x2))),
8.  pad0(nil)                    --> cons(b0, nil),
9.  pad0(cons(b0, y))            --> cons(b0, pad0(y)),
10. pad0(cons(b1, y))            --> cons(b1, pad0(y)).
```

The function pad0 is compatible with ntob; bton is compatible with pad0 as well as ntob. However, ntob is not compatible with bton (since there is no way to rewrite ntob(x + x) and ntob(s(x + x))).

A proof attempt of (C3) leads to two basis and two step subgoals based on the cover set of ntob. In the first basis subgoal where m <- 0, bton(pad0(ntob(0))) = 0, ntob(0) rewrites using the definition of ntob to cons(b0, nil), then pad0(cons(b0, nil))) rewrites to cons(b0, cons(b0, nil))), and finally, bton(pad0(ntob(0))) rewrites, giving the formula 0 + 0 + 0 + 0 = 0, which is valid in Presburger arithmetic. The second basis subgoal is similar.

Consider the first induction step subgoal. The conclusion is bton(pad0(ntob(s(s(x2 + x2))))) = s(s(x2 + x2)) with the hypothesis being bton(pad0(ntob(s(x2)))) = s(x2). Subterm ntob(s(s(x2 + x2))) in the conclusion rewrites to cons(b0, ntob(s(x2))) by the definition of ntob; the subterm pad0(cons(b0, ntob(s(x2)))) then rewrites to cons(b0, pad0(ntob(s(x2)))). Term bton(pad0(ntob(s(s(x2 + x2))))) thus rewrites to bton(pad0(ntob(s(x2)))) + bton(pad0(ntob(s(x2)))) which on the hypothesis application, gives a valid formula s(x2) + s(x2) = s(s(x2 + x2)) in Presburger arithmetic. It can be shown that the second step subgoal also simplifies similarly to a valid formula in Presburger arithmetic. Hence, the validity of (C3) can be decided. The reader would have noticed that the compatibility requirement ensures that all the function symbols are pushed over interpreted symbols for the induction hypothesis to be applicable.

Theorem 2. *A conjecture $l = r$, where l is a composed term and r is interpreted in $\mathcal{T}$, can be decided by the cover set method if the sequence $\langle f_1, f_2, \cdots, f_{d-1}, f_d \rangle$ of function symbols from the outermost function symbol f_1 of l to the basic subterm $f_d(x_1, \cdots, x_m)$ is compatible.*

Acknowledgement: Most of this work has been done jointly with my former student Mahadevan Subramaniam. I thank Jürgen Giesl for helpful comments.

References

Gupta, A. (1994): *Inductive Boolean Function Manipulation: A Hardware Verification Methodology for Automatic Induction.* Ph.D. Thesis, CMU, Pittsburgh.

Boyer, R.S., and Moore, J S. (1988): *A Computational Logic Handbook.* Academic Press.

Boyer, R.S., and Moore, J S. (1988): "Integrating decision procedures into heuristic theorem provers: A case study of linear arithmetic", *Machine Intelligence 11*, P. Hayes, D.Mitchie and J. Richards (eds).

Bryant, R.E. (1986): "Graph-based algorithms for boolean function manipulation," *IEEE Trans. on Computers*, C-35(8).

Dershowitz, N. (1986): "Termination of rewriting," *J. Symbolic Computation*, 3, 69-116.

Enderton, H. (1972): *A Mathematical Introduction to Logic.* Academic Press.

Kapur, D. (1994): "Automated tools for analyzing completeness of specifications," Proc. *1994 Intl. Symp. on Software Testing and Analysis (ISSTA)*, Seattle, WA. 28-43.

Kapur, D. (1997a): "Shostak's congruence closure as completion," Proc. *Intl. Conf. on Rewriting Techniques and Applications (RTA-97)*, Barcelona, Spain.

Kapur, D. (1997b): "Rewriting, decision procedures and lemma speculation for automated hardware verification," Proc. *Theorem Provers in Higher-order Logics*, (ed. Gunter and Felty), Springer LNCS 1275, Murray Hill, 171-182.

Kapur, D., Narendran, P., Rosenkrantz, D., and Zhang, H. (1991): "Sufficient-completeness, quasi-reducibility and their complexity," *Acta Informatica*, 28, 311-350.

Kapur, D., and Nie, X. (1994): "Reasoning about numbers in Tecton," Proc. *8th Intl. Symp. Methodologies for Intelligent Systems*, Charlotte, North Carolina, 57-70.

Kapur, D., and Subramaniam, M. (1996a): "New uses of linear arithmetic in automated theorem proving for induction," *J. Automated Reasoning*, 16(1-2), 39-78.

Kapur, D., and Subramaniam, M. (1996b): "Mechanically verifying a family of multiplier circuits," Proc. *CAV*, (eds. Alur and Henzinger), LNCS 1102, 135-146.

Kapur, D., and Subramaniam, M. (1997): "Mechanizing reasoning about arithmetic circuits: SRT division," Proc. *17th FSTTCS*, Springer LNCS 1346, Kharagpur, India.

Kapur, D., and Subramaniam, M. (1998a): "Mechanical verification of adder circuits using powerlists," *J. Formal Methods in System Design,* 13(2), 127-158.

Kapur, D., and Subramaniam, M. (1998b): "Mechanizing reasoning about large finite tables in a rewrite based theorem prover,", Proc. *ASIAN-98,* LNCS 1538, Manila.

Kapur, D., and Subramaniam, M. (2000a): "Extending decision procedures with induction schemes" Proc. *CADE-17* (ed. McAllester) Springer LNAI 1831, Pittsburgh.

Kapur, D., and Subramaniam, M. (2000b): D. Kapur and M. Subramaniam, "Using an induction prover for verifying arithmetic circuits," *J. Software Tools for Technology Transfer,* Springer Verlag, Sep 2000.

Kapur, D., and Zhang, H. (1995): "An overview of Rewrite Rule Laboratory (RRL)," *J. of Computer and Mathematics with Applications,* 29, 2, 91-114.

Nelson, G., and Oppen, D.C. (1979): "Simplification by cooperating decision procedures," *ACM Trans. on Programming Languages and Systems* 1 (2), 245-257.

Shostak, R.E. (1984): "Deciding combination of theories," *J. ACM* 31 (1), 1-12.

Subramaniam, M. (1997): *Failure Analyses in Inductive Theorem Provers.* Ph.D. Thesis, Department of Computer Science, University of Albany, New York.

Zhang, H. (1992): "Implementing contextual rewriting," Proc. *3rd Intl. Workshop on Conditional Term Rewriting Systems,* (eds. Remy and Rusinowitch), Springer LNCS 656, France, 363-377.

Zhang, H., Kapur, D., and Krishnamoorthy, M.S. (1988): "A mechanizable induction principle for equational specifications," Proc. *9th Intl. Conf. Automated Deduction (CADE),* (eds. Lusk and Overbeek), Springer LNCS 310, Chicago, 250-265.

This article was processed using the LaTeX macro package with LLNCS style

Derivative-Based Subdivision in Multi-Dimensional Verified Gaussian Quadrature

B. Lang

1 Summary

An implementation of verified Gaussian quadrature involves algorithmic parameters whose correct setting is crucial for adequate performance. We identify several of these control parameters and, based on experimental data, we try to give advice for choosing suitable parameter values in order to obtain reasonable average performance for a "black-box" quadrature routine.

2 d-dimensional verified Gauss–Legendre quadrature

One-dimensional Gauss–Legendre quadrature approximates the value of an integral

$$I := \int_{-1}^{1} f(x)\, dx$$

with an n-point formula

$$A^{(n)} = \sum_{i=1}^{n} \omega_i^{(n)} \cdot f(x_i^{(n)}) .$$

Golub and Welsch (1969) showed that the nodes $x_i^{(n)} \in (-1, 1)$ and the weights $\omega_i^{(n)} > 0$ in the formula can be obtained by solving a suitable symmetric tridiagonal eigenvalue problem. The remainder term $R^{(n)} := I - A^{(n)}$ can be written in the form

$$R^{(n)} = \underbrace{\frac{2^{2n+1}}{2n+1} \cdot \binom{2n}{n}^{-2} \cdot \frac{1}{(2n)!}}_{=: \, e^{(n)}} f^{(2n)}(\xi) \quad \text{for some } \xi \in (-1, 1).$$

Other representations of the remainder term involve the Peano kernel, see Davis and Rabinowitz (1984).

The corresponding d-dimensional $(n_1 \times \ldots \times n_d)$-point product formula,

$$A^{(n_1 \times \ldots \times n_d)} = \sum_{i_1=1}^{n_1} \cdots \sum_{i_d=1}^{n_d} \omega_{i_1}^{(n_1)} \cdots \omega_{i_d}^{(n_d)} \cdot f(x_{i_1}^{(n_1)}, \ldots, x_{i_d}^{(n_d)}) , \tag{1}$$

satisfies

$$I = \int\limits_{-1}^{1} \cdots \int\limits_{-1}^{1} f(x_1, \ldots, x_d)\, dx_d \cdots dx_1 = A^{(n_1 \times \ldots \times n_d)} + R^{(n_1 \times \ldots \times n_d)} \qquad (2)$$

with the following bound for the remainder term:

$$\left| R^{(n_1 \times \ldots \times n_d)} \right| \leq 2^{d-1} \cdot \sum_{j=1}^{d} e^{(n_j)} \cdot \max_{x \in [-1,1]^d} \left| \frac{1}{(2n_j)!} \cdot \frac{\partial^{2n_j} f(x)}{\partial x_j^{2n_j}} \right|, \qquad (3)$$

see Stroud (1971).

Equations (1)–(3) lead easily to an enclosure for the interval value,

$$I = \int\limits_{-1}^{1} \cdots \int\limits_{-1}^{1} f(x_1, \ldots, x_d)\, dx_d \cdots dx_1 \in [A^{(n_1 \times \ldots \times n_d)}] + [R^{(n_1 \times \ldots \times n_d)}],$$

with

$$[A^{(n_1 \times \ldots \times n_d)}] = \sum_{i_1=1}^{n_1} \cdots \sum_{i_d=1}^{n_d} [\omega_{i_1}^{(n_1)}] \cdots [\omega_{i_d}^{(n_d)}] \cdot f([x_{i_1}^{(n_1)}], \ldots, [x_{i_d}^{(n_d)}]) \qquad (4)$$

and

$$[R^{(n_1 \times \ldots \times n_d)}] = [-1,1] \cdot 2^{d-1} \cdot \sum_{j=1}^{d} e^{(n_j)} \cdot \left| \frac{1}{(2n_j)!} \cdot \frac{\partial^{2n_j} f}{\partial x_j^{2n_j}} [-1,1]^d \right|. \qquad (5)$$

Here, $\|[a]\| := \max\{|\alpha| : \alpha \in [a]\}$ is the maximum modulus of an element in a given interval $[a]$, and $(\partial^{2n_j} f)/((2n_j)!\, \partial x_j^{2n_j})[-1,1]^d$ denotes the range of f's Taylor coefficient over the cube $[-1,1]^d$. This range can be enclosed via standard automatic differentiation techniques. The enclosures $[x_{i_j}^{(n_k)}]$ and $[\omega_{i_j}^{(n_k)}]$ for the nodes and weights, respectively, of the *one-dimensional* n_k-point formulas may be pre-computed with verified eigensolvers, cf. Storck (1993).

3 The algorithm

The following recursive algorithm relies on Equations (4) and (5) to compute an enclosure of width $\leq \varepsilon$ for I (if width$[A^{(n_1 \times \ldots \times n_d)}]$ is not negligible then the width of the enclosure can slightly exceed ε; this might be prevented with additional checks, which we have omitted for brevity). The algorithm is called with $[x] = [-1,1]^d$.

for $j = 1 : d$ and $k = 1 : 2 \cdot n_{\max}$, compute intervals $[f_{k,j}]$ enclosing the ranges

$\qquad f_{k,j}[x]$ of the Taylor coefficients $f_{k,j} := \dfrac{\partial^k f}{k!\, \partial x_j^k}$ over $[x]$

146

if there is some combination $(n_1 \times \cdots \times n_d)$ such that

$$\text{width}[R^{(n_1 \times \cdots \times n_d)}] = 2^d \cdot \sum_{j=1}^{d} e^{(n_j)} |[f_{2n_j,j}]| \leq \varepsilon$$

then
  minimize $n_1 \cdot n_2 \cdots n_d$ subject to $\text{width}[R^{(n_1 \times \cdots \times n_d)}] \leq \varepsilon$
  compute $[A^{(n_1 \times \cdots \times n_d)}]$ and return $[I] := [R^{(n_1 \times \cdots \times n_d)}] + [A^{(n_1 \times \cdots \times n_d)}]$
otherwise
  split the current box $[x]$ into $m \geq 2$ subboxes $[x^{(\ell)}]$
  recursively compute enclosures $[I^{(\ell)}]$ over the $[x^{(\ell)}]$ with $\text{width}[I^{(\ell)}] \leq \dfrac{\varepsilon}{m}$
  return $[I] := [I^{(1)}] + \ldots + [I^{(m)}]$

4   Algorithmic parameters

The algorithm contains several parameters whose setting heavily influences the performance. First, the order of the Gauss formulas must be limited by some $n_{\max}$. Then we must choose among several ways to enclose $f_{k,j}$ over $[x]$. And if the termination criterion cannot be met at the current recursion level we must select some dimension(s) along which the current box $[x]$ is split for further recursion, and we must decide on the number of subboxes to be generated.

4.1   Enclosing the Taylor coefficients

We have investigated three techniques for enclosing the $f_{k,j}[x]$.
  Plain interval evaluation: The simplest way to obtain enclosures for the $f_{k,j}[x]$ is to use Taylor arithmetic with interval evaluation, i.e.,

$$f_{k,j}[x] \subseteq [f_{k,j}] := f_{k,j}([x]).$$

This method is simple to implement, but it tends to severely over-estimate the actual range if the function is somewhat complicated.
  Union: Subdivide $[x]$ into subboxes $[x^{(\ell)}]$ and compute

$$f_{k,j}[x] \subseteq [f_{k,j}] := \bigcup_{\ell} f_{k,j}([x^{(\ell)}]) \,,$$

i.e., use the **plain** approach on each subbox. This method typically yields sharper enclosures, but it requires significantly more work per box.
  Note that subdivision for enclosing Taylor coefficients is *not* made redundant by the subdivision already present in the recursive quadrature algorithm because a sharper termination criterion must be met at the next recursion level.
  Centered: For any $\tilde{x} \in [x]$ (e.g., $\tilde{x} = \text{mid}[x]$),

$$f_{k,j}[x] \subseteq [f_{k,j}] := f_{k,j}(\tilde{x}) + \sum_{i=1}^{d} \frac{\partial f_{k,j}}{\partial x_i}([x]) \cdot ([x_i] - \tilde{x}_i) \,. \tag{6}$$

Table 1: Work per box in multiples of a function evaluation.

Taylor arithmetic	$\mathcal{O}(d \cdot (2n_{\max} + 1)^2)$
Taylor+gradient arithmetic	$\mathcal{O}(d \cdot (d+2) \cdot (2n_{\max} + 1)^2)$

For narrow boxes $[x]$ this technique gives much sharper enclosures than the other methods, and it also provides information that may be used in the recursive subdivision, as described in Section 4.2. On the other hand, computing the gradients of the Taylor coefficients significantly increases the overall work. In addition, it is not obvious how to obtain the mixed partial derivatives at reasonable cost.

Griewank et al. (1997) described a method for obtaining any higher-order mixed partial derivative from suitably chosen uni-directional Taylor series. While this technique is certainly applicable, it is too general and too expensive for our restricted needs.

Instead we used operator overloading to implement a combination of Taylor arithmetic and gradient arithmetic. Indeed, the required gradients of the Taylor coefficients can be obtained either

1. by performing Taylor arithmetic on gradients instead of scalars, or

2. by performing gradient arithmetic on Taylor series instead of scalars.

Both approaches increase the amount of work by a factor of $d+2$ as compared to Taylor arithmetic alone (evaluation of the Taylor series itself at $[x]$ and at $\tilde{x}$, plus evaluation of its gradients), see Table 1. As the order $k_{\max} = 2n_{\max}$ of the Taylor series typically exceeds the dimension d, the second implementation requires less administrative overhead (creation of intermediate objects, etc.).

Finally it should be noted that the **centered** and **union** approaches can be used together to obtain even tighter enclosures, albeit at further increased cost.

4.2 Choice of the subdivision direction(s) for recursion

At the two extremes, the **full** subdivision strategy cuts the box along *each* axis into ν pieces, resulting in $m = \nu^d$ subboxes, and the **largest** strategy cuts only along a single direction, yielding $m = \nu$ subboxes.

While the **full** subdivision effectively minimizes the recursion depth (and therefore also the number of intermediate boxes that are considered without contributing to the final result), the last subdivision step tends to produce more (and smaller) boxes than would be required to fulfill the termination criterion, thus unnecessarily increasing the total work. The **largest** strategy, on the other hand, tries to minimize the number of boxes at the lowest recursion levels, at the cost of having to handle more intermediate boxes.

The **largest two** strategy is a compromise between these two strategies. It subdivides the current box along two axis, resulting in $m = \nu^2$ subboxes.

The **largest** and **largest two** strategies select the direction(s) j that yield the largest value (the two largest values) for some functions v_j associated with the coordinate axes.

A commonly used criterion is $v_j([x]) = \text{width}[x_j]$, meaning that the box is split along its longest edge(s). As the recursion is started with the cube $[x] = [-1, 1]^d$, this criterion is equivalent to **round-robin**, where the direction of the cut varies cyclically with the recursion depth. This criterion is very cheap to implement, and it produces boxes with an aspect ratio of at most 2, which tends to reduce the over-estimation.

The **worst-of-best** criterion uses

$$v_j([x]) = \min_{n_j=1}^{n_{\max}} e^{(n_j)} \left| [f_{2n_j,j}] \right| ,$$

i.e., by Equation (5), it selects the direction that contributes most to the best error bound $\text{width}[R^{(n_1 \times \cdots \times n_d)}]$ that can be attained with the current box. This is a purely heuristic criterion because subdivision in the direction of the largest contribution does not necessarily give an above-average reduction of $\text{width}[R]$.

The **gradient-based** criterion takes a more systematic approach to effectively reducing $\text{width}[R]$: Here, we first determine the "worst-of-best" remainder coefficient

$$(j^*, n_j^*) \longleftarrow \max_{j=1}^{d} \min_{n_j=1}^{n_{\max}} e^{(n_j)} \left| [f_{2n_j,j}] \right|$$

and then we set

$$v_i([x]) = \left| [f_{2n_j^*,j^*,i}] \right| \cdot \text{width}[x_i] ,$$

where $[f_{2n_j^*,j^*,i}]$ is an enclosure for the range of $(\partial f^{2n_j^*+1})/(\partial x_i \partial x_{j^*}^{2n_j^*})$ over $[x]$. By Equation (6) this means that we reduce the largest contribution to the largest (of the best) remainder coefficients. Of course, this criterion is available only with centered Taylor evaluation. A similar criterion is used in global optimization, cf. Ratz and Csendes (1995).

The **full**, **largest** and **largest two** subdivision strategies, in conjunction with the **worst-of-best** or **gradient-based** criterion, may also be used for the Taylor subdivision discussed in Section 4.1. Then the values of the current box' *father* are the basis for determining the direction(s) for subdivision.

4.3   Number of segments for subdivision

Together with the number of cutting direction(s) the number ν of segments in each direction determines the number of resulting subboxes. For Taylor subdivision we can use any $\nu_T \geq 1$, with larger values leading to sharper enclosures at a higher cost per box. In the recursive subdivision $\nu_R \geq 2$ is required. Here, larger values reduce the maximum recursion depth but may produce an excessive number of boxes at the deepest recursion levels.

4.4   Maximum Gauss order

The parameter $n_{\max}$ influences the recursion depth reached by the adaptive algorithm and the amount of work spent on each box. Higher-order formulas allow the termination criterion to be met after a few recursion steps, whereas small orders may force very deep recursion. On the other hand, the work per box increases with n_{max} in two ways: First, the approximation $[A^{(n_1 \times \cdots \times n_d)}]$ may require up to n_{max}^d function evaluations, and the computation of the Taylor coefficients is more expensive, too; see Table 1.

5 Experimental results

The goal of our investigations was to find heuristics and settings for the algorithmic parameters that will yield good average performance in a "black-box" quadrature routine, which has no specific a priori knowledge about the integrand. In practice this means that the maximum recursion depth, the number of boxes generated during the adaptive quadrature, the total number of function evaluations, and in particular the overall execution time should be as small as possible.

In our experiments, various settings for the parameters discussed in Section 4 were tested with integrals of dimension up to four. The test functions were:

- $f_1(x) = e^{-\|x\|_2^2}$ with $x \in [-4, 4]^d$, for $d = 1, 2, 3, 4$. This function is not very difficult.

- $f_2(x) = N(x) \cdot \sin(e^{N(x)})$ with $N(x) = \frac{1}{d} \cdot \sum_{j=1}^{d} (x_j - \beta)^2$ and $x \in [\alpha, \beta]^d$, for $d = 1, 2, 3, 4$. This function is somewhat harder in the vicinity of α^d.

- $f_3(x) = N(x) \cdot \sin(e^{N(x)})$ with $N(x) = \frac{1}{d} \cdot \sum_{j=1}^{d} (x_j - \xi_j)^2$, $\xi_j = \alpha + \frac{d-j}{d-1} \cdot (\beta - \alpha)$, and $x \in [\alpha, \beta]^d$, for $d = 2, 4$. Here the partial derivatives depend heavily on the direction.

- f_4 is as f_3, but before evaluating the function, the coordinates are twice rotated to and fro in the (x_1, x_2) and (x_3, x_4) planes. Thus f_4 is identical to f_3 for point arguments, but the rotations cause severe over-estimation for larger interval arguments.

Nearly 4000 runs were made with the recursive quadrature algorithm from Section 3, implemented in Pascal-XSC. The results of the experiments can be summarized as follows.

- Taylor evaluation with the **union** approach was able to reduce the recursion depth. On average, the number of recursive calls went down by roughly 10%, and the total number of function evaluations for the approximations $[A^{(n_1 \times \cdots \times n_d)}]$ was also reduced by some 5%. (This gain being somewhat smaller indicates that with **union** the termination criterion could be reached at earlier recursion levels, but with formulas of slightly higher average order.)

  Unfortunately, these reductions did not carry over to execution time. In fact, subdivision into $\nu_T = 2$ subboxes along the **largest** axis caused the time to increase by an average of approximately 50%. A more detailed analysis reveals that the loss is highest for "easy" problems (up to 80%) and smaller for the harder higher-dimensional problems (roughly 20% for $d = 4$). Thus, really hard problems might actually benefit from the **union** approach; indeed, for a sextuple integral arising in a physical experiment (see Holzmann et al. (1996)) subdivision along two axes even halved the total time.

- **Centered** Taylor evaluation could again slightly reduce the number of boxes and of function evaluations, but in terms of execution time it *did not pay in a single test case*. This is due to the significantly higher amount of work per box that is induced by the Taylor-with-gradients or gradient-with-Taylors arithmetic. As the time goes up by a factor of five or more, further optimizations of the Taylor+gradient arithmetics (like avoiding the generation of intermediate objects) can only slightly alleviate this problem. Even a speedup of two would not be sufficient, so substantially different approaches for obtaining the same information are needed.

  The performance gap decreased when the problems became harder. Therefore, the situation might change for very difficult problems because at deeply nested recursion levels the **centered** evaluation gives much sharper enclosures for the $f_{k,j}[x]$. So it might become the only feasible way to fulfill the termination criterion at all.

- Subdivision strategies: The **largest** subdivision strategy, based on the **worst-of-best** criterion, was able to reduce the number of boxes, function evaluations, and the execution time by 25%, 10%, and 20%, respectively, as compared to **round-robin**. (The gain in time being much larger than the reduction of function evaluations indicates that the overall time is dominated by the Taylor evaluations.) For the four-dimensional integrals, the speedup reached 35%. Thus, the harder problems gained more from the heuristic strategy. Switching to the **largest two** gave another 10–20% improvement in the number of boxes and the execution time, whereas *full subdivision never paid* for $d > 2$.

  As mentioned above, no run with **centered** Taylor evaluation was competitive. Even considering only such runs, the heuristic **worst-of-best** criterion performed almost as well as the **gradient-based** criterion. Given the former's greater flexibility (it does not require **centered** Taylor evaluation), it should be preferred.

- The best subdivision factors were $\nu = 2$ or $\nu = 3$, with bisection being superior in 80% of the cases.

- The harder the problem, the higher the maximum Gauss order should be; $n_{max} = 12$ proved adequate in most cases. For the simplest problems $n_{max} = 6$ or $n_{max} = 8$ performed slightly better since they require less work for Taylor evaluation, but for $d = 4$ they were slower by a factor of 5 (3, respectively) because they lead to deeper recursions.

6   Conclusions

The results from the experiments suggest the following multi-stage strategy:

- Above a certain recursion level ($2d$, say) the control parameters are set to achieve *minimum work per box*, that is, low Gauss order, no centered form, etc. With these settings, easy problems can be handled efficiently.

- Below this recursion level we assume that the problem is harder and switch to parameters that yield sharper enclosures and allow earlier termination (e.g., higher Gauss order, Taylor evaluation with moderate subdivision).

- If the recursion reaches much further down then the problem is really hard, and we should try to limit the recursion depth by further increasing the Gauss order and using the centered form, together with moderate subdivision, for the Taylor evaluation.

Even with well-tuned control parameters, hard problems may take several hours to complete. This is also true if the dimension is increased beyond $d = 4$. A straight-forward master–slave type parallelization might help to reduce the computing time in these cases.

Two other directions for further research are the development of more efficient ways to compute the mixed partials in the **gradient-based** Taylor evaluation, and investigating non-product verified quadrature formulas in order to reduce the number function evaluations.

References

Davis, P.J., Rabinowitz, P. (1984): Methods of Numerical Integration. Academic Press, New York

Golub, G.H., Welsch, J.H. (1969): Calculation of Gauss quadrature rules. Math. Comp. 23: 221–230

Griewank, A., Utke, J., Walther, A. (1997): Evaluating higher derivative tensors by forward propagation of univariate Taylor series. Preprint IOKOMO-09-971, TU Dresden, Inst. of Scientific Computing. Revised June 1998

Holzmann, O., Lang, B., Schütt, H. (1996): Newton's constant of gravitation and verified numerical quadrature. Reliable Comput. 2: 229–239

Ratz, D., Csendes, T. (1995): On the selection of subdivision directions in interval branch-and-bound methods for global optimization. Journal of Global Optimization 7: 183–207

Storck, U. (1993): Verified Calculation of the nodes and weights for Gaussian quadrature formulas. Interval Computations 4: 114–124

Stroud, A.H. (1971): Approximative Calculation of Multiple Integrals. Prentice-Hall, Englewood Cliffs

On the Shape of the Fixed Points of $[\mathbf{f}]([\mathbf{x}]) = [\mathbf{A}][\mathbf{x}] + [\mathbf{b}]$

G. Mayer, I. Warnke

Dedicated to Prof. Dr. Lothar Berg on the occasion of his 70th birthday.

1 Introduction

When solving linear systems of equations

$$Cx = b \tag{1}$$

with a real $n \times n$ matrix $C = (c_{ij})$ and a real vector b with n components one often uses iterative methods – particularly when C is a large sparse matrix. Probably the most elementary iterative method can be derived from the so-called Richardson splitting $C = I - A$ of C, where I is the identity matrix and $A := I - C$. This splitting induces the equivalent fixed point formulation $x = Ax + b$ of (1) which leads to the iterative method

$$x^{k+1} = Ax^k + b. \tag{2}$$

It converges for any starting vector x^0 if and only if the spectral radius $\rho(A)$ of A is less than one. In this case the limit x^* is the same for any x^0 and fulfills $x^* = Ax^* + b$, i.e., it is the unique fixed point of the function $f : \mathbb{R}^n \to \mathbb{R}^n$ with $f(x) = Ax + b$. Interval linear systems

$$[C]x = [b], \quad [C] \in I(\mathbb{R}^{n \times n}), \ [b] \in I(\mathbb{R}^n) \tag{3}$$

$(I(\mathbb{R}^{n \times n}) = $ set of real $n \times n$ interval matrices, $I(\mathbb{R}^n) = $ set of real interval vectors with n components) are, by definition, the collection of linear systems $Cx = b$ with $C \in [C]$, $b \in [b]$. Solving (3) means computing the solution set

$$S := \{\, x \in \mathbb{R}^n \mid Cx = b, \ C \in [C], \ b \in [b] \,\}$$

which is the union of at most 2^n intersections of finitely many half spaces (cf. Oettli and Prager (1964), Beeck (1974), Hartfiel (1980), or Alefeld et al. (1997), e.g.). Unfortunately, the determination of S is complicated. Therefore, one often confines oneself to computing an enclosure $[x]^*$ of S by an interval vector – preferably the interval hull $[x]^S \in I(\mathbb{R}^n)$ which is defined as the tightest of these enclosures. An interval enclosure $[x]^*$ of S can, for instance, be obtained by using the analogue of (2) which reads

$$[x]^{k+1} = [A][x]^k + [b], \quad k = 0, 1, \ldots, \tag{4}$$

with $[A] := I - [C]$. O. Mayer (1968) showed that (4) is convergent to a unique interval vector $[x]^* \supseteq S$ if and only if the spectral radius $\rho(|[A]|)$ of the absolute value $|[A]|$ of $[A]$ (cf. Section 2) is less than one. It is easy to see that in this case $[x]^*$ is the unique solution of the interval equation

$$[x] = [A][x] + [b] \tag{5}$$

or, equivalently, $[x]^*$ is the unique fixed point of the interval function $[f]$ defined by

$$[f]([x]) := [A][x] + [b]. \tag{6}$$

For completeness we mention that in general $[x]^*$ is not an algebraic solution of the interval system $[C][x] = [b]$ with $[C] = I - [A]$. For algebraic solutions we refer, e.g., to Ratschek and Sauer (1982). Although O. Mayer's results on (4) and (6) are nearly exhaustive they do not indicate how $[x]^*$ looks like. It is this aspect to which our paper is devoted. For particular classes of matrices $[A]$ with $\rho(|[A]|) < 1$ and particular classes of vectors $[b]$, respectively, we are able to derive expressions for $[x]^*$ which depend only on the interval bounds or – equivalently – on the midpoint and the radius of the input data $[A]$ and $[b]$. It is well–known that $[x]^*$ normally overestimates not only S but also the interval hull $[x]^S$. For symmetric interval matrices $[A]$ (which are defined by $[A] = -[A]$) we will show that in each entry of $[x]^*$ at least one bound is optimal in the sense that it coincides with the corresponding bound of $[x]^S$. This result turns out to be completely analogous to that in a paper of Mayer and Rohn (1998) in which the interval vector $[x]^G$ from the interval Gaussian algorithm was compared with $[x]^S$ for matrices of the form $[C] = I - [A]$ with symmetric $[A]$.

Before we present our results in Section 3 we list some notations which we use throughout the paper.

2 Notations

In addition to the notations $I(\mathbb{R}^n)$, $I(\mathbb{R}^{n \times n})$, $\rho(\cdot)$ in Section 1 we denote the set of real compact intervals $[a] = [\underline{a}, \overline{a}]$ by $I(\mathbb{R})$. We use $[A] = [\underline{A}, \overline{A}] = ([a]_{ij}) = ([\underline{a}_{ij}, \overline{a}_{ij}]) \in I(\mathbb{R}^{n \times n})$ simultaneously without further reference, and we apply a similar notation for interval vectors. If $[a] \in I(\mathbb{R})$ contains only one element a then, trivially, $\underline{a} = \overline{a} = a$. In this case we identify $[a]$ with its element writing $[a] \equiv a$ and calling $[a]$ degenerate or a point interval. Analogously, we define degenerate interval vectors / point vectors and degenerate interval matrices / point matrices, respectively. We call $[a] \in \mathbb{R}$ symmetric if $[a] = -[a]$, i.e., if $[a] = [-a, a]$ with some real number $a \geq 0$. For intervals $[a]$, $[b]$ we introduce the midpoint $\breve{a} := (\underline{a} + \overline{a})/2$, the absolute value $|[a]| := \max\{|\underline{a}|, |\overline{a}|\}$, the diameter $d[a] := \overline{a} - \underline{a}$, the radius $\mathrm{rad}[a] := d[a]/2$ and the (Hausdorff) distance $q([a], [b]) := \max\{|\underline{a} - \underline{b}|, |\overline{a} - \overline{b}|\}$. For interval vectors and interval matrices these quantities are defined entrywise, for instance $|[A]| := (|[a]_{ij}|) \in \mathbb{R}^{n \times n}$. We assume some familiarity when working with these definitions and when applying interval arithmetic. For details see, e.g., the introductory chapters in the book of Alefeld and Herzberger (1983) or Neumaier (1990). As usual we call $A \in \mathbb{R}^{n \times n}$ non-negative if $a_{ij} \geq 0$ for $i, j = 1, \ldots, n$, writing $A \geq O$ in this case. Non-negative vectors $b \geq 0$ and non-positive vectors $b \leq 0$ are defined analogously. We call $A \in \mathbb{R}^{n \times n}$ an M matrix if it is regular with all its off-diagonal entries being non-positive and with its inverse being non-negative. We use this term also for interval matrices $[A] \in \mathrm{I}(\mathbb{R}^{n \times n})$ if each matrix $A \in [A]$ is an M matrix.

3 Results

We first recall O. Mayer's result mentioned in Section 1 assuming here and in the sequel $[A] \in \mathrm{I}(\mathbb{R}^{n \times n})$, $[b] \in \mathrm{I}(\mathbb{R}^n)$.

Theorem 1

For every starting vector $[x]^0 \in \mathrm{I}(\mathbb{R}^n)$ the iteration (4) converges to the same vector $[x]^* \in \mathrm{I}(\mathbb{R}^n)$ if and only if $\rho(|[A]|) < 1$. In this case $[x]^*$ contains the solution set

$$S := \{\, x \in \mathbb{R}^n \mid \exists A \in [A],\ b \in [b] : (I - A)x = b \,\} \tag{7}$$

and is the unique fixed point of $[f]$ from (6).

$\square$

The proof of Theorem 1 which we do not want to repeat here is essentially based on Banach's fixed point theorem: Since $q([f]([x]), [f]([y])) \leq |[A]| q([x], [y])$, the function $[f]$ is a so-called $|[A]|$-contraction (Alefeld and Herzberger 1983), therefore, there is a nonsingular diagonal matrix $D \in \mathbb{R}^{n \times n}$ with positive entries d_{ii} such that $\|Dq([f]([x]), [f]([y]))\|_\infty$ with the maximum norm $\|\cdot\|_\infty$ is a contraction on $\mathrm{I}(\mathbb{R}^n)$. For details see the book of Alefeld and Herzberger (1983).

We now list our classes of input data $[A]$ and $[b]$ for which we were able to represent $[x]^*$ by a *finite* expression. We skip the trivial case where $[A]$ and $[b]$ both are degenerate and start with a degenerate interval matrix $[A]$, i.e., $[A] \equiv A$, and an arbitrary interval vector $[b]$.

Theorem 2

Let $[A] \equiv A \in \mathbb{R}^{n \times n}$ with $\rho(|[A]|) < 1$ and let $[b] \in \mathrm{I}(\mathbb{R}^n)$. Then

$$[x]^* = (I - A)^{-1}\breve{b} + (I - |A|)^{-1}\mathrm{rad}[b] \cdot [-1, 1] \tag{8}$$

is the unique fixed point of $[f]$ from (6).

Proof.

From

$$[x]^* = A[x]^* + [b] \tag{9}$$

we get $\check{x}^* = A\check{x}^* + \check{b}$ whence $\check{x}^* = (I - A)^{-1}\check{b}$. In addition (9) implies $\text{rad}[x]^* = |A|\text{rad}[x]^* + \text{rad}[b]$. Hence $\text{rad}[x]^* = (I - |A|)^{-1}\text{rad}[b]$, and (8) follows from $[x]^* = \check{x}^* + \text{rad}[x]^*[-1, 1]$.

□

As a second particular case we consider an interval matrix $[A]$ whose entries all are symmetric intervals, i.e., $[A] = -[A]$.

Theorem 3

Let $[A] = -[A] \in I(\mathbb{R}^{n \times n})$ with $\rho(|[A]|) < 1$ and let $[b] \in I(\mathbb{R}^n)$. Then

$$\begin{aligned}
[x]^* &= [b] + (I - |[A]|)^{-1}[A][b] \\
&= \check{b} + (I - |[A]|)^{-1}|[A]||\check{b}| \cdot [-1, 1] + (I - |[A]|)^{-1}\text{rad}[b] \cdot [-1, 1]
\end{aligned} \tag{10}$$

is the unique fixed point of $[f]$ from (6).

Proof.

We first remark that $(I - |[A]|)^{-1} \geq O$ holds by virtue of $\rho(|[A]|) < 1$ and the Neumann series for $|[A]|$. By the symmetry of the entries of $[A]$ we get at once from (5) $\check{x}^* = \check{b}$ and

$$\begin{aligned}
\text{rad}[x]^* &= |[A]| \cdot |[x]^*| + \text{rad}[b] = |[A]|(|\check{x}^*| + \text{rad}[x]^*) + \text{rad}[b] \\
&= |[A]|(|\check{b}| + \text{rad}[x]^*) + \text{rad}[b]
\end{aligned}$$

whence $(I - |[A]|)\text{rad}[x]^* = |[A]||\check{b}| + \text{rad}[b]$. This implies

$$\text{rad}[x]^* = (I - |[A]|)^{-1}(|[A]||\check{b}| + \text{rad}[b])$$

and proves the last expression in (10). From

$$\begin{aligned}
[b] + (I &- |[A]|)^{-1}[A][b] = \check{b} + \text{rad}[b] \cdot [-1, 1] + (I - |[A]|)^{-1}|[A]| \cdot |[b]|[-1, 1] \\
&= \check{b} + \{(I - |[A]|)^{-1}(|[A]| \cdot |[b]|) + (I - |[A]|)\text{rad}[b])\} \cdot [-1, 1] \\
&= \check{b} + (I - |[A]|)^{-1}\{|[A]|(|[b]| - \text{rad}[b]) + \text{rad}[b]\} \cdot [-1, 1] \\
&= \check{b} + (I - |[A]|)^{-1}(|[A]||\check{b}| + \text{rad}[b]) \cdot [-1, 1]
\end{aligned}$$

we get the first representation of $[x]^*$ in (10).

□

Remark 1

a) Although – by virtue of the missing distributivity in $I(\mathbb{R})$ – the multiplication of interval matrices is not associative we could omit brackets in (10) because of symmetric entries and point matrices, respectively.

b) Theorem 3 follows immediately from Theorem 4.4.10 in the book of Neumaier (1990), p. 150: In order to avoid a duplicate notation denote Neumaier's matrix $[A]$ by $[B]$. Then $[B] = I - [A]$ with $[A]$ as in Theorem 2. Hence $\check{B} = I$, in particular $\check{B}$ is diagonal. Since $\rho(|[A]|) < 1$ the matrix $[B]$ is an H matrix and $[x]^* = [b] + \{(I - |[A]|)^{-1} \cdot |[b]| - |[b]|\}[-1,1]$ according to Neumaier's Theorem 4.4.10. Using $[A] = [-|[A]|, |[A]|]$ we get

$$
\begin{aligned}
[x]^* &= [b] + \{(I - |[A]|)^{-1}(|[b]| - (I - |[A]|)|[b]|)\}[-1,1] \\
&= [b] + (I - |[A]|)^{-1}|[A]||[b]|[-1,1] = [b] + (I - |[A]|)^{-1}[A][b]
\end{aligned}
$$

as in (10).

$\square$

In the two preceding cases we could isolate the term $(I - |[A]|)^{-1}\mathrm{rad}[b] \cdot [-1,1]$ which only depends on the diameter of $[b]$ and which influences the diameter of $[x]^*$ but not its midpoint $\check{x}^*$. In our next theorem we prove that such a splitting of $[x]^*$ always exists for arbitrary $[A]$ with $\rho(|[A]|) < 1$ and arbitrary $[b]$. In addition, we present bounds for $[x]^*$.

Theorem 4

Let $[A] \in I(\mathbb{R}^{n \times n})$ with $\rho(|[A]|) < 1$ and let $[b] \in I(\mathbb{R}^n)$. Choose any $\tilde{A} \in [A]$ and define

$$
\begin{aligned}
[u]^* &= (I - \tilde{A})^{-1}\check{b} + (I - |\tilde{A}|)^{-1}\mathrm{rad}[b] \cdot [-1,1] &\text{(cf. (8))} \\
[v]^* &= \check{b} + (I - |[A]|)^{-1}|[A]||\check{b}| \cdot [-1,1] + (I - |[A]|)^{-1}\mathrm{rad}[b] \cdot [-1,1] \\
& &\text{(cf. (10))}
\end{aligned}
$$

Then the fixed point $[x]^*$ of $[f]$ from (6) satisfies

$$
[u]^* \subseteq [x]^* \subseteq [v]^* \tag{11}
$$

and can be represented as

$$
[x]^* = [\check{x}]^* + (I - |[A]|)^{-1}\mathrm{rad}[b] \cdot [-1,1] \tag{12}
$$

with some (unknown) interval vector $[\check{x}]^*$.

If $\tilde{A}$ satisfies $|\tilde{A}| = |[A]|$ then

$$
(I - \tilde{A})^{-1}\check{b} \in [\check{x}]^* \subseteq \check{b} + (I - |[A]|)^{-1}|[A]||\check{b}| \cdot [-1,1], \tag{13}
$$

in particular,

$$
\check{b} + |\check{b}| - (I - |[A]|)^{-1}|\check{b}| \le \underline{\check{x}}^* \le (I - \tilde{A})^{-1}\check{b} \le \overline{\check{x}}^* \le \check{b} - |\check{b}| + (I - |[A]|)^{-1}|\check{b}|. \tag{14}
$$

Proof.

Start (4) with $[x]^0 := [u]^*$. Then

$$[u]^* = \tilde{A}[u]^* + [b] \subseteq [A][u]^* + [b] = [A][x]^0 + [b] = [x]^1,$$

and by induction one obtains $[u]^* \subseteq [x]^k$, $k = 0, 1, \ldots$, hence $[u]^* \subseteq \lim_{k \to \infty} [x]^k = [x]^*$. This proves the left inclusion of (11). The right one follows analogously, this time starting (4) with $[x]^0 := [v]^*$ and taking into account $[A] \subseteq [-|[A]|, |[A]|] =: [B]$, $\rho(|[B]|) = \rho(|[A]|) < 1$. The representation (12) is possible since (5) implies $\mathrm{rad}[x]^* \geq |[A]|\mathrm{rad}[x]^* + \mathrm{rad}[b]$ whence $\mathrm{rad}[x]^* \geq (I - |[A]|)^{-1}\mathrm{rad}[b]$. The inclusion (13) follows from (11) and (12), the inequalities (14) result from (13) combined with $(I - |[A]|)^{-1}|[A]||\check{b}| = -|\check{b}| + (I - |[A]|)^{-1}|\check{b}|$.

□

In order to compare the solution $[x]^*$ in Theorem 3 with the interval hull $[x]^S$ of the solution set S in (2) we first cite a theorem due to Hansen and Rohn which expresses $[x]^S$ explicitly in terms of the solution of a single linear system of equations.

Theorem 5   *(Hansen 1992), (Rohn 1993)*

Let $[A] = -[A] \in I(\mathbb{R}^{n \times n})$, $[C] := I - [A]$, $\rho(|[A]|) < 1$, $M := (I - |[A]|)^{-1}$, $[b] \in I(\mathbb{R}^n)$. Denote by $[x]^S$ the interval hull of the solution set $S := \{x \mid \exists \tilde{C} \in [C], \tilde{b} \in [b] : \tilde{C}x = \tilde{b}\}$. Then for each $i \in \{1, \ldots, n\}$ we have

$$\underline{x}_i^S := \min\{\underline{x}_i^S, \nu_i \tilde{\underline{x}}_i^S\}, \qquad \overline{x}_i^S := \max\{\tilde{x}_i^S, \nu_i \tilde{x}_i^S\}, \tag{15}$$

where

$$\underline{x}_i^S := -x_i^* + m_{ii}(\check{b} + |\check{b}|)_i, \quad \tilde{x}_i^S := x_i^* + m_{ii}(\check{b} - |\check{b}|)_i, \quad x_i^* := \left(M(|\check{b}| + \mathrm{rad}[b])\right)_i, \tag{16}$$

and

$$\nu_i := \frac{1}{2m_{ii} - 1} \in (0, 1].$$

□

Comparing (10) with (15) reveals that in the situation of Theorem 3 at least one bound of each component $[x]_i^*$ coincides with the corresponding one of the interval hull $[x]^S$ for the interval linear system $(I - [A])x = [b]$. This is stated and proved in the subsequent Theorem 6. That the second bounds can differ is illustrated by Example 1.

Theorem 6

Let $[A] = -[A] \in I(\mathbb{R}^{n \times n})$ with $\rho(|[A]|) < 1$ and let $[b] \in I(\mathbb{R}^n)$. Denote by $[x]^*$ the solution (10) of (5) and by $[x]^S$ the interval hull of the solution set S associated with the interval linear system $(I - [A])x = [b]$. Then $\underline{x}_i^* = \underline{x}_i^S$ if $\breve{b}_i \leq 0$ and $\overline{x}_i^* = \overline{x}_i^S$ if $\breve{b}_i \geq 0$ for each $i \in \{1, \ldots, n\}$. In particular,

$$[x]_i^* = [x]_i^S = \left((I - |[A]|)^{-1} \mathrm{rad}[b] \right)_i \cdot [-1, 1]$$

if $\breve{b}_i = 0$.

Proof.

Fix $i \in \{1, \ldots, n\}$ and assume $\breve{b}_i \geq 0$. Then $\tilde{x}_i^S = x_i^*$ where we used (16) and the notation of Theorem 5. From (15) we therefore obtain

$$\overline{x}_i^S = \tilde{x}_i^S = x_i^* = \left(M(|\breve{b}| + \mathrm{rad}[b]) \right)_i.$$

From (10) we get

$$\begin{aligned}
\overline{x}^* &= \breve{b} + M|[A]| \cdot |\breve{b}| + M\mathrm{rad}[b] \\
&= \breve{b} + M\{(|[A]| - I) \cdot |\breve{b}| + |\breve{b}| + \mathrm{rad}[b]\} \\
&= \breve{b} - |\breve{b}| + M\{|\breve{b}| + \mathrm{rad}[b]\},
\end{aligned}$$

whence $\overline{x}_i^* = \overline{x}_i^S$.

If $\breve{b}_i \leq 0$ an analogous argumentation yields to $\underline{x}_i^S = -\underline{x}_i^* = -\left(M(|\breve{b}| + \mathrm{rad}[b]) \right)_i = \underline{x}_i^*$.

From these two cases the remaining part of the theorem follows now trivially.

□

Remark 2

When applying the interval Gaussian algorithm (Alefeld and Herzberger 1983) to $I - [A]$ and $[b]$ with $[A]$, $[b]$ from Theorem 6 then the resulting vector $[x]^G$ shows a similar behaviour as $[x]^*$: According to Theorem 3.4 in a paper of Mayer and Rohn (1998) one has $\underline{x}_i^G = \underline{x}_i^S$ if $\breve{b}_i \leq 0$ and $\overline{x}_i^G = \overline{x}_i^S$ if $\breve{b}_i \geq 0$ for each $i \in \{1, \ldots, n\}$. In particular, $[x]_i^G = [x]_i^S$ holds in the case $\breve{b}_i = 0$.

□

Example 1

Let

$$[A] = \begin{pmatrix} 0 & [-1, 1] \\ [-\frac{1}{2}, \frac{1}{2}] & 0 \end{pmatrix}, \qquad [b] \equiv b = \begin{pmatrix} -1 \\ 1 \end{pmatrix}.$$

Then $\rho(|[A]|) = \dfrac{1}{\sqrt{2}} < 1$ and $[x]^* = ([-4,2],[-1,3])^T$. By virtue of Theorem 5 with

$$[C] = I - [A] = \begin{pmatrix} 1 & [-1,1] \\ [-\frac{1}{2}, \frac{1}{2}] & 1 \end{pmatrix} \quad \text{and} \quad M = (I - |[A]|)^{-1} = \begin{pmatrix} 2 & 2 \\ 1 & 2 \end{pmatrix},$$

we obtain the interval hull $[x]^S = ([-4,0],[\frac{1}{3},3])^T$ which confirms Theorem 6 and which shows $[x]^* \neq [x]^S$.

The vector $[x]^G$ resulting from the interval Gaussian algorithm for $[C]$ and b reads $[x]^G = ([-4,2],[\frac{1}{3},3])^T$. Thus it satisfies $[x]^S \subseteq [x]^G \subseteq [x]^*$ but differs from $[x]^S$ and $[x]^*$.

$\blacksquare$

In our third particular case we restrict $[b]$ to be symmetric, i.e., $[b] = -[b]$.

Theorem 7

Let $[A] \in I(\mathbb{R}^{n \times n})$ with $\rho(|[A]|) < 1$ and let $[b] = -[b] \in I(\mathbb{R}^n)$. Then

$$[x]^* = (I - |[A]|)^{-1}[b] = (I - |[A]|)^{-1}\mathrm{rad}[b] \cdot [-1,1] \qquad (17)$$

is the unique fixed point $[x]^*$ of $[f]$ from (6).

Proof.

Start the iteration (4) with $[x]^0 := [b]$. Then $[x]^1 = [A][b] + [b] = (|[A]| + I)[b]$ which is symmetric. Therefore,

$$\begin{aligned} [x]^2 &= [A]\{(I + |[A]|)[b]\} + [b] = \{|[A]|(I + |[A]|)\}[b] + [b] \\ &= (I + |[A]| + |[A]|^2)[b], \end{aligned}$$

and (17) follows by induction using the Neumann series for $|[A]|$.

$\blacksquare$

Our fourth case is very specific because we assume that $I - [A]$ has degenerate columns and $\check{b}$ is contained in the linear space spanned by these columns.

Theorem 8

Let $[b] \in I(\mathbb{R}^n)$ and $[A] \in I(\mathbb{R}^{n \times n})$ with $\rho(|[A]|) < 1$. Denote by M the set of all column indices j for which at least one of the entries $[a]_{ij}$, $i = 1, \ldots, n$, is not degenerate. Denote by $\hat{A}$ the matrix which arises from $[A]$ by replacing the columns with index $j \in M$ by the corresponding column of the identity matrix I. Assume that the equation

$$x = \hat{A}x + \check{b} \qquad (18)$$

160

is solvable. Then there is exactly one solution x^* of (18) which satisfies

$$x_i^* = 0 \quad \text{for all } i \in M, \tag{19}$$

and

$$[x]^* = x^* + (I - |[A]|)^{-1}\text{rad}[b] \cdot [-1, 1] \tag{20}$$

is the unique fixed point of $[f]$ from (6).

Proof.

If $\tilde{x}^* \in \mathbb{R}^n$ differs from a solution of (18) only in those components $\tilde{x}_i^*$ for which $i \in M$ then $\tilde{x}^*$ is also a solution of (18). This results immediately from the particular shape of $\hat{A}$. Therefore, since we assumed (18) to be solvable, there is at least one solution x^* of (18) which satisfies (19). If $y^* \neq x^*$ is another solution of (18), (19) then

$$[A](x^* - y^*) = \hat{A}(x^* - y^*) = x^* - y^*,$$

i.e., $x^* - y^*$ is a solution of the interval equation $[x] = [A][x]$. Since $\rho(|[A]|) < 1$ the solution $[z]^*$ of this system is unique by Theorem 1 and thus reads $[z]^* = 0 = x^* - y^*$. Hence the solution x^* of (18), (19) is unique. By virtue of the zero components of x^* we get with $[x]^*$ from (20)

$$
\begin{aligned}
[A][x]^* + [b] &= [A]\{x^* + (I - |[A]|)^{-1}\text{rad}[b] \cdot [-1, 1]\} + [b] \\
&= [A]x^* + [A]\{(I - |[A]|)^{-1}\text{rad}[b] \cdot [-1, 1]\} + [b] \\
&= \hat{A}x^* + |[A]|\{(I - |[A]|)^{-1}\text{rad}[b] \cdot [-1, 1]\} + \check{b} + \text{rad}[b][-1, 1] \\
&= x^* + (I - |[A]|)^{-1}\text{rad}[b] \cdot [-1, 1] = [x]^*.
\end{aligned}
$$

Note that the distributivity needed for the second equality holds because $[a]_{ij}$ is degenerate if x_j^* differs from zero. $\qquad \square$

For completeness we mention without proof the following result on matrices $[A] \in I(\mathbb{R}^{n \times n})$ with $\rho(|[A]|) < 1$ and $\underline{A} \geq O$ which is essentially contained in a paper of Barth and Nuding (1974) and – in parts – in a paper of Kulisch (1969). Note that the assumptions on $[A]$ imply that $[C] = I - [A]$ is an M matrix (Berman and Plemmons 1994).

Theorem 9

Let $[A] \in I(\mathbb{R}^{n \times n})$ with $\rho(|[A]|) < 1$ and $\underline{A} \geq O$, and let $[b] \in I(\mathbb{R}^n)$ fulfill $\overline{b} \leq 0$ or $\underline{b} \leq 0 \leq \overline{b}$ or $0 \leq \underline{b}$. Then

$$[x]^* = \begin{cases} [(I - \overline{A})^{-1}\underline{b}, (I - \underline{A})^{-1}\overline{b}], & \text{if } \overline{b} \leq 0, \\ [(I - \overline{A})^{-1}\underline{b}, (I - \overline{A})^{-1}\overline{b}], & \text{if } \underline{b} \leq 0 \leq \overline{b}, \\ [(I - \underline{A})^{-1}\underline{b}, (I - \overline{A})^{-1}\overline{b}], & \text{if } 0 \leq \underline{b} \end{cases}$$

is the unique fixed point $[x]^*$ of $[f]$ from (6). It equals the interval hull $[x]^S$ of the solution set S associated with the interval linear system $(I - [A])x = [b]$. $\qquad \square$

In agreement with a referee we finally remark that our results are primarily of theoretical character since they involve exact solutions z of real linear systems, for instance $z = (I - |[A]|)^{-1} \mathrm{rad}[b]$, $z = (I - A)^{-1} \breve{b}$, ...

References

Alefeld, G., Herzberger, J. (1983): Introduction to Interval Computations. Academic Press, New York

Alefeld, G., Kreinovich, V., Mayer, G. (1997): On the Shape of the Symmetric, Persymmetric, and Skew–Symmetric Solution Set. SIAM J. Matrix Anal. Appl. 18: 693–705

Barth, W., Nuding, E. (1974): Optimale Lösung von Intervallgleichungssystemen. Computing 12: 117–125

Beeck, H. (1974): Zur scharfen Außenabschätzung der Lösungsmenge bei linearen Intervallgleichungssystemen. Z. Angew. Math. Mech. 54, T208–T209

Berman, A., Plemmons, R. J. (1994): Nonnegative Matrices in the Mathematical Sciences. Classics in Applied Mathematics 9, SIAM, Philadelphia

Hansen, E. R. (1992): Bounding the solution of linear interval equations. SIAM J. Numer. Anal. 29: 1493–1503

Hartfiel, D. J. (1980): Concerning the Solution Set of $Ax = b$ where $P \leq A \leq Q$ and $p \leq b \leq q$. Numer. Math. 35: 355–359

Kulisch, U. (1969): Grundzüge der Intervallrechnung. In: Laugwitz, L. (ed.): Überblicke Mathematik 2. Bibliographisches Institut, Mannheim, pp. 51–98

Mayer, G., Rohn, J. (1998): On the Applicability of the Interval Gaussian Algorithm. Reliable Computing 4: 205–222

Mayer, O. (1968): Über die in der Intervallrechnung auftretenden Räume und einige Anwendungen. Ph.D. Thesis, Universität Karlsruhe, Karlsruhe

Neumaier, A. (1990): Interval Methods for Systems of Equations. Cambridge University Press, Cambridge

Oettli, W., Prager, W. (1964): Compatibility of Approximate Solution of Linear Equations with Given Error Bounds for Coefficients and Right–hand Sides. Numer. Math. 6: 405–409

Ratschek, H., Sauer, W. (1982): Linear Interval Equations. Computing 28: 105–115

Rohn, J. (1993): Cheap and Tight Bounds: The Recent Result by E. Hansen Can Be Made More Efficient. Interval Computations 4: 13–21

Exact Computation with `leda_real` — Theory and Geometric Applications

Kurt Mehlhorn and Stefan Schirra

1 Introduction

The number type `leda_real` provides exact computation for a subset of real algebraic numbers: Every integer is a `leda_real`, and `leda_reals` are closed under the basic arithmetic operations $+, -, *, /$ and k-th root operations. `leda_reals` guarantee correct results in all comparison operations. The number type is available as part of the LEDA C++ software library of efficient data types and algorithms (LEDA, Mehlhorn and Näher 2000). `leda_reals` provide user-friendly exact computation. All the internals are hidden to the user. A user can use `leda_reals` just like any built-in number type. The number type is successfully used to solve precision and robustness problems in geometric computing (Burnikel et al. 2000, Seel). It is particularly advantageous when used in combination with the computational geometry algorithms library CGAL.

2 Theory and practice in geometric computing

For the sake of better understanding underlying mathematical principles, theory of algorithm design often makes simplifying assumptions. In practice of geometric computing such simplifying assumptions cause notorious problems, if they don't hold, see for example Fortune (1993), Hoffmann (1989), Mehlhorn and Näher (1994), Schirra (2000), and Yap (1997b). Besides the assumption of so-called *general position* which excludes special cases in the input for an algorithm like three points lying on a line, the most puzzling assumption the implementor of a geometric algorithm is left to deal with is the assumption of *exact arithmetic* over the reals. The ubiquitous model of computation in computational geometry is the so-called *real RAM*. A real RAM can hold a single real number in each of its storage locations and perform basic arithmetic operations $+, -, *, /$, comparison between two real numbers, and $\sqrt{}$ and exp, log, if needed (Preparata and Shamos 1985). All these operations are assumed to give the correct result at unit cost.

With standard floating-point arithmetic, the default substitution for the real numbers in scientific computing, this assumption does not hold. Due to rounding and cancellation errors, the operations above might give incorrect results. Implementations simply using floating-point arithmetic as a substitute for exact real arithmetic typically "work" most of the time, but sometimes produce catastrophic errors, i.e., they crash, compute useless output, or even loop forever, although the implemented algorithms are theoretically correct. Such behavior can be observed also with currently available CAD-systems.

The root cause of such errors is that rounding errors in the computation lead to incorrect values resulting in incorrect and contradictory decisions. For example, a program using imprecise floating-point arithmetic might detect two different points of intersection between two non-identical straight lines. Of course, such a situation can not arise in real geometry, so the algorithm is not designed to handle such situations and it is not surprising that the program crashes.

Geometric computing goes beyond numerical computing, since it also has a discrete combinatorial component. Whereas in numerical computing one can often argue that the numerical results computed with floats are correct for some (small) perturbation of the numerical values in the input, such reasoning is intrinsically much harder in geometric computing. Computing the orientation of three points p, q, and r in the plane corresponds to computing the sign of a 3×3 determinant. Even if the orientation computed using floating-point arithmetic is not correct the three points can always be perturbed a little bit to points p', q', and r' such that the computed orientation is the orientation of p', q', and r'. However, in a geometric program points p, q, and r are involved in many orientation computations, and although for each single orientation computation there is always a correcting perturbation of the points, it is not guaranteed at all that there is a valid perturbation for all points making all computed orientation computations correct. Now, deciding whether there is a set of points realizing given orientation information is a hard problem. In fact, it is at least as hard as deciding the existential theory over the reals (Mnev 1989).

There are two obvious approaches to close the gap between theory and practice of geometric computing with respect to precision problems. Either change theory or change practice, i.e., take imprecision into account when designing a geometric algorithm or compute "exactly". The latter approach is known as the "exact geometric computation paradigm" (Yap and Dubé 1995, Yap 1997a). It assures that all decisions, i.e., all comparison operations, made in a program are correct and thereby assures that a program behaves as its theoretical counterpart. leda_reals are an extremely useful tool to implement an algorithm according to the exact geometric computation paradigm. Redesigning algorithms such that imprecision is taken into account, is considered not very attractive as the many algorithms developed in theory so far under the real RAM model would get lost.

3   Verified computation with leda_reals

In this section we describe how the leda_reals assure correct comparisons. The leda_reals record computation history in an expression dag (i.e., a directed acyclic graph) in order to allow for re-evaluation. The leaves of the dag are integers, and each internal node of the dag is labeled with a unary or binary operation and points to the operands from which the subexpression was computed. Fig. 1 shows an example. Furthermore, during construction of the expression dag, the leda_reals already compute rough approximations. Here, one can maintain a value and an error bound or, alternatively, use interval arithmetic.

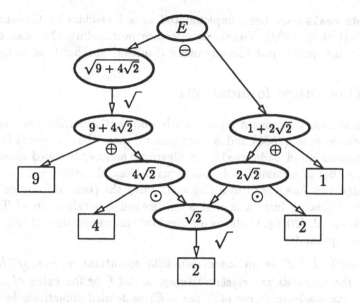

Fig. 1. Recorded computation history for the code fragment

```
leda_real root_two = sqrt(leda_real (2));
leda_real E = sqrt(9 + 4 * root_two) - (1 + 2 * root_two);
```

The actual work, however, is done in the comparison operations which are re-duced to sign computations. Let E be the expression whose sign we are interested in, and let ξ denote the value of the expression E. If, in a sign computation for E, the current approximation $\tilde{\xi}$ is not sufficient to verify the sign, the expression dag is used to re-compute better approximations. Approximations are computed as leda_bigfloats, a software floating-point number type that allows you to choose the mantissa length. We compute approximations $\tilde{\xi}$ of ξ with increasing quality until either the sign of ξ is equal to the sign of $\tilde{\xi}$ or until we can conclude that ξ is zero. We use separation-bounds to check for zero. A separation bound for an expression E is a number sep_E such that

$$|\xi| < sep_E \Rightarrow \xi = 0.$$

In the next section we prove such a separation bound for expressions. With each approximation $\tilde{\xi}$ we compute an error bound $\Delta_{\text{error}} \geq |\xi - \tilde{\xi}|$. If

$$|\xi - \tilde{\xi}| \leq \Delta_{\text{error}} < |\tilde{\xi}|$$

we have $\text{sign}(\tilde{\xi}) = \text{sign}(\xi)$. If, on the other hand,

$$|\tilde{\xi}| + \Delta_{\text{error}} < sep_E$$

we know that ξ must be zero. Since the quality of the approximations increases and hence Δ_{error} decreases, we eventually get the correct sign. The sign compu-tation is adaptive. The running time depends on how small the value $|\xi|$ of E is.

leda_reals have been implemented as a C++-class by Christoph Burnikel (Burnikel et al. 1996). Thanks to operator overloading, they can be used with the natural syntax, just like any other (built-in) number type in C++.

4 Theoretical foundations

In this section we prove a separation bound that generalizes the bound proved by Burnikel et al. (2000) and is more general than what is needed for the current implementation of leda_reals. An algebraic number is called *algebraic integer*, if it is a root of a monic[1] polynomial with integral coefficients. It is well known that algebraic integers form a ring containing the (rational) integers.

The following lemma is a straightforward generalization of Theorem 1 of Burnikel et al. (2000). It allows for algebraic integer instead of (rational) integer as basic operands.

Lemma 1. *Let E be an expression with operations $+, -, *, \sqrt[k]{\ }$ for integral k where the operands are algebraic integers. Let ξ be the value of E and $\deg(\xi)$ denote the algebraic degree of ξ. Let $u(E)$ be defined inductively by the structure of E by the rules shown in the table below.*

E	$u(E)$
algebraic integer γ	U_γ
$E_1 \pm E_2$	$u(E_1) + u(E_2)$
$E_1 \cdot E_2$	$u(E_1) \cdot u(E_2)$
$\sqrt[k]{E_1}$	$\sqrt[k]{u(E_1)}$

Here U_γ is chosen such that $|\varrho| \leq U_\gamma$ for all roots ϱ of monic $P \in \mathbb{Z}[X]$ with $P(\gamma) = 0$. We have

$$\left(u(E)^{\deg(\xi)-1}\right)^{-1} \leq |\xi| \leq u(E).$$

Algebraic integers are closed under the operations $+, -, *$ and $\sqrt[k]{\ }$. More generally, if we have a monic polynomial whose coefficients are algebraic integers, then the roots of this polynomial are algebraic integers again. We briefly prove that there is $P_E(X) = \prod(X - \gamma_\ell) \in \mathbb{Z}[X]$ with $P_E(\xi) = 0$, such that for all roots γ_ℓ of $P_E(X)$:

$$|\gamma_\ell| \leq u(E)$$

By the theorem of elementary symmetric functions, any polynomial which is symmetric in $z_1, \ldots, z_n$, can be written as a polynomial in the z_is' elementary symmetric functions

$$\sigma_1 = z_1 + z_2 + \cdots + z_n$$

$$\sigma_2 = z_1 z_2 + z_1 z_3 + \cdots + z_{n-1} z_n$$

$$\vdots$$

$$\sigma_n = z_1 z_2 \cdots z_n$$

---

[1] a polynomial is called monic, if its leading coefficient is 1.

We prove the claim by structural induction. In the base case, the claim holds by definition of an algebraic integer and $u(E)$. Note the special case, that E is an integer N. In this case, $P(X) = X - N \in \mathbb{Z}[X]$ and $u(E) = |N|$.

For the induction step we distinguish unary and binary operations.

$E_1 \overset{\pm}{*} E_2$:
By induction hypothesis we have

$$P_{E_1}(X) = \prod_{i=1}^{n}(X - \alpha_i) = \sum_{s=0}^{n} a_s X^s \in \mathbb{Z}[X],$$

$$P_{E_2}(X) = \prod_{j=1}^{m}(X - \beta_j) = \sum_{t=0}^{m} b_t X^t \in \mathbb{Z}[X].$$

By the theorem of elementary symmetric functions, the polynomial

$$P_E(X) := \prod_{i=1}^{n}\prod_{j=1}^{m}(X - (\alpha_i \overset{\pm}{*} \beta_j))$$

has integral coefficients. Furthermore, we have $|\alpha_i \overset{\pm}{*} \beta_j| \leq u(E_1) \overset{+}{*} u(E_2)$ by induction hypothesis and definition of u.

$E = \sqrt[k]{E_1}$:
By induction hypothesis, we have

$$P_{E_1}(X) = \prod_{i=1}^{n}(X - \alpha_i) = \sum_{s=0}^{n} a_s X^s \in \mathbb{Z}[X].$$

Then

$$P_E(X) = P_{E_1}(X^k) = \prod_{j=1}^{k}\prod_{i=1}^{n}(X - \zeta_{(k)}^j \sqrt[k]{\alpha_i}) \in \mathbb{Z}[X]$$

where $\zeta_{(k)}$ is a primitive k-th root of unity. We have $|\zeta_{(k)}^j \sqrt[k]{\alpha_i}| = |\sqrt[k]{\alpha_i}| \leq |\sqrt[k]{u(E_1)}|$ by induction hypothesis. Thus, $|\zeta_{(k)}^j \sqrt[k]{\alpha_i}| \leq u(E)$ by definition of u.

To complete the proof of Lemma 1, let $\xi \neq 0$ and $M_E(X)$ be the minimal polynomial of ξ.
Since $M_E(X)$ divides $P_E(X) = \prod(X - \gamma_\ell)$, we have

$$M_E(X) = \prod_{t=1}^{\deg(\xi)} (X - \gamma_{\ell_t}) \in \mathbb{Z}[X].$$

W.l.o.g. $\xi = \gamma_{\ell_1}$. Since all coefficients are integral, we have $\left|\prod_{t=1}^{\deg(\xi)} \gamma_{\ell_t}\right| \geq 1$.
Hence $\left(\prod_{t=2}^{\deg(\xi)} |\gamma_{\ell_t}|\right)^{-1} \leq |\xi|$ and $(u(E)^{\deg(\xi)-1})^{-1} \leq |\xi|$. $\qquad\square$

Lemma 2. *Let E be an expression with operations $+, -, *, /, \sqrt[k]{\ }$ for integral k where the operands are roots of univariate polynomials with integral coefficients. Let ξ be the value of E and $\deg(\xi)$ denote the algebraic degree of ξ. Let $u(E)$ and $l(E)$ be defined inductively by the structure of E by the rules shown in the table below. Let $K(E)$ be the product of the indices of the radical operations in E. Furthermore, let $D(E)$ be the product of the degree of the polynomials defining the operands.*

	$u(E)$	$l(E)$
algebraic number α	U_α	L_α
$E_1 \pm E_2$	$u(E_1) \cdot l(E_2) + l(E_1) \cdot u(E_2)$	$l(E_1) \cdot l(E_2)$
$E_1 \cdot E_2$	$u(E_1) \cdot u(E_2)$	$l(E_1) \cdot l(E_2)$
E_1 / E_2	$u(E_1) \cdot l(E_2)$	$l(E_1) \cdot u(E_2)$
$\sqrt[k]{E_1}$	$\sqrt[k]{u(E_1)}$	$\sqrt[k]{l(E_1)}$

Here $P = \sum_{i=0}^{d} a_i X^i \in \mathbb{Z}[X]$ with $P(\alpha) = 0$, not necessarily monic, $L_\alpha = a_d$ and U_α such that $|\varrho| \leq U_\alpha / a_d$ for all roots ϱ of P. Then we have

$$\left(l(E) u(E)^{K(E)^2 D(E)^2 - 1} \right)^{-1} \leq |\xi| \leq u(E) l(E)^{K(E)^2 D(E)^2 - 1}$$

Lemma 2 follows by Lemma 1, if we postpone division operations, i.e. replace an expression by a quotient of two division-free expressions whose values are algebraic integers. The size of the numerator is bounded by u, while l bounds the size of the denominator of this quotient. Note that if α is a root of $P = \sum_{i=0}^{d} a_i X^i \in \mathbb{Z}[X]$, then $\gamma = a_d \alpha$ is an algebraic integer. γ is a root of $a_d^{d-1} P(X/a_d)$. So we can replace α by $\frac{\gamma}{a_d}$.

There is no need to compute the l- and u-values exactly. It suffices to compute upper bounds on these values in order to derive separation bounds. The current implementation of leda_reals computes such upper bounds logarithmically. We plan to extend the implementation of leda_reals to include algebraic integer operands as discussed above.

5 Geometric applications

Exact computation with radicals is frequently required in geometric computations involving distances. Fig. 2 shows a Voronoi diagram of line segments, computed with an algorithm that uses leda_reals (Seel). In contrast to a previous implementation, the new program never crashes due to rounding errors.

Parametric search (Megiddo 1983) is a very nice technique to derive algorithms for solving certain optimization problems from algorithms for related decision problems (Salowe 1997). Parametric search can be applied if the problem can be phrased as an optimization problem parameterized by a real-valued parameter r where the goal is to compute a unique optimal value r^* for a given input parameterized by r. The related decision problem is to decide whether a given concrete r_0 is at least as large as r^*. Parametric search simulates the

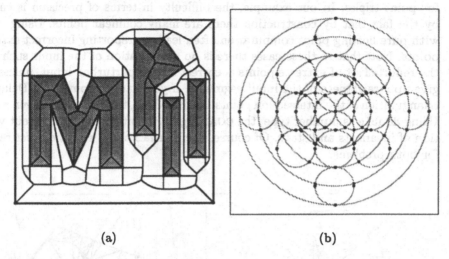

(a)   (b)

Fig. 2. Two geometric problems involving square roots: (a) Voronoi diagram of line segments, (b) arrangement of circles (Halperin and Hanniel 2000). The points shown are intersection points and points of vertical tangency.

decision problem for the unknown r^* and computes r^* during the simulation. Applying the parametric search technique involves solving the decision problem for zeros of functions in r that arise during the parameterized simulation. In almost all geometric applications, these functions are polynomials in r. Schwerdt et al. (1997) provide, to the best of our knowledge, the first actual implementation of parametric search for the problem of computing the point in time where a set of points moving with constant velocity has minimum diameter. It uses leda_reals to exactly solve decision problems for roots of polynomials of degree 2.

In combination with the CGAL framework (CGAL, Fabri et al. 1998, Overmars 1996) for geometric computation, the number type leda_real is particularly fruitful. The geometry kernels of CGAL, a computational geometry algorithms library, are ready to use leda_reals. CGAL provides kernels with Cartesian coordinate representation as well as a kernel based on homogeneous coordinates. Both are parameterized with the number type used for coordinates and arithmetic. The use of an exact number type yields an exact kernel. Using leda_reals with the CGAL kernels yields easy-to-write, correct and still reasonably efficient geometric programs. Fig. 2 (b) shows an arrangement of circles computed with the CGAL arrangement algorithm using leda_reals as number type.

Fig. 3 (a) shows a basic geometric problem that is challenging in terms of precision because of degenerate configurations. The task is to compute the extreme points among the intersection points of a set of line segments. A point is called *extreme* with respect to a set of points P, if it is a corner point of the convex hull of P. Extreme point computation involves orientation computation for point triples. In our example, the difficulty in terms of precision is caused by the fact that by construction there are many collinear points. Using CGAL with pure floating point computation often leads to reporting incorrect extreme points. Note that in these cases there is no perturbation of the input such that the reported set of extreme points is correct for the perturbed input, unless you give up straightness of the input segments. If you want to compute a Delaunay triangulation of the intersection points, see Fig. 3 (b), the situation is even worse: Using `double` as number type, the CGAL algorithm often complains about violation of invariants and stops. Of course, using `leda_reals`, we get correct results for both problems.

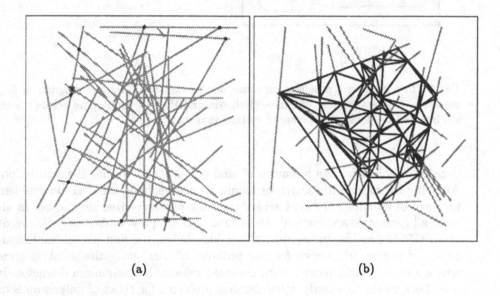

(a) (b)

Fig. 3. Extreme point computation (a) and Delaunay triangulation (b) for intersection points of random line segments with `double` coordinates. Points incorrectly classified as extreme by floating-point computation with `double` precision are encircled.

`leda_reals` are reasonably efficient in geometric computing whenever the expressions defining the algebraic numbers do have bounded size. In these cases the use of `leda_reals` slows down the computation by a constant factor. However, exact computation has its costs. In our experience, the slow down factor with respect to pure floating-point computation is between 5 and 20. For challenging problem instances it might be even more. However, here we compare apples and oranges. While the float based algorithm produces catastrophic errors sometimes, the `leda_real` based algorithm always gives a correct result. The comparison of running times is with respect to those cases only, where we get a result with both number types. If we would compare the time it takes a program to compute the correct result, a `leda_real`-based program would be "much faster", because it computes the correct result in finite time whereas the float-based program never reaches this goal. For further discussion on running times with **leda_reals** we refer the interested reader to the experimental results presented by Burnikel et al. (1999) and Schirra (1999).

leda_reals are not a panacea. If there are algebraic numbers involved whose expressions do not have bounded size, the use of **leda_reals** might not lead to satisfactory solutions anymore.

6 Conclusions

leda_reals offer a convenient and (reasonably) efficient way of computing with expressions involving roots. For geometric computing, **leda_reals** are especially useful in combination with the CGAL kernels.

A number type with similar functionality is available as part of the CORE-package developed at NYU (Karamcheti et al. 1999). The work presented in this paper is furthermore related to algebra systems and algebra toolkits like those presented by Keyser et al. (1999) and Rege (1996).

References

Brönniman, H., Kettner, L., Schirra, S., Veltkamp, R. (1998): Applications of the generic programming paradigm in the design of CGAL. Research Report MPI-I-98-1-030, Max-Planck-Insitut für Informatik

Burnikel, C., Fleischer, R., Mehlhorn, K., Schirra, S. (1999): Companion page to 'Efficient exact geometric computation made easy'. http://www.mpi-sb.mpg.de/~stschirr/exact/made_easy

Burnikel, C., Fleischer, R., Mehlhorn, K., Schirra, S. (2000): A strong and easily computable separation bound for arithmetic expressions involving radicals. Algorithmica 27:87-99

Burnikel, C., Mehlhorn, K., Schirra, S. (1996): The LEDA class **real** number. Research Report MPI-I-96-1-001, Max-Planck-Institut für Informatik. A more recent documentation of the implementation is available at http://www.mpi-sb.mpg.de/~burnikel/reports/real.ps.gz.

CGAL: http://www.cgal.org

Fabri, A., Giezeman, G.-J., Kettner, L., Schirra, S., Schönherr, S. (1998): On the design of CGAL, the computational geometry algorithms library. Research Report MPI-I-98-1-007, Max-Planck-Institut für Informatik

Fortune, S. (1993): Progress in computational geometry. In: Martin R., (ed.): Directions in Geometric Computing, pp. 81 - 128. Information Geometers Ltd.

Hanniel, I., Halperin, D. (2000): Two-dimensional arrangements in CGAL and adaptive point location for parametric curves. Manuscript. Tel-Aviv University, Israel

Hoffmann, C. (1989): The problem of accuracy and robustness in geometric computation. IEEE Computer 3:31-41

Karamcheti, V., Li, C., Pechtanski, I., Yap, C.K. (1999): A core library for robust numeric and geometric computation. In: Proc. 15th Annu. ACM Sympos. Comput. Geom., pp. 351-359

Keyser, J., Culver, T., Manocha, D., Krishnan, S. (1999): MAPC: A library for efficient and exact manipulation of algebraic points and curves. In: Proc. 15th Annu. ACM Sympos. Comput. Geom., pp. 360-369. See also http://www.cs.unc.edu/~geom/MAPC

LEDA: http://www.mpi-sb.mpg.de/LEDA

Megiddo, N. (1983): Applying parallel computation algorithms in the design of serial algorithms. Journal of the ACM 30:852-865

Mehlhorn, K., Näher, S. (1994): The implementation of geometric algorithms. In: Proceedings of the 13th IFIP World Computer Congress, vol. 1, pp. 223-231. Elsevier Science B.V. North-Holland, Amsterdam

Mehlhorn, K., Näher, S. (2000): LEDA – A Platform for Combinatorial and Geometric Computing. Cambridge University Press

Mnev, M.E. (1989): The universality theorems on the classification problem of configuration varieties and convex polytopes varieties. In: Topology and Geometry - Rohlin Seminar, Springer, pp. 527–544 (Lecture Notes Math., vol. 1346)

Overmars, M. (1996): Designing the computational geometry algorithms library CGAL. In Lin, M.C., Manocha, D., (eds.): Applied Computational Geometry: Towards Geometric Engineering (WACG96), Springer, pp. 53-58 (Lecture Notes in Computer Science, vol. 1148)

Preparata, F.P., Shamos, M.I. (1985): Computational Geometry: An Introduction. Springer

Rege, A. (1996): APU User Manual – Version 2.0. http://www.cs.berkeley.edu/~rege/apu/apu.html

Salowe, J. (1997): Parametric search. In: Goodman, J.E., Rourke, J.O. (eds.): Handbook of Discrete and Computational Geometry, pp. 683–698, CRC Press

Schirra, S. (2000): Robustness and Precision Issues in Geometric Computation. In: Sack, J.-R., Urrutia, J. (eds.): Handbook of Computational Geometry, Elsevier, pp. 597-632

Schirra, S. (1999): A case study on the cost of geometric computing. In: McGeoch C.G, Goodrich M. (eds.): Algorithm Engineering and Experimentation (ALENEX'99), Springer pp. 156-176 (Lecture Notes in Computer Science, vol. 1619)

Seel, M.: Abstract Voronoi Diagrams. http://www.mpi-sb.mpg.de/LEDA/friends/avd.html

Schwerdt, J., Smid, M., Schirra, S. (1997): Computing the Minimum Diameter for Moving Points: An Exact Implementation using Parametric Search. In: Proc. of the 13th ACM Symp. on Computational Geometry, ACM Press, pp. 466-468.

Yap, C.K. (1997a): Towards exact geometric computation. Comput. Geom. Theory Appl., 7:3–23

Yap, C.K. (1997b): Robust geometric computation. In: Goodman, J.E., Rourke, J.O. (eds.): Handbook of Discrete and Computational Geometry, pp. 653–668, CRC Press

Yap, C.K., Dubé, T. (1995): The exact computation paradigm. In: Du, D., Hwang, F. (eds.), Computing in Euclidean Geometry, 2nd edition, World Scientific Press, pp. 452–492

Numerical Verification Method for Solutions of Nonlinear Hyperbolic Equations

Teruya Minamoto

1 Introduction

Recently, several methods to the computer-assisted existence proof of solutions for various differential equations have been developed. However, there are very few approaches for partial differential equations. As far as we know, there are only two methods, that is, Nakao's method (e.g. Nakao 1993) using C^0 finite element and explicit error estimates, and Plum's method (e.g. Plum 1994) using C^1-class approximate solution with high accuracy and an exact eigenvalue enclosure for a linearized operator. Almost all papers by these authors deal with elliptic equations.

In this article, we consider a numerical method to verify the existence and uniqueness of the solutions of nonlinear hyperbolic problems with guaranteed error bounds. This verification method is based on Plum's formulation of verification methods and weak formulation for determining a bound on the inverse norm of the linearized operator, and the idea contained in Nakao's method to ensure existence and uniqueness. More precisely, using a C^1 finite element solution and an inequality constituting a bound on the norm of the inverse operator of the linearized operator, we numerically construct a set of functions which satisfies the hypothesis of Banach's fixed point theorem for a map on L^p-space in a computer.

In the following section, we introduce the problem considered and its fixed point formulation. In Section 3, a fundamental theorem which contains the verification conditions of our method is presented. In Section 4, we estimate the inverse norm of the linearized operator and give the algorithm for our method using a weak formulation, Some numerical examples are illustrated in the last part of this article.

2 Problem and a Fixed Point Formulation

Consider the problem of finding a function u that satisfies the following relations:

$$u \in L^2(J; H_0^1(\Omega)), u_t \in L^2(J; L^2(\Omega)),$$

$$\frac{d^2}{dt^2}(u, v) + (\nabla u, \nabla v) = (-f(\cdot, u), v), \quad v \in H_0^1(\Omega), \quad t \in J := (0, T),$$

$$u(\cdot, 0) = 0, \tag{1}$$

where Ω is a bounded open interval on $\mathbf{R}$ or a bounded rectangular domain in $\mathbf{R}^2$, $(\cdot, \cdot)$ is the usual $L^2(\Omega)$ inner product, and f is a function on $Q \times \mathbf{R}$ with $Q = \Omega \times J$. Also, suppose that $\hat{f}$ defined by $(\hat{f}(u))(x, t) := f(x, t, u(x, t))$ maps $L^p(Q)$ into $L^2(Q)$ for some p satisfying $2 \le p \le 6$.

To be precise, the derivative d^2/dt^2 in (1) is treated as the generalized derivative of real functions on $(0, T)$, that is,

$$\int_0^T (u(\cdot, t), v)\varphi''(t)dt + \int_0^T (\nabla u(\cdot, t), \nabla v)\varphi(t)dt = \int_0^T (-f(\cdot, t, u(\cdot, t)), v)\varphi(t)dt,$$
$$\forall \varphi \in C_0^\infty[0, T).$$

Next, we define the time-dependent Sobolev space H by

$$H \equiv L^2(J; H_0^1(\Omega)) \cap H^1(J; L^2(\Omega))$$

with norm

$$\|u\|_H^2 = \int_J \|\nabla u(\cdot, t)\|_{L^2(\Omega)}^2 dt + \int_J \|u_t(\cdot, t)\|_{L^2(\Omega)}^2 dt,$$

where $\|\cdot\|_{L^p(\Omega)}$ is the usual $L^p(\Omega)$ norm.

Let $\tilde{H} := \{\phi \in H | \phi(\cdot, 0) = 0\}$ and let $u_h \in \tilde{H}$ be an approximate solution of (1). It is most common to think of such a solution as some finite element solution depending on h. Then suppose the following conditions hold for the nonlinear map f in (1):

(A1) $\hat{f} : L^p(Q) \to L^2(Q)$ is continuous and maps bounded sets into bounded sets.

(A2) $\hat{f}$ is Fréchet differentiable in $L^p(Q)$.

(A3) $f'(\cdot, u_h)(x, t) \in C^1([0, T]; L^\infty(\Omega))$.

Here $f'(\cdot, u_h)(x, t) := \frac{\partial f}{\partial u}(x, t, u_h(x, t))$.

Now, as well known (e.g. Lions 1971), for each $g \in L^2(Q)$, if $a \in C^1([0, T]; L^\infty(\Omega))$, the following problem has a unique solution $\phi \in \tilde{H}$:

$$\frac{d^2}{dt^2}(\phi, v) + (\nabla \phi, \nabla v) + (a\phi, v) = (g, v), \qquad v \in H_0^1(\Omega), \quad t \in J. \qquad (2)$$

We denote the above correspondence by $Ag = \phi$. Moreover, assuming that $a = f'(\cdot, u_h)$, we define the fixed-point operator T by

$$Tu \equiv A[f'(\cdot, u_h)u - f(\cdot, u)]. \qquad (3)$$

Then, from (A1),(A2),(A3) and the fact that the operator $A : L^2(Q) \to \tilde{H}$ and the injection $H \hookrightarrow L^p(Q)$ are continuous and bounded(See Lemma 1 and Lemma 2), the operator $T : L^p(Q) \to L^p(Q)$ is Fréchet differentiable in $L^p(Q)$.

The map $f(\cdot, u) = gu^m$ is an example that satisfies assumption (A1) and (A2), where $g \in L^\infty(Q)$, and m is an arbitrary nonnegative integer satisfying $1 \leq m \leq 3$. In this case, $f'(\cdot, u_h) = mgu_h^{m-1}$, and if $u_h \in C^1([0, T]; L^\infty(\Omega))$, the operator T in (3) is well-defined.

Remark 1. In the one-dimensional case we can choose $2 \leq p < \infty$, which implies $1 \leq m < \infty$ in the above example. In any case, we assume that the nonlinearity of f has a polynomial form with respect to u.

3 Verification Condition

We derive a verification condition by using a a "residual-form". Defining $v = u - u_h$, we introduce the operator $\tilde{T} : L^p(Q) \to L^p(Q)$ defined by

$$\tilde{T}v \equiv T(u_h + v) - u_h. \tag{4}$$

To construct a set V which includes the error of a solution to (1), taking some real number α, we set

$$V \equiv \{v \in L^p(Q) | \|v\|_{L^p(Q)} \le \alpha\}. \tag{5}$$

Moreover, we choose the positive real numbers β and γ such that

$$\|\tilde{T}(0)\|_{L^p(Q)} \le \beta, \tag{6}$$
$$\|\tilde{T}'(v_1)v_2\|_{L^p(Q)} \le \gamma \quad \forall v_1, v_2 \in V, \tag{7}$$

and define the set $K \subset L^p(Q)$ by

$$K \equiv \{v \in L^p(Q) | \|v\|_{L^p(Q)} \le \beta + \gamma\}. \tag{8}$$

Then, our verification condition is described in the following theorem, which is similar to those in Nagatou et al. (1999) and is proved in Minamoto (2000) by virtue of Banach's Fixed Point Theorem.

Theorem 1. If $K \subset V$ holds for V(that is, if $\beta + \gamma \le \alpha$), then there exists a solution to

$$v = \tilde{T}(v)$$

in K, and this solution is unique within the set V.

In actual computing, we carry out the following procedures to construct the number α.

i. Compute a constant β and set $\alpha = \beta$.

ii. Compute γ.

iii. Check $\beta + \gamma \le \alpha$. If this condition is satisfied, then stop. Otherwise, replace α by $(1 + \delta)\alpha$ for a certain positive number δ and return to ii.

4 Constants in the Verification Condition

In this section we describe how to estimate β and γ introduced in the previous section. We assume that there exist the constants C_1 and C_2 satisfying

$$\|Ar\|_H \le C_1\|r\|_{L^2(Q)} \quad \forall r \in L^2(Q), \tag{9}$$
$$\|Ar\|_{L^p(Q)} \le C_2\|Ar\|_H. \tag{10}$$

If we then compute an approximate solution $u_h \in \tilde{H}$ so as to satisfy $d[u_h] \equiv u_{htt} - \triangle u_h + f(\cdot, u_h) \in L^2(Q)$, the following relations holds:

$$
\begin{aligned}
\|\tilde{T}(0)\|_{L^p(Q)} &= \|Ad[u_h]\|_{L^p(Q)} \\
&\leq C_2\|Ad[u_h]\|_H \\
&\leq C_1 C_2\|d[u_h]\|_{L^2(Q)}.
\end{aligned} \tag{11}
$$

Similarly, we can obtain

$$
\begin{aligned}
\|\tilde{T}'(v_1)v_2\|_{L^p(Q)} &\leq C_1 C_2\|\hat{f}'(u_h + v_1)v_2 - \hat{f}'(u_h)v_2\|_{L^2(Q)} \\
&\leq C_1 C_2 G_\alpha \quad \forall v_1, v_2 \in V.
\end{aligned} \tag{12}
$$

Here G_α is a constant that depends on α such that $\sup\{\|\hat{f}'(u_h + v_1)v_2 - \hat{f}'(u_h)v_2)\|_{L^2(Q)}; v_1, v_2$ $V\} \leq G_\alpha$.

In what follows, we consider the open intervals $I_{x_1} = (a_{x_1}, b_{x_1})$ and $I_{x_2} = (a_{x_2}, b_{x_2})$ for real numbers $a_{x_1} < b_{x_1}$ and $a_{x_2} < b_{x_2}$ and define $\Omega = I_{x_1} \times I_{x_2}$ and $d = \max\{b_{x_1} - a_{x_1}, b_{x_2} - a_{x_2}\}$. The following two lemmas give the values of C_1 and C_2. The proofs of these lemmas can be found in Minamoto (2000).

Lemma 1. Let $\underline{a}$ and $\bar{a}$ denote constants satisfying $\underline{a} \leq a(x, t) \leq \bar{a}$ for almost all $(x, t) \in Q$. Then C_1 in (9) is given by

$$
C_1 = \sqrt{\frac{1}{c}(e^{cT} - 1)},
$$

where $c = \max(1 - \underline{a}T, \frac{d^2}{n_0 \pi^2}\|a_t\|_{L^\infty(Q)})$ for $\underline{a} < 0$ and $c = \max(1, \frac{d^2}{n_0 \pi^2}\|a_t\|_{L^\infty(Q)})$ for $\underline{a} \geq 0$ and n_0 is the dimension.

Lemma 2. C_2 in (10) is given by

$$
C_2 = \begin{cases} \frac{p}{4}|Q|^{\frac{1}{p}} & (\Omega \text{ is one dimensional}) \\ (\frac{1}{2})^{\frac{2}{3}}\frac{2\sqrt{3}}{9}p|Q|^{\frac{6-p}{6p}} & (\Omega \text{ is two dimensional}). \end{cases}
$$

5 Verification Procedures

In this section, we describe the computation of the approximate solution u_h and defect $\|d[u_h]\|_{L^2(Q)}$ in (11).

Let S_h be a finite-dimensional subspace of $H_0^1(\Omega) \cap H^2(\Omega)$ depending on h and let N be the dimension of S_h. Then we can represent u_h by

$$
u_h(x, t) = \sum_{i=1}^{N} u_i(t)\hat{\phi}_i(x),
$$

where $\hat{\phi}_i$ are base functions in S_h. The function $u_i(t)$ constitutes the time-dependent coefficient of the base function $\hat{\phi}_i(x)$.

Now u_h is computed by the following Newton-iteration:

$$(u_{htt}^{(n)}, \hat{\phi}_j) + (\nabla u_h^{(n)}, \nabla \hat{\phi}_j) + (f'(u_h^{(n-1)})u_h^{(n)}, \hat{\phi}_j) = (f'(u_h^{(n-1)})u_h^{(n-1)} - f(u_h^{(n-1)}), \hat{\phi}_j),$$
(13)

where n is the iteration number.

For the discretization of time, we take equal time steps of length Δt and define

$$t_k = k\Delta t, \qquad k = 0, 1, 2, \cdots.$$

We used the Newmark method (e.g. Quarteroni and Valli 1994), which generates the following scheme:

$$u_i^{(n)}(t + \Delta t) \approx u_i^{(n)}(t) + \Delta t \dot{u}_i^{(n)}(t) + \Delta t^2 [\beta \ddot{u}_i^{(n)}(t + \Delta t) + (\frac{1}{2} - \beta)\ddot{u}_i^{(n)}(t)] \quad (14)$$

$$\dot{u}_i^{(n)}(t + \Delta t) \approx \dot{u}_i^{(n)}(t) + \Delta t[\theta \ddot{u}_i^{(n)}(t + \Delta t) + (1 - \theta)\ddot{u}_i^{(n)}(t)], \qquad (15)$$

where $\dot{u}_i = \dfrac{du_i}{dt}$ and $\ddot{u}_i = \dfrac{d^2 u_i}{dt^2}$, and θ and β are some non-negative parameters.

We compute an approximate solution by combining the Newton iteration and the Newmark method.

From (14),

$$\ddot{u}_i^{(n)}(t + \Delta t) \approx \frac{1}{\beta \Delta t^2}[u_i^{(n)}(t + \Delta t) - u_i^{(n)}(t)] - \frac{1}{\beta \Delta t}\dot{u}_i^{(n)}(t) - (\frac{1}{2\beta} - 1)\ddot{u}_i^{(n)}(t) \quad (16)$$

We approximate the iteration (13) by carrying out the following procedure (i)-(iii), and get an approximate solution in each time step on condition that $u_i^{(n)}(t)$, $\dot{u}_i^{(n)}(t)$, $\ddot{u}_i^{(n)}(t)$ are already known.

(i):Substitute (16) to the next equation:

$$\begin{aligned}(u_{htt}^{(n)}(t + \Delta t), \hat{\phi}_j) \quad &+ \quad (\nabla u_h^{(n)}(t + \Delta t), \nabla \hat{\phi}_j) + (f'(u_h^{(n-1)}(t + \Delta t))u_h^{(n)}(t + \Delta t), \hat{\phi}_j) \\ &= \quad (f'(u_h^{(n-1)}(t + \Delta t))u_h^{(n-1)}(t + \Delta t) - f(u_h^{(n-1)}(t + \Delta t)), \hat{\phi}_j), \quad (1\end{aligned}$$

and compute $u_i^{(n)}(t + \Delta t)$.

(ii): Substitute $u_i^{(n)}(t + \Delta t)$ in (i) to (16), and compute $\ddot{u}_i^{(n)}(t + \Delta t)$.

(iii): Substitute $\ddot{u}_i^{(n)}(t + \Delta t)$ in (ii) to (15), and compute $\dot{u}_i^{(n)}(t + \Delta t)$.

For initial value of this scheme, we find $u_i(0) = \dot{u}_i(0) = 0$ from initial condition in (1) for each i, and $\ddot{u}_i(0)$ is computed by solving (17) in $t + \Delta t = 0$ with respect to $\ddot{u}_i(0)$, i.e. by solving the linear system $\sum_{i=1}^{N}(\hat{\phi}_j, \hat{\phi}_i)\ddot{u}_i(0) = -(f(0), \hat{\phi}_j)(j = 1, 2, \ldots, N)$. In particular, $u_i(0), \dot{u}_i(0), \ddot{u}_i(0)$ are independent of n.

Since $u_{htt} - \triangle u_h + f(\cdot, u_h) \in L^2(Q)$ is required, we use the piecewise cubic Hermite function in one-dimensional case and the piecewise bi-cubic Hermite function in two-dimensional case as the base function in space, and also use the piecewise cubic Hermite interpolation in time. Moreover, since f has a polynomial restriction with respect to u and piecewise polynomials are used in space and time, we can compute $\|d[u_h]\|_{L^2(Q)}$ directly, elementwise in each time step in some cases. In case that we can't compute $\|d[u_h]\|_{L^2(Q)}$ directly, we use some overestimations for it.

6 Numerical Examples

One-dimensional case

We consider a nonlinear Klein-Gordon Equation

$$\begin{cases} u_{tt} - u_{xx} = -u - u^3 & (x,t) \in Q, \\ u(x,t) = 0 & x \in \partial\Omega \times J, \\ u(x,0) = \psi(x) & x \in \Omega, \\ u_t(x,0) = 0 & x \in \Omega, \end{cases} \tag{18}$$

where $\partial\Omega$ stands for the boundary of Ω, and $\psi(x)$ is a smooth function defined in a finite interval that vanishes on $\partial\Omega$. This equation is important to analyze the behavior of some waves appearing in mathematical physics, and is investigated by some authors (Abolowitz et al. 1979, Jiménez 1990, Li and Guo 1997).

Defining $\tilde{u}(x,t) := u(x,t) - \psi(x)$ and noting that the equation (1) is a generalized problem corresponding to (18), we can apply our method to the equation (18) after replacing u by $\tilde{u}$.

We let $\Omega = (0,1)$, $T = 2$ and $\psi(x) = 0.05(1 - \cos(2\pi x))$. Figures 1 and 2 display the numerical solutions of the equation (18) up to t=5.

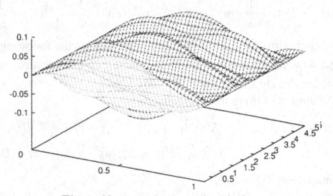

Fig.1. Numerical solutions of (18)

If we take $p = 6$, then $(A1)$ and $(A2)$ are satisfied. Since we have

$$\|\hat{f}'(u_h + v_1)v_2 - \hat{f}'(u_h)v_2\|^2_{L^2(Q)} = \int\int (|6u_h v_1 v_2|^2 + |3v_1^2 v_2|^2 + 36|u_h v_1^3 v_2^2|)dxdt$$

$$\leq 36\|u_h\|_{L^\infty(Q)}|Q|^{\frac{1}{6}}\|v_1\|^2_{L^6(Q)}\|v_2\|^2_{L^6(Q)}$$

$$\times (\|u_h\|_{L^\infty(Q)}|Q|^{\frac{1}{6}} + \|v_1\|_{L^6(Q)}) + 9\|v_1\|^4_{L^6(Q)}\|v_2\|^2_{L^6(Q)}$$

$$\leq 36\|u_h\|_{L^\infty(Q)}|Q|^{\frac{1}{6}}\alpha^4(\|u_h\|_{L^\infty(Q)}|Q|^{\frac{1}{6}} + \alpha) + 9\alpha^6,$$

(12) is satisfied for $G_\alpha = \sqrt{36\|u_h\|_{L^\infty(Q)}|Q|^{\frac{1}{6}}\alpha^4(\|u_h\|_{L^\infty(Q)}|Q|^{\frac{1}{6}} + \alpha) + 9\alpha^6}$, where we regard u_h as an approximation of $\tilde{u}$.

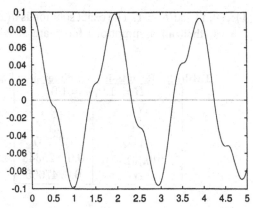

Fig.2. Numerical solutions of (18) at $x = 0.5$

We illustrate our method with the numerical results, where NS is the partition numbers of space. Generally speaking, it is difficult to describe the stability of the Newmark method for nonlinear problems, but according to Quarteroni and Valli (1994), if $\theta = \frac{1}{2}$ and $\beta = \frac{1}{4}$, the Newmark method is unconditionally stable for linear hyperbolic equations. Thus we choose $\theta = \frac{1}{2}$ and $\beta = \frac{1}{4}$ in these examples.

In our numerical examples we used $u_h((k-1)\Delta t)(k \geq 1)$ as the initial value of Newton iteration at $t = k\Delta t$. When we compute $\|d[u_h]\|_{L^2(Q)}$, we must compute the term $\int_{Q_{ij}} u_h^6 dx dt$, where $Q_{ij} = [(i-1)h, ih] \times [t_{j-1}, t_j](i = 1, \ldots, NS, j = 1, 2, \ldots, h = 1/NS)$. For this term, we use the estimation $\int_{Q_{ij}} u_h^6 dx dt \leq |Q_{ij}|\|u_h\|_{L^\infty(Q_{ij})}^6$ to save computing time. Noting that $u_h(x,t) = \sum_{k,l=1}^{4} u_{kl}\psi_k(x,h)\psi_l(t,\Delta t)$ in Q_{ij}, where $\psi_1(a,b) = 2(a-a_{i-1})^3/b^3 - 3(a-a_{i-1})^2/b^2 + 1$, $\psi_2(a,b) = -(a-a_{i-1})^3/b^3 + 3(a-a_{i-1})^2/b^2$, $\psi_3(a,b) = (a-a_{i-1})(a_i-a)^2/b^2$ and $\psi_4(a,b) = (a-a_{i-1})^2(a-a_i)/b^2$ for $a \in [a_{i-1}, a_i]$, the inequality $\|u_h\|_{L^\infty(Q_{ij})} \leq U_1 + hU_2/4 + \Delta t U_3/4 + h\Delta t U_4/16$ is obtained by using $|\psi_1(a,b)| + |\psi_2(a,b)| = 1$ and $|\psi_3(a,b)| + |\psi_4(a,b)| \leq b/4$. Here $U_1 = \max_{k,l=1,2} |u_{kl}|, U_2 = \max_{k=3,4,l=1,2} |u_{kl}|, U_3 = \max_{k=1,2,l=3,4} |u_{kl}|$, and $U_4 = \max_{k,l=3,4} |u_{kl}|$ and then $\|u_h\|_{L^\infty(Q)} \leq \max_{i,j} \|u_h\|_{L^\infty(Q_{ij})}$.

In the following, α represents the verified error bound in Theorem 1.

Table 1. Results in one dimensional Klein-Gordon equation

NS	200
Δt	0.001
$C_1 C_2$	4.8660324
$\|d[u_h]\|_{L^2(Q)}$	0.003066
α	0.01491926

Two dimensional case

In (1), we set

$$f(x,t,u) = f(x_1, x_2, t, u) = Bu^2 - k\sin \pi x_1 \sin \pi x_2(2 + 2t^2\pi^2 + Bkt^4 \sin \pi x_1 \sin \pi x_2),$$

and let $\Omega = (0,1) \times (0,1)$ and $T = 1$. The exact solution is $u(x_1, x_2, t) = kt^2 \sin \pi x_1 \sin \pi x_2$. If we choose $p = 4$, then all assumptions for f are satisfied and we may choose $G_\alpha = 2|B|\alpha^2$.

Table 2. Results in two dimensional case

$B = 1.0, k = 1.0$	
NS	8
Δt	$\frac{1}{64}$
$C_1 C_2$	1.27137
$\|d[u_h]\|_{L^2(Q)}$	0.01825348
α	0.0247665

Remark 2. In these computations, we used the usual floating-point number system with double precision. Therefore, the above results may include some unknown rounding errors. From the author's experiences, however, the order of magnitude of the effect of round-off errors is smaller than 10^{-10}. With this observation, we can assume that the numerical results are sufficiently reliable to at least six digits or so. Of course, we need to use arithmetic system with guaranteed accuracy for a really rigorous verification.

References

Ablowitz, M. J., Kruskal, M. D., and Ladik, J. F. (1979): Solitary wave collisions. SIAM J. Appl. Math. 36(3):428-437

Adams, R. A. (1978): Sobolev Spaces. Academic Press, Boston San Diego New York London Sydney Tokyo Toronto

Jiménez, S. (1990): Analysis of four numerical schemes for a nonlinear Klein-Gordon equation. Appl. Math. Comp. 35:61-94

Johnson, C., Schatz, A. H. and Wahlbin, L. B. (1987): Crosswind Smear and Pointwise Errors in Streamline Diffusion Finite Element Methods. Math. Comp. 49:25-38

Li, X. and Guo, B. Y. (1997): A Legendre pseudospectral method for solving nonlinear Klein-Gordon equation. J. Comput. Math. 15(2):105-126

Lions, J. L. (1971): Optimal Control of Systems Governed by Partial Differential Equations. Springer-Verlag, Berlin Heidelberg New York

Minamoto, T. and Nakao, M.T. (1997): Numerical verifications of solutions for nonlinear parabolic equations in one-space dimensional case. Reliable Computing 3:137-147

Minamoto, T. (1997): Numerical verifications of solutions for nonlinear hyperbolic equations. Appl. Math. Lett. 10:91-96

Minamoto, T.(2000): Numerical existence and uniqueness proof for solutions of nonlinear hyperbolic equations. J. Comp. Appl. Math.(to appear)

Nagatou, K., Yamamoto, N. and Nakao, M. T. (1999): An approach to the numerical verification of solutions for nonlinear elliptic problems with local uniqueness. Numer. Funct. Anal. Optimz. 20(5&6):543-565

Nakao, M. T. (1993): Solving nonlinear elliptic problems with result verification using an H^{-1} residual iteration. Computing Suppl. 9:161-173

Nakao, M. T. (1994): Numerical Verifications of Solutions for Nonlinear Hyperbolic Equations. Interval Computations 4:64-77

Plum, M. (1994): Inclusion Methods for Elliptic Boundary Value Problems. In:Herzberger, J.(eds.):Topics in validated computation. North-Holland, Amsterdam Lausanne New York Oxford Shannon Tokyo, pp.323-379(Studies in Computational Mathematics, vol.5)

Quarteroni, A. and Valli, A. (1994): Numerical Approximation of Partial Differential Equations. Springer-Verlag, New York Berlin Heidelberg

Yamamoto, N. (1998): A numerical verification method for solutions of boundary value problems with local uniqueness by Banach's fixed point theorem. SIAM J. Numer. Anal. 35:2004-2013

Zeidler, E. (1990): Nonlinear Functional Analysis and its Applications II/A. Springer, New York Berlin Heidelberg

Geometric Series Bounds for the Local Errors of Taylor Methods for Linear n-th Order ODEs

Markus Neher

1 Introduction

Interval Taylor methods for the validated solution of initial value problems for ODEs were introduced by Moore (1965a, 1965b, 1966). Lohner (1987, 1988, 1992) developed a comprehensive software package of an advanced interval Taylor method, which he applied successfully to many linear and nonlinear problems. But, as Lohner (1988) remarked, the step size of his method is limited by the step size for the explicit Euler method. Lohner (1995), Corliss & Rihm (1996) and Nedialkov (1999) have proposed modified versions of Lohner's algorithm that remove this restriction. However, to the author's knowledge these alternatives have not been extensively tested and have not been implemented in software for general IVPs.

In this paper, a new enclosure method for linear n-th order ODEs is proposed. It is also based on Taylor series expansions, but the foundation of the error bound is completely different from Moore's or Lohner's methods. Its main advantage is that it works with very large step sizes. Recently, (Neher 1999), the method was presented for linear ODEs with polynomial coefficient functions. Here it is extended to linear ODEs with analytic coefficient functions.

We assume that the reader is familiar with interval arithmetic (Moore 1966, Alefeld and Herzberger 1983) and its implementation on a computer (Kulisch and Miranker 1981). Our presentation is concentrated on the analytic foundations of the method rather than on customary interval arithmetic modifications that are necessary when the calculations are performed in floating point arithmetic.

In the next two sections, the model IVP is introduced and an error estimation scheme for a single integration step is developed. After that, we comment on the implementation of the method and on the continuation of the integration of the ODE. In the last section, numerical examples are given.

2 IVP for linear n-th order ODE

For $n \geq 2$, we consider the following normalized initial value problem for a linear ODE with analytic coefficient functions on some interval $[0, r)$:

$$y^{(n)} = \sum_{i=0}^{n-2} p_i(x)\, y^{(i)} + p_{-1}(x),$$

$$y^{(i)}(0) = y_{io}, \quad i = 0, \ldots, n-1,$$

(1)

where $p_i(x)$, $i = -1, \ldots, n-2$ are assumed to be analytic functions with Taylor series

$$p_i(x) = \sum_{j=0}^{\infty} b_{ij} x^j, \quad x \in [0, r). \tag{2}$$

For the numbers b_{ij}, we assume that there are constants $m_i \in \mathbb{N}_0$ and $B_i \geq 0$ such that

$$|b_{ij}| \leq \frac{B_i}{r^j}, \quad \text{for } j > m_i, \quad i = -1, \ldots, n - 2 \tag{3}$$

(the computation of such bounds will be discussed in Section 5).

Remark: The normalization of the IVP applies for technical reasons, without loss of generality. If the initial values are given at some $x_0 \neq 0$, then let $u(x) := y(x - x_0)$ and reformulate the problem for u. If the differential equation contains also $p_{n-1}(x) y^{(n-1)}$ then the transformation $w(x) = y(x) e^{\frac{1}{n} \int p_{n-1}(x) dx}$ yields an ODE of the form (1) for w (cf. Heuser 1989, p. 257).

It is well known (Heuser 1989, p. 260) that the solution of (1) can be written as a power series

$$y(x) := \sum_{k=0}^{\infty} a_k x^k, \tag{4}$$

and that this series converges at least for all x with $|x| < r$. Inserting the series into (1), we have

$$a_k = \frac{y_{k0}}{k!} \quad \text{for } k = 0, \ldots, n - 1, \tag{5}$$

$$a_{k+n} = \sum_{i=0}^{n-2} \sum_{j=0}^{k} \frac{P(k-j,i) b_{ij}}{P(k,n)} a_{k+i-j} + \frac{b_{-1,k}}{P(k,n)} \quad \text{for } k = 0, 1, \ldots, \tag{6}$$

where

$$P(k, i) := (k+1) \cdots (k+i), \quad P(k, 0) := 1 \quad \text{for } i \in \mathbb{N}, k \in \mathbb{N}_0.$$

3 Error estimation for the IVP

We use (4) to construct lower and upper bounds of $y(h)$ for some $h \in [0, r)$. A pair of such bounds is called an *enclosure* of $y(h)$.

First, we compute finitely many, say κ, coefficients a_k from (5) and (6) and then use the truncated series

$$s(\kappa) := \sum_{k=0}^{\kappa-1} a_k h^k \tag{7}$$

as an approximation to $y(h)$. The approximation error $y(h) - s(\kappa)$ is then the truncation error given by the infinite series

$$t(\kappa) := \sum_{k=\kappa}^{\infty} a_k h^k.$$

Our goal is to bound $t(\kappa)$ by a geometric series. As a prerequisite, we recall an elementary summation formula. For $i \in \mathbf{N}_0$, the proof is with induction with respect to k.

Lemma 1 *For all* $i, k \in \mathbf{N}_0$: $\displaystyle\sum_{j=0}^{k} P(j, i) = \frac{P(k, i+1)}{i+1}$.

Now suppose that (3) holds. Then we can deduce the following theorem:

Theorem 1 *The sequence* $\{a_k r^k\}_{k=0}^{\infty}$ *with* a_k *from (5) and (6) is bounded.*

Proof: Let $m := \max\limits_{i=-1}^{n-2} m_i$. Then for $k > m$,

$$|a_{k+n}\, r^{k+n}| \overset{(6)}{\leq} \sum_{i=0}^{n-2} \sum_{j=0}^{m_i} \frac{r^{n-i}\, P(k-j,i)\, |b_{ij}|\, r^j}{P(k,n)}\, |a_{k+i-j}|\, r^{k+i-j}$$

$$+ \sum_{i=0}^{n-2} \sum_{j=m_i+1}^{k} \frac{r^{n-i}\, P(k-j,i)\, |b_{ij}|\, r^j}{P(k,n)}\, |a_{k+i-j}|\, r^{k+i-j} + \frac{|b_{-1,k}|\, r^{k+n}}{P(k,n)}$$

$$\overset{(3)}{\leq} \max_{\nu=k-m}^{k+n-2} |a_\nu\, r^\nu| \sum_{i=0}^{n-2} \sum_{j=0}^{m_i} \frac{r^{n-i+j}\, P(k-j,i)\, |b_{ij}|}{P(k,n)}$$

$$+ \max_{\nu=0}^{k+n-2} |a_\nu\, r^\nu| \sum_{i=0}^{n-2} \frac{B_i\, r^{n-i}}{P(k,n)} \sum_{j=m_i+1}^{k} P(k-j,i) + \frac{B_{-1}\, r^n}{P(k,n)}$$

$$\overset{\text{Lemma 1}}{=} \max_{\nu=k-m}^{k+n-2} |a_\nu\, r^\nu| \sum_{i=0}^{n-2} \sum_{j=0}^{m_i} \frac{r^{n-i+j}\, P(k-j,i)\, |b_{ij}|}{P(k,n)}$$

$$+ \max_{\nu=0}^{k+n-2} |a_\nu\, r^\nu| \sum_{i=0}^{n-2} \frac{B_i\, r^{n-i}\, P(k-m_i-1,i+1)}{(i+1)\, P(k,n)} + \frac{B_{-1}\, r^n}{P(k,n)}$$

$$\leq \max_{\nu=0}^{k+n-1} |a_\nu\, r^\nu| \left\{ \frac{\max\limits_{\nu=k-m}^{k+n-1} |a_\nu\, r^\nu|}{\max\limits_{\nu=0}^{k+n-1} |a_\nu\, r^\nu|} \underbrace{\sum_{i=0}^{n-2} \sum_{j=0}^{m_i} \frac{r^{n-i+j}\, P(k-j,i)\, |b_{ij}|}{P(k,n)}}_{\searrow\, 0\ (k\to\infty)} \right.$$

$$\left. + \underbrace{\sum_{i=0}^{n-2} \frac{B_i\, r^{n-i}\, P(k-m_i-1,i+1)}{(i+1)\, P(k,n)}}_{\searrow\, 0\ (k\to\infty)} + \underbrace{\frac{B_{-1}\, r^n}{P(k,n)}}_{\searrow\, 0} \cdot \underbrace{\frac{1}{\max\limits_{\nu=0}^{k+n-1} |a_\nu\, r^\nu|}}_{\text{nonincreasing}} \right\}.$$

(In the last inequality, the upper indexes on the maxima have been increased on purpose.)

Since the (finitely many) summands within the brackets tend to zero for $k \to \infty$, there is a number κ so that the *recess condition*

$$
\frac{\displaystyle \max_{v=k-m}^{k+n-1} |a_v \, r^v|}{\displaystyle \max_{v=0}^{k+n-1} |a_v \, r^v|} \; \sum_{i=0}^{n-2} \sum_{j=0}^{m_i} \frac{r^{n-i+j} \, P(k-j,i) \, |b_{ij}|}{P(k,n)}
$$

$$
+ \sum_{i=0}^{n-2} \frac{B_i \, r^{n-i} \, P(k-m_i-1, i+1)}{(i+1) \, P(k,n)} + \frac{B_{-1} \, r^n}{P(k,n) \displaystyle \max_{v=0}^{k+n-1} |a_v \, r^v|} \le 1 \tag{8}
$$

holds for all $k \ge \kappa$. Hence,

$$
|a_{\kappa+n} \, r^{\kappa+n}| \; \le \; \max_{v=0}^{\kappa+n-1} |a_v \, r^v|,
$$

and, by induction,

$$
|a_k \, r^k| \; \le \; A := \; \max_{v=0}^{\kappa+n-1} |a_v \, r^v| \quad \text{for all } \; k \in \mathbb{N}_0. \tag{9}
$$

$\square$

For a given IVP (1), the smallest number κ that fulfills (8) and the number A defined in (9) depend on r. This dependency can be the basis of a step size and order control strategy in practical computations, as will be outlined in Section 5.

As a direct consequence of Theorem 1, $t(\kappa)$ is bounded by a geometric series for $0 \le h < r$:

Corollary 1 *Under the above assumptions, for $h = \omega r$, $0 \le \omega < 1$, and all $k \ge 0$,*

$$
\left| y(h) - \sum_{v=0}^{k-1} a_v \, h^v \right| \; \le \; \sum_{v=k}^{\infty} |a_v \, r^v| \, \omega^v \; \le \; A \sum_{v=k}^{\infty} \omega^v \; = \; \frac{A \omega^k}{1-\omega},
$$

where A is defined by (9).

Continuous enclosures of the solution, that is lower and upper function bounds for $y(x)$ on $[0, r)$, follow immediately:

Corollary 2 *Under the above assumptions, for $x \in [0, r)$ and all $k \ge 0$,*

$$
\left| y(x) - \sum_{v=0}^{k-1} a_v \, x^v \right| \; \le \; A \sum_{v=k}^{\infty} \left(\frac{x}{r} \right)^v \; = \; A \cdot \frac{\left(\dfrac{x}{r} \right)^k}{1 - \dfrac{x}{r}}.
$$

186

4 Continuation of the integration

So far, we have only described one step of the integration of the given IVP. When the integration domain is split into subintervals, then initial values for the first $n - 1$ derivatives are required on subsequent integration intervals. Also, since all intermediate errors must be enclosed in the course of computation, instead of real initial values one has to deal with interval initial values that include the set of all possible real initial values.

4.1 Enclosures of derivatives

Enclosures for derivatives at $x = h$ are gained by an estimation of the truncation error similar to the estimation according to Corollary 1. From (4),

$$y^{(i)}(h) = \sum_{\nu=i}^{\infty} P(\nu - i, i) a_\nu h^{\nu-i}.$$

If $A = \max_{\nu=0}^{\infty} |a_\nu r^\nu|$ and if κ and q are such that (12) and (13) hold, then for $k \geq \kappa$,

$$\left| y^{(i)}(h) - \sum_{\nu=i}^{k-1} P(\nu - i, i) a_\nu h^{\nu-i} \right| \leq \sum_{\nu=k}^{\infty} P(\nu - i, i) |a_\nu| r^\nu r^{-i} \omega^{\nu-i}$$

$$\leq \frac{qA}{r^i} \sum_{\nu=k}^{\infty} P(\nu - i, i) \omega^{\nu-i} = \frac{qA}{r^i} \frac{d^i}{d\omega^i} \sum_{\nu=k}^{\infty} \omega^\nu = \frac{qA}{r^i} \frac{d^i}{d\omega^i} \frac{\omega^k}{1-\omega}.$$

We conclude that error bounds for derivatives of the solution y of (1) follow immediately from error bounds for y.

4.2 Interval initial values

Let us assume that the integration domain $[0, x_J]$ is split into J subintervals $I_j = [x_{j-1}, x_j]$, $j = 1, \ldots, J$, and that in each subinterval I_j, $j \geq 1$, interval initial values appear. Real initial value problems are then used to compute tight enclosures of y. In the case of a homogeneous differential equation, n initial value problems

$$u_\nu^{(n)} = \sum_{i=0}^{n-2} p_i(x) u_\nu^{(i)}, \quad x \in I_j,$$

$$u_\nu^{(\mu)}(x_{j-1}) = \delta_{\mu\nu}, \quad \mu, \nu = 0, 1, \ldots, n - 1,$$

replace one interval initial value problem. If interval matrices $[A_j] = ([a_{\mu\nu}^j])$ are built of intervals $[a_{\mu\nu}^j] \ni u_\nu^{(\mu)}(x_j)$ for $j = 1, \ldots, J$, then $y(x_J)$ is contained in the first component of the matrices–vector product

$$[A_J]C_J\left(C_J^{-1}[A_{J-1}]C_{J-1}\left(\cdots\left(C_3^{-1}[A_2]C_2\left(C_2^{-1}[A_1]\begin{pmatrix} y(0) \\ \vdots \\ y^{(n-1)}(0) \end{pmatrix}\right)\right)\cdots\right)\right), \quad (10)$$

where the C_j are arbitrary nonsingular matrices. Using $n+1$ IVPs on each subinterval, a similar representation can be derived for nonhomogeneous ODEs.

The matrices C_j are used to rule out the wrapping effect that appears when intermediate results of successive matrix-vector multiplications are stored in interval vectors. Lohner (1987, 1988, 1992) discussed several choices for the matrices C_j.

5 Practical calculation of the enclosure

By Corollary 1, the enclosure of $y(h)$ is built from the Taylor polynomial approximate solution $s(\kappa)$ in (7) and a geometric series error bound. For the construction of $s(\kappa)$, only finitely many Taylor coefficients b_{ij} in (2) are needed. They are obtained by automatic differentiation (Rall 1981) of the functions p_i. The numbers a_k in (7) are then calculated recursively from (5) and (6).

The geometric series bound relies on the bounds B_i in (3) and on a number κ such that (8) holds for all $k \geq \kappa$. The calculation of these quantities will be described in two subsections. The number A then follows from (9).

5.1 Estimates for Taylor coefficients of analytic functions

A well known bound for Taylor coefficients of analytic functions is Cauchy's estimate (Conway 1973, p. 73). For an analytic function

$$p(z) = \sum_{j=0}^{\infty} b_j z^j, \quad |z| \leq r,$$

it holds that

$$|b_j| \leq \frac{\max\limits_{|z|=r} |p(z)|}{r^j}, \quad j \in \mathbb{N}.$$

Now let $T_{m_i}(z; p_i)$ denote the Taylor polynomial of order m_i for p_i in (1). Then we can use

$$B_i := \max_{|z|=r} |p_i(z) - T_{m_i}(z; p_i)|, \quad i = -1, \ldots, n-2 \quad (11)$$

in (3). The practical computation of B_i has thus become the problem of determining the range of a complex function.

For many compositions of complex standard functions, the real and the imaginary parts can be expressed by compositions of real standard functions. Braune and Krämer (1987) used these decompositions for the construction of high–accuracy complex interval standard functions that are easily implemented in modern programming languages, and that yield tight range bounds.

5.2 Determination of κ

If κ is a given integer, then we can check the recess condition (8) for $k = \kappa$. However, the fulfillment of (8) for $k = \kappa$ does not imply that (8) holds for all $k \geq \kappa$, because the left of (8) is not monotonically decreasing with k in general. Monotonicity only holds for some of the terms and for sufficiently large k (the proof of Lemma 2 follows from straightforward calculations):

Lemma 2 *For* $k \geq n(m+1)$, $\dfrac{P(k-j,i)}{P(k,n)}$ *and* $\dfrac{P(k-m_i-1,i+1)}{P(k,n)}$ *are monotonically decreasing with* k.

$$\frac{\max\limits_{\nu=k-m}^{k+n-1} |a_\nu r^\nu|}{\max\limits_{\nu=0}^{k+n-1} |a_\nu r^\nu|} \text{ is not necessarily monotonic with } k, \text{ not even for large } k.$$

Hence, we modify the recess condition slightly, to obtain an enclosure criterion that is verifiable by simple practical calculations. Suppose that for some $k \geq n(m+1)$ and some q, $0 < q < 1$, it holds that

$$\frac{\max\limits_{\nu=k-m}^{k+n-1} |a_\nu r^\nu|}{\max\limits_{\nu=0}^{k+n-1} |a_\nu r^\nu|} \leq q \tag{12}$$

and that

$$q \cdot \sum_{i=0}^{n-2} \sum_{j=0}^{m_i} \frac{r^{n-i+j} P(k-j,i) |b_{ij}|}{P(k,n)}$$

$$+ \sum_{i=0}^{n-1} \frac{B_i r^{n-i} P(k-m_i-1,i+1)}{(i+1) P(k,n)} + \frac{B_{-1} r^n}{P(k,n)} \leq q. \tag{13}$$

Then

$$|a_{k+n} r^{k+n}| \leq q \cdot \max_{\nu=0}^{k+n-1} |a_\nu r^\nu|,$$

so that

$$|a_{k+j} r^{k+j}| \leq q \cdot \max_{\nu=0}^{k+n-1} |a_\nu r^\nu| \text{ for all } j \geq n.$$

Due to Lemma 2, the fulfillment of (12) and (13) for $k = \kappa \geq n(m+1)$ is a sufficient condition that (12) and (13) also hold for all $k > \kappa$. Instead of A we can then use qA in Corollary 1 or Corollary 2, to obtain pointwise or continuous enclosures of the solution of the given IVP (1).

5.3 Step size and order control strategy

In many numerical examples, the following step size and order control strategy has been found to be effective: If the interval of integration is given by $[0, x_J]$,

189

then start the enclosure algorithm with $h = x_J$. Choose $\omega \in (0,1)$ (often, $\omega \in (0.5, 0.9)$ proved to be a good choice) and let $r := \frac{h}{\omega}$. Compute the numbers B_i from (11), provided that the functions p_i are analytic for $|z| \leq r$. Check the modified recess condition (13) for $q = 1$ and a suitable value of $k = \kappa$ (the approximate order of the method). If (13) is fulfilled, then compute $s(\kappa)$ and enclose $t(\kappa)$ according to Corollary 1. Increase κ until $t(\kappa)$ becomes small enough for an accurate enclosure of $y(h)$. If any of the above steps failed, then bisect the integration domain and apply the same procedure recursively to both subintervals.

6 Numerical examples

We have implemented our method in a computer program written in PASCAL–XSC (Klatte et al. 1992), a PASCAL extension with a machine interval arithmetic. In such an arithmetic, all of the above calculations can be performed virtually unchanged, by replacing all operations between real numbers by the corresponding operations between machine intervals, with the automatic enclosure of all roundoff errors in the result of any computation.

The program includes a multiple precision interval staggered correction arithmetic (Stetter 1984), to handle the cancellations that occur when the calculations are performed in floating arithmetic. The complete code of the program is available via Internet at

http//www.uni-karlsruhe.de/~Markus.Neher/livptayp.html

Example 1: $y'' = \alpha e^x y + e^{-x} - \alpha$, $y(0) = 1$, $y'(0) = -1$.

The exact solution of this problem is $y(x) = e^{-x}$. The IVP gets numerically unstable when α is increasing. In Table 1 we show the computed enclosures $[y(h)]$ for several values of α and for the largest values of h that could be reached with one single integration step (for $\omega = 0.82$ and $\kappa \leq 300$). The computation times (in seconds) were obtained on a PC with a Pentium II processor with 266 MHz.

In the last column the maximum integration domains that were reached with Lohner's program AWA (that runs with single precision) are given. Long before the integration with AWA aborted, the enclosures had already become inaccurate.

Table 1. Integrations for different values of α.

α	x	κ	$[y](x)$	Time	$x_{\max}$ with AWA
100	3.25	288	$3.877\ 420\ 783\ 172\ 20^{3}_{0}\text{E-}02$	54	2.231
1000	2	261	$1.353\ 352\ 832\ 366\ 12^{8}_{6}\text{E-}01$	47	0.981
10000	1	279	$3.678\ 794\ 411\ 714\ ^{588}_{354}\text{E-}01$	47	0.356

Example 2: Computation of eigenvalues of

$$-u'' + \cos(2x)\,u = \lambda u$$
$$u(0) = u(\pi) = 0. \tag{14}$$

A real number λ is called an eigenvalue of the boundary value problem (14) if there is a nontrivial solution $u(x)$ of (14). Following the well–known Sturm–Liouville theory, we compute eigenvalues of (14) with a shooting method with shooting parameter λ, using the initial value problem

$$y'' = (\cos x - \lambda)\,y$$
$$y(0) = 0,\; y'(0) = 1. \tag{15}$$

If $y(\pi; \lambda) = 0$ then λ is an eigenvalue of (14). The index of the eigenvalue is given by the number of zeros of $y(\cdot; \lambda)$ in the interval $(0, \pi)$. Due to the symmetry of the differential equation and the boundary conditions, eigenvalues can also be computed by testing $y(\frac{\pi}{2}) = 0$ (for eigenvalues with even indexes) or $y'(\frac{\pi}{2}) = 0$ (for eigenvalues with odd indexes). Eigenvalue bounds for specific eigenvalues are computed by solving (15) for different values of λ and $x \in [0, \pi]$, and by determining the number $N(\lambda)$ of zeros of the respective solutions $y(\cdot; \lambda)$ within $(0, \pi)$.

The latter task involves validated rootfinding. The zeros are all simple and isolated. Uniqueness of a root in a given interval $X = [x_1, x_2] \subset [0, \pi]$ can be proved with a continuous enclosure of the derivative of $y(\cdot; \lambda)$ (Neher 1992). Let U be an interval such that $y'(x, \lambda) \in U$ for all $x \in X$, then $y(x, \lambda)$ has a unique zero in X if $y(x_1, \lambda) \cdot y(x_2, \lambda) < 0$ and $0 \notin U$ holds.

Table 2. Integrations of (15) for different values of λ.

λ	$N(\lambda)$	$[y(\frac{\pi}{2})]$
16.008 310 459 709 47	3	$-3.87^4_5 \text{E} - 16$
16.008 310 459 709 48	4	$1.36^8_6 \text{E} - 16$

λ	$N(\lambda)$	$[y'(\frac{\pi}{2})]$
121.001 041 672 579 0	10	$-1.00^7_8 \text{E} - 16$
121.001 041 672 579 1	11	$^{7.27}_{6.79} \text{E} - 15$

In Table 2, the enclosures of $y(\frac{\pi}{2}; \lambda)$ from (15) are listed for four different values of λ. They were computed with only one integration step. From these enclosures, the eigenvalues λ_4 and λ_{10} of (14) are determined with 16 decimal digits of accuracy:

$$\lambda_4 \in 16.008\ 310\ 459\ 709\ 4^8_7,$$
$$\lambda_{10} \in 121.001\ 041\ 672\ 579\ ^1_0.$$

Conclusion

We have presented a new enclosure method for linear n-th order ODEs with analytic coefficients. Our numerical examples demonstrate that the method can be successfully implemented on a computer, and that it works with very large step sizes.

Future work will concentrate on the utilization of the method to nonlinear ODEs.

Acknowledgement. The author thanks the referee for his helpful comments.

References

Alefeld, G., Herzberger, J. (1983): Introduction to interval computations. Academic Press, New York

Braune, K., Krämer, W. (1987): High–accuracy standard functions for real and complex intervals. In: Kaucher, E., Kulisch, U., Ullrich, Ch. (eds.): Computerarithmetic: Scientific computation and programming languages. Teubner, Stuttgart, pp. 81–114

Conway, J. B. (1973): Functions of one complex variable. Springer, New York

Corliss, G. F., Rihm, R. (1996): Validating an a priori enclosure using high-order Taylor series. In: Alefeld, G., Frommer, A., Lang, B. (eds.): Scientific Computing and Validated Numerics. Akademie–Verlag, Berlin, pp. 228–238

Heuser, H. (1989): Gewöhnliche Differentialgleichungen. Teubner, Stuttgart

Klatte, R., Kulisch, U., Neaga, M., Ratz, D., Ullrich, Ch. (1992): PASCAL–XSC — Language reference with examples. Springer, Berlin

Kulisch, U., Miranker, W. L. (1981): Computer arithmetic in theory and practice. Academic Press, New York

Lohner, R. (1987): Enclosing the solutions of ordinary initial- and boundary-value problems. In: Kaucher, E., Kulisch, U., Ullrich, Ch. (eds.): Computerarithmetic: Scientific computation and programming languages. Teubner, Stuttgart, pp. 255–286

Lohner, R. (1988): Einschließung der Lösung gewöhnlicher Anfangs- und Randwertaufgaben und Anwendungen. Dissertation, Universität Karlsruhe

Lohner, R. (1992): Computation of guaranteed solutions of ordinary initial and boundary value problems. In: Cash, J. R., Gladwell, I. (eds.): Computational ordinary differential equations. Clarendon Press, Oxford, pp. 425–435

Lohner, R. (1995): Step size and order control in the verified solution of ordinary initial value problems. Presented at the SciCADE 95 International Conference on Scientific Computation and Differential Equations, Stanford, California

Moore, R.E. (1965a): The automatic analysis and control of error in digital computation based on the use of interval numbers. In: Rall, L. B. (ed.): Error in digital computation, vol. I. John Wiley and Sons, New York, pp. 61–130

Moore, R.E. (1965b): Automatic local coordinate transformations to reduce the growth of error bounds in interval computation of solutions of ordinary differential equations. In: Rall, L. B. (ed.): Error in digital computation, vol. II. John Wiley and Sons, New York, pp. 103–140

Moore, R.E. (1966): Interval analysis. Prentice Hall, Englewood Cliffs

Nedialkov, N. (1999): Computing rigorous bounds on the solution of an IVP for an ODE. Ph. D. Thesis, University of Toronto

Neher, M. (1992): Inclusion of eigenvalues and eigenfunctions of the Sturm-Liouville problem. In: Atanassova, L., Herzberger, J. (eds.): Computer arithmetic and enclosure methods. Elsevier, Amsterdam, pp. 401–408

Neher, M. (1999): An enclosure method for the solution of linear ODEs with polynomial coefficients. Numer. Funct. Anal. and Optimiz. 20: 779–803

Rall, L. B. (1981): Automatic differentiation: Techniques and applications. Lecture Notes in Computer Science, Vol. 120, Springer, Berlin

Stetter, H. J. (1984): Sequential defect correction for high-accuracy floating-point arithmetic. In: Numerical analysis. Springer, Berlin, pp. 186–202 (Lecture notes in mathematics, vol. 1066)

Safe Numerical Error Bounds for Solutions of Nonlinear Elliptic Boundary Value Problems

Michael Plum

1 Introduction

This paper is concerned with nonlinear elliptic boundary value problems of the form

$$-\Delta u + F(x, u) = 0 \qquad \text{on } \Omega,$$
$$u = 0 \qquad \text{on } \partial\Omega, \qquad (1)$$

where $\Omega \subset \mathbb{R}^n$ is a bounded domain with Lipschitz-continuous boundary $\partial\Omega$, and F is a given nonlinearity on $\overline{\Omega} \times \mathbb{R}$ with values $F(x, y) \in \mathbb{R}$. Further regularity assumptions will follow below in appropriate places.

We will consider two concepts of solutions to problem (1):

i) *strong* solutions $u \in H_2^B(\Omega) := \text{closure}_{H_2(\Omega)}\{v \in C_2(\overline{\Omega}) : v|_{\partial\Omega} = 0\}$;

ii) *weak* solutions $u \in H_1^0(\Omega)$; the term Δu in (1) is then to be understood in the distributional sense.

Our goal is to present a computer-assisted method for proving, under appropriate conditions, the *existence* of a solution to problem (1) within a "close" and explicitly given neighborhood of some approximate solution. In slightly different words, the method provides safe and verified *error bounds* for approximate solutions. The conditions, under which such an existence and enclosure result can be stated, shall moreover be testable in an automatic way on a computer.

The usual general way of proceeding in order to derive existence and enclosure statements of the desired kind is to transform problem (1) into some equivalent fixed-point equation

$$u \in X, \quad u = T(u), \qquad (2)$$

and to apply some fixed-point theorem. Under appropriate conditions on the space X and the operator $T : X \to X$ (e.g., continuity, compactness, contractivity), which usually have to be verified by theoretical means, the fixed-point theorem yields the existence of a solution u of problem (2) (and thus, of problem (1)) in some suitable set $U \subset X$, provided that

$$T(U) \subset U. \qquad (3)$$

The statement "$u \in U$" constitutes the desired enclosure result. In order to compute an explicit enclosure, one must therefore construct U *explicitly*. Moreover, U should provide tight bounds, i.e., it should be "*small in diameter*" in an appropriate sense. These requirements can usually be satisfied only by numerical means. For the numerical verification of condition (3), one hat to use *interval-analysis* on many levels between basic interval arithmetic and functional analysis, as we will discuss in more detail in later sections.

The differences between existing existence and enclosure methods for problems like (1) are characterized by different choices of fixed-point operators T and different types of sets U. The "classical" choice, initialized by L. Collatz already in the 1950's (Collatz 1952) and developed in detail later by J. Schröder (1980), (1991), uses *function intervals* U and a *monotone* fixed-point operator T. Since it involves certain inherent difficulties concerning applicability to wide ranges of problems (it requires the linearization of problem (1) at the expected solution to be positive definite), we will propose another method which avoids these difficulties, as will be shown in several examples. A different approach also avoiding these problems has been proposed by Nakao (see e.g. Nakao (1993), Nakao and Yamamoto (1995)).

2 A method based on norm bounds - Abstract formulation

Here, we describe our method first on a more abstract level, which will later be made more concrete in different ways.

Let X, Y, Z denote three Banach spaces, $X \subset Y$, and let $L_0 \in \mathcal{B}(X, Z)$ (the space of bounded linear operators from X to Z). Moreover, let $\mathcal{F} : Y \to Z$ denote a Fréchet-differentiable operator. We consider the problem

$$u \in X, \quad L_0[u] + \mathcal{F}(u) = 0, \tag{4}$$

again aiming at *existence* and *enclosure* statements. The enclosing set U will now be a *norm ball* with known center and radius (the latter being moreover "small"). We make the following abstract *regularity* assumptions (in particular the first one can be weakened):

A) The embedding $E_X^Y : X \to Y$ is compact,

B) For some $\sigma \in \mathcal{B}(Y, Z)$, $L_0 + \sigma E_X^Y : X \to Z$ is one-to-one and onto.

As a consequence we obtain, for *every* $\rho \in \mathcal{B}(Y, Z)$, the implication

$$\text{If } L_0 + \rho E_X^Y : X \to Z \text{ is one-to-one, then it is also onto,}$$
$$\text{and } (L_0 + \rho E_X^Y)^{-1} \in \mathcal{B}(Z, X). \tag{5}$$

(For proof, let $r \in Z$ and rewrite the equation $(L_0 + \rho E_X^Y)[u] = r$ as $u = (L_0 + \sigma E_X^Y)^{-1}[(\sigma - \rho)E_X^Y u + r]$, so that Fredholm's Alternative and the Open Mapping Theorem provide the assertion.)

Now let $\omega \in X$ denote some approximate solution to problem (4), and denote by

$$d := L_0[\omega] + \mathcal{F}(\omega) \in Z \tag{6}$$

its *defect* (residual). Simple calculations show that the following equation for the *error* $v = u - \omega$ is equivalent to (4):

$$v \in X, \; L_0[v] + \mathcal{F}(\omega + v) - \mathcal{F}(\omega) = -d. \tag{7}$$

Now, with $\mathcal{F}'(\omega) \in \mathcal{B}(Y, Z)$ denoting the Fréchet derivative of $\mathcal{F}$ at ω, let

$$g(v) := \mathcal{F}(\omega + v) - \mathcal{F}(\omega) - \mathcal{F}'(\omega)[v] \quad \text{for } v \in Y, \tag{8}$$

$$L := L_0 + \mathcal{F}'(\omega)E_X^Y. \tag{9}$$

Assuming that $L : X \to Z$ is one-to-one, so that it is also onto and L^{-1} is bounded according to (5), we can therefore rewrite (7) as

$$v \in X, \; v = -L^{-1}[d + g(E_X^Y v)] =: T(v) \tag{10}$$

and apply Schauder's Fixed-Point-Theorem: Since L^{-1} is bounded, g is continuous, and E_X^Y is compact, we conclude that $T : X \to X$ is continuous and compact. We are therefore left to find a closed, bounded, and convex set $V \subset X$ such that $T(V) \subset V$. Here, we aim in particular at a norm ball V centered at 0 with radius α (to be constructed). For this purpose, suppose that constants $\delta, C,$ and K, as well as some monotonically nondecreasing function $G : [0, \infty) \to [0, \infty)$ are known which satisfy

$$\|L_0[\omega] + \mathcal{F}(\omega)\|_Z \le \delta, \tag{11}$$

$$\|u\|_Y \le C\|u\|_X \quad \text{for all } u \in X, \tag{12}$$

$$\|u\|_X \le K \, \|L[u]\|_Z \quad \text{for all } u \in X, \tag{13}$$

$$\|\mathcal{F}(\omega + u) - \mathcal{F}(\omega) - \mathcal{F}'(\omega)[u]\|_Z \le G(\|u\|_Y) \quad \text{for all } u \in Y, \tag{14}$$

$$G(t) = o(t) \quad \text{for } t \to 0+ \tag{15}$$

(regard that (14) and (15) are consistent due to the Fréchet differentiability of $\mathcal{F}$ at ω). Using (6), (8), and (11) to (14), we obtain from (10) that

$$\|T(v)\|_X \le K[\delta + G(C\|v\|_X)] \quad \text{for each } v \in X,$$

so that the norm ball $V = \{v \in X : \|v\|_X \le \alpha\}$ is mapped into itself by T if $K[\delta + G(C\alpha)] \le \alpha$, i.e., if

$$\delta \le \frac{\alpha}{K} - G(C\alpha). \tag{16}$$

We have therefore proved the following

Existence and Enclosure Theorem

If (16) holds for some $\alpha \geq 0$, there exists a solution $u \in X$ to problem (4) satisfying

$$\|u - \omega\|_X \leq \alpha. \tag{17}$$

An important observation is that, due to (15), the crucial condition (16) is indeed satisfied for some "small" α if the constant δ is sufficiently small, which means according to (11) that the approximate solution ω of problem (4) must be computed with *sufficient accuracy*, and (16) tells *how* accurate the computation has to be. This meets the general philosophy of computer-assisted proofs: The "hard work" of the proof is left to the computer!

We are left to describe how δ, C, K, and G satisfying (11) to (15) can be computed. We will do so in the next sections in our different realizations of the abstract operator setting.

We remark that the requirement of L being one-to-one (which was stated after (9)) is contained in (13). Observe moreover that (13) does not require any restriction on the sign of the eigenvalues of the linearization L, in contrast to the classical monotonicity methods.

3 Strong solutions

Here we describe the application of our abstract results to the elliptic boundary value problem (1), choosing

$$L_0 := -\Delta, \quad \mathcal{F}(u)(x) := F(x, u(x)) \tag{18}$$

$$X := H_2^B(\Omega) := \text{closure}_{H_2(\Omega)}\{v \in C_2(\overline{\Omega}) : v|_{\partial\Omega} = 0\}, \quad Z := L_2(\Omega), \tag{19}$$

i.e., we are aiming at existence and enclosure results for *strong* solutions of problem (1) now. For the choice of the Banach space Y we have to ensure that E_X^Y is compact and that $\mathcal{F} : Y \to Z$ is Fréchet differentiable; these two requirements point into opposite directions concerning the "strength" of the norm in Y. A good "balance" is achieved for the choice

$$Y = C(\overline{\Omega}),$$

if $n \leq 3$. If $n = 1$ (so that Ω is a bounded open real interval and $\Delta u = u''$), also $Y := C_1(\overline{\Omega})$ is appropriate. The case $n \geq 4$ will not be considered here.

The regularity assumption B) (see Section 2) requires here that the Poisson equation

$$-\Delta u = r \text{ on } \Omega, \quad u = 0 \text{ on } \partial\Omega \tag{20}$$

has a unique solution $u \in H_2^B(\Omega)$ for each $r \in L_2(\Omega)$. In fact, this is the usual H_2-regularity condition on the domain Ω resp. on its boundary $\partial\Omega$. It is satisfied e.g. for C_2-smoothly bounded domains Ω or for convex polygonal domains $\Omega \subset \mathbb{R}^2$; it excludes e.g. domains with reentrant corners.

For the computation of an approximate solution ω, any numerical method providing an approximation in $H_2^B(\Omega)$ is suitable. In particular, ω has to satisfy the Dirichlet boundary conditions exactly. (However, a generalization of our method is able to handle approximations ω satisfying also the boundary conditions only approximately.) Using Finite Elements for computing ω, one must choose C_1-elements (e.g., triangular Argyris- or Bell-elements, or rectangular Bogner-Schmidt-Fox-elements), in order to meet the required H_2-property of ω. This is certainly a disadvantage, but on the other hand a rather natural condition when one looks for *strong* solutions.

We now comment on the computation of the terms δ, C, K, and G satisfying (11) to (15). Condition (11) reads, in the present context,

$$|| - \Delta\omega + F(\cdot, \omega)||_{L_2} \leq \delta, \tag{21}$$

i.e., a bound for an *integral* is required. Depending on F and on the concrete representation of ω, such a bound can be computed either by explicit integration using a computer algebra package (e.g., if F is a polynomial function and ω is piecewise polynomial), or by use of a quadrature formula and a bound for its remainder term (Storck n.n.); the latter can often be obtained by automatic differentiation techniques. In any case, *interval arithmetic* (Klatte et al. 1993) has to be used in all numerical evaluations here (but not during the computation of ω described above!), in order to take rounding errors into account.

For the computation of C, K, and G, we restrict ourselves to the case $n \in \{2, 3\}$ (for $n = 1$, see Plum (1991)), and choose $Y := C(\overline{\Omega})$. For convex Ω we proved (Plum 1992) that

$$||u||_\infty \leq C_0||u||_{L_2} + C_1||\nabla u||_{L_2} + C_2||u_{xx}||_{L_2} \quad \text{for all} \ u \in H_2(\Omega), \tag{22}$$

with u_{xx} denoting the Hessian matrix of u, and with

$$C_j = \frac{\gamma_j}{\text{vol}(\Omega)} \left[\max_{x_0 \in \overline{\Omega}} \int_\Omega |x - x_0|^{2j} dx \right]^{\frac{1}{2}} \quad (j = 0, 1, 2),$$

where $\gamma_0, \gamma_1, \gamma_2$ are fixed numbers less than 2. For non-convex Ω (satisfying, however, a boundary Lipschitz condition), the formulas for C_0, C_1, C_2 look a bit more complicated.

Due to (22), the inequality (12) holds with $C := 1$ if we choose the right-hand side of (22) as norm $||u||_X$, and $||u||_Y := ||u||_\infty$. Now (13) reads

$$C_0||u||_{L_2} + C_1||\nabla u||_{L_2} + C_2||u_{xx}||_{L_2} \leq K||L[u]||_{L_2} \quad \text{for all} \ u \in H_2^B(\Omega), \tag{23}$$

where

$$L[u] := -\Delta u + cu, \quad c(x) := \frac{\partial F}{\partial y}(x, \omega(x)). \tag{24}$$

Clearly, (23) is satisfied for $K := C_0 K_0 + C_1 K_1 + C_2 K_2$ where K_0, K_1, K_2 are constants such that, for all $u \in H_2^B(\Omega)$,

$$
\begin{array}{lll}
\text{(a)} & \|u\|_{L_2} \le K_0 \||L[u]|\|_{L_2}, & \text{(b)} \quad \|\nabla u\|_{L_2} \le K_1 \||L[u]|\|_{L_2}, \\[2mm]
\text{(c)} & \|u_{xx}\|_{L_2} \le K_2 \||L[u]|\|_{L_2}.
\end{array}
\tag{25}
$$

Expansion into eigenfunction series shows that (25a) holds for

$$K_0 := [\min \{|\lambda| : \lambda \text{ eigenvalue of } L \text{ on } H_2^B(\Omega)\}]^{-1},$$

so that *eigenvalue bounds* are needed to compute K_0. This is again carried out by computer assisted methods of their own (which we cannot describe in detail here), namely the Rayleigh-Ritz-method, the Temple-Lehmann-Goerisch method, and a homotopy method connecting the given eigenvalue problem to a "simple" one with known eigenvalues; see Behnke and Goerisch (1994) and Plum (1997) for an overview on eigenvalue bounds.

The remaining constants K_1 and K_2 satisfying (25b,c) can now be computed rather directly, using (25a) and the a priori bounds

$$\|\nabla u\|_{L_2} \le [\|u\|_{L_2} \||L[u]|\|_{L_2} - \underline{c}\|u\|_{L_2}^2]^{\frac{1}{2}} \quad (u \in H_2^B(\Omega))$$

(which is obtained by partial integration, with $\underline{c} \le \min_{x \in \overline{\Omega}} c(x)$), and

$$\|u_{xx}\|_{L_2} \le \||L[u]|\|_{L_2} + \|c\|_\infty \|u\|_{L_2} \quad (u \in H_2^B(\Omega))$$

which holds true in case of a convex domain Ω. If Ω is not convex (but has a piecewise C_2-smooth boundary), the latter estimate must be replaced by a more involved one; see Plum (1992) for details.

Finally, (14) is satisfied here if, for all $x \in \overline{\Omega}$ and $y \in \mathbb{R}$,

$$\sqrt{\text{vol}(\Omega)} \, |F(x, \omega(x) + y) - F(x, \omega(x)) - c(x)y| \le G(|y|), \tag{26}$$

with c defined in (24). A monotonically nondecreasing function G satisfying (26) and (15) can in most cases be calculated directly, using e.g. Taylor expansion and constant upper and lower bounds for ω.

To illustrate our strong solution approach, we consider the problem

$$\Delta u + u^2 = \lambda \sin(\pi x_1)\sin(\pi x_2) \text{ on } \Omega := (0,1)^2, \quad u = 0 \text{ on } \partial\Omega. \tag{27}$$

The results presented here are joint work with P.J. McKenna and B. Breuer and will be reported in detail in a paper which is presently in preparation; see also Breuer (1998).

In the PDE-community it has been an open question since many years if problem (27) has at least *four solutions* for sufficiently large values of λ. Apparently, this question could not be answered by purely analytical means. By our existence and enclosure method, combined with a numerical mountain pass method (see Choi and McKenna (1993)), we could give a positive answer, at least for the particular value $\lambda = 800$.

The numerical mountain pass method was used to find four essentially different *approximate* solutions, which were then improved by a Newton iteration, where the linear problems in the Newton steps were treated by a collocation method with ~ 16000 trigonometric basis functions. In this way, we arrived at highly accurate approximations w_i ($i = 1, \ldots, 4$) with defect bounds δ_i (see (11), (21)) in order of magnitude of 0.001 to 0.01. Since the eigenvalues of $L_i = -\Delta - 2w_i$ turned out to be well separated from zero for all four approximations w_i, our existence and enclosure method was successful in proving the existence of four solutions u_i ($i = 1, \ldots, 4$) of problem (27) such that $\|u_i - w_i\|_\infty \leq \alpha_i$ ($i = 1, \ldots, 4$), with error bounds $\alpha_1, \ldots, \alpha_4$ between $5 \cdot 10^{-4}$ and $5 \cdot 10^{-2}$. Since simple computations show $\|w_i - w_j\|_\infty > \alpha_i + \alpha_j$ for $i \neq j$, the four solutions $u_1, \ldots, u_4$ are indeed pairwise different.

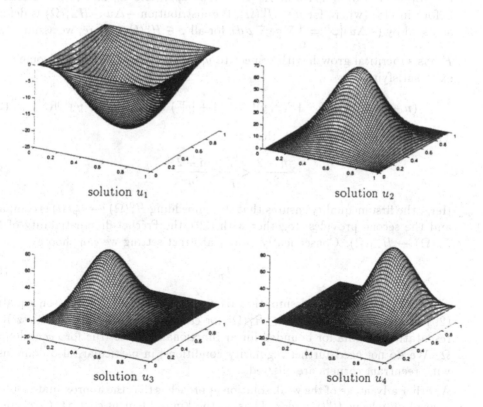

solution u_1

solution u_2

solution u_3

solution u_4

Figure 1: Solution plots for problem (27), $\lambda = 800$

It should be remarked that the linearization L_1 in the approximation ω_1 has only positive eigenvalues, while $L_2, L_3,$ and L_4 have also negative eigenvalues, so that "monotonicity" methods could not be used for our purpose.

Figure 1 shows plots of the four solutions. u_1 and u_2 are "fully" symmetric (i.e., with respect to reflection at the axes $x_1 = \frac{1}{2}$, $x_2 = \frac{1}{2}$, $x_1 = x_2$, and $x_1 = 1 - x_2$), while u_3 is symmetric only with respect to $x_2 = \frac{1}{2}$, and u_4 only with respect to $x_1 = x_2$. (Of course, further approximations resp. solutions arise from ω_3 and ω_4 resp. u_3 and u_4 by rotations.)

4 Weak solutions

Besides providing the strong solution approach, our abstract operator setting is also appropriate for yielding existence and enclosure results for *weak* solutions to problem (1). For this purpose, we choose now

$$X := H_1^0(\Omega), \; Z := H_{-1}(\Omega) \tag{28}$$

(where $H_1^0(\Omega)$ is endowed with the norm $\|u\|_X = \|\nabla u\|_{L_2}$, and $H_{-1}(\Omega)$ denotes the topological dual space of $H_1^0(\Omega)$). The operators L_0 and $\mathcal{F}$ are chosen as before in (18) (where for $u \in H_1^0(\Omega)$, the distribution $-\Delta u \in H_{-1}(\Omega)$ is defined as usual by $(-\Delta u)[\varphi] := \int_\Omega \nabla u \cdot \nabla \varphi \, dx$ for all $\varphi \in H_1^0(\Omega)$). Here, we assume that

F has subcritical growth with respect to y, i.e., that some $\widehat{C} > 0$ and some $r \geq 1$ exist satisfying

$$(n-2)r < n + 2, \quad |F(x,y)| \leq \widehat{C}(1 + |y|^r) \text{ for all } x \in \overline{\Omega}, \; y \in \mathbb{R}. \tag{29}$$

Then, some $p \geq 2$ exists such that

$$\frac{n-2}{2n} < \frac{1}{p} < \frac{n+2}{2nr}. \tag{30}$$

Here, the first inequality ensures that the embedding $H_1^0(\Omega) \hookrightarrow L_p(\Omega)$ is compact, and the second provides, together with (29), the Fréchet-differentiability of $\mathcal{F} : L_p(\Omega) \to H_{-1}(\Omega)$. Consequently, in our abstract setting we can choose

$$Y := L_p(\Omega). \tag{31}$$

The abstract regularity assumption B) requires here that the Poisson equation (20) has a unique solution $u \in H_1^0(\Omega)$ for each $r \in H_{-1}(\Omega)$. By the Riesz Representation Lemma for bounded linear functionals, this is true for *every* domain Ω. We do not need further regularity conditions; in particular, also domains Ω with reentrant corners are allowed.

Another advantage of the weak solution approach is that the approximate solution ω need only be in $H_1^0(\Omega)$ here. Thus, in the Finite Element context, C_0-elements are sufficient!

Again, we comment now on the computation of the terms δ, C, K, and G satisfying (11) to (15). The defect $-\Delta\omega + F(\cdot, \omega)$ has to be bounded, according to (11) and (28), in the H_{-1}-norm, which under some aspects is a bit more involved than the simple L_2-bound needed for strong solutions. The direct definition of the dual space norm gives a *supremum* over $H_1^0(\Omega) \setminus \{0\}$ and is therefore not well-suited for the computation of an upper bound. However, it can be shown (see Plum (1994)) that the following complementary variational characterization holds true:

$$\| -\Delta\omega + F(\cdot, \omega)\|_{H_{-1}} = \min\{\|\nabla\omega - v\|_{L_2} : v \in L_2(\Omega)^n, \operatorname{div} v = F(\cdot, \omega)\} \quad (32)$$

(with div v defined in the distributional sense), and moreover, that the minimum in (32) is attained at the unique $v \in L_2(\Omega)^n$ satisfying div $v = F(\cdot, \omega)$ and

$$\int_\Omega v \cdot w \, dx = 0 \quad \text{for all } w \in L_2(\Omega)^n, \quad \operatorname{div} w = 0 \quad (33)$$

(the minimizing v is given by $v = \nabla u$, where $u \in H_1^0(\Omega)$ solves $\Delta u = F(\cdot, \omega)$). Thus, a defect bound δ can be obtained by computing a $v \in L_2(\Omega)^n$ satisfying the side condition div $v = F(\cdot, \omega)$ exactly, but solving equation (33) only approximately, and then computing an upper bound δ for $\|\nabla\omega - v\|_{L_2}$ (by verified integration), which is the desired defect bound according to (32).

To satisfy (12) we need a constant C satisfying

$$\|u\|_{L_p} \leq C \|\nabla u\|_{L_2} \quad \text{for all } u \in H_1^0(\Omega). \quad (34)$$

There exist several approaches to this problem in the literature. A constant C which is (in general) not optimal but simple is the following one (see Plum (1994), Lemma 5.1):

$$C = \frac{1}{2\sqrt{2}} \operatorname{vol}(\Omega)^{\frac{1}{p}} \left[\prod_{j=0}^{\nu-1} \left(\frac{p}{2} - j \right) \right]^{\frac{1}{\nu}} \quad \text{if } n = 2,$$

$$\tag{35}$$

$$C = \frac{n-1}{\sqrt{n(n-2)}} \operatorname{vol}(\Omega)^{\frac{1}{p}-\frac{1}{2}+\frac{1}{n}} \quad \text{if } n \geq 3,$$

where ν is the largest integer $\leq \frac{p}{2}$.

For the analysis of (13) we first observe that the linearization L is here (formally) again given by (24), and that (13) now reads

$$\|\nabla u\|_{L_2} \leq K \|L[u]\|_{H_{-1}} \quad \text{for all } u \in H_1^0(\Omega). \quad (36)$$

By rather straightforward calculations (see Plum (1994)) one arrives at the eigenvalue problem

$$u \in H_1^0(\Omega), \quad L[u] = \lambda(-\Delta u) \quad (37)$$

and obtains that (36) holds for

$$K := [\inf\{|\lambda| : \lambda \text{ eigenvalue of problem (37)}\}]^{-1},$$

so that again *eigenvalue bounds* are required to compute K. Observe that the eigenvalues of (37) accumulate at 1 (if there are infinitely many), which may complicate the computation of eigenvalue bounds. In the particular case $c \geq 0$ (which is, roughly speaking, the "maximum principle case"), the Rayleigh quotient of problem (37) is always ≥ 1, so that all eigenvalues are ≥ 1, which provides $K = 1$ without any further computation. This may be regarded as reflection of the "simplicity" of the "maximum principle case", which in turn is closely related to the cases where the classical monotonicity methods are successful.

Finally, a monotonically nondecreasing function G satisfying (15) and (14), i.e.,

$$\|F(\cdot, \omega + u) - F(\cdot, \omega) - cu\|_{H_{-1}} \leq G(\|u\|_{L_p}) \quad \text{for all } u \in L_p(\Omega), \tag{38}$$

is obtained by first calculating a monotonically nondecreasing function $\widetilde{G}$ such that $\widetilde{G}(t) = o(t)$ for $t \to 0+$ and the pointwise inequality (26) (with $\widetilde{G}$ in place of G, and without the factor $\sqrt{\text{vol}(\Omega)}$ on the left-hand side) is fulfilled, and moreover, $\widetilde{G}(t^{\frac{1}{r}})$ is a *concave* function of t, where r is the growth exponent in (29). Then, using the embedding constant $C = C(p)$ satisfying (34), we obtain, for all $u \in L_p(\Omega)$,

$$\|F(\cdot, \omega + u) - F(\cdot, \omega) - cu\|_{H_{-1}} \leq \sup_{\varphi \in H_1^0(\Omega) \setminus \{0\}} \left\{ \|\nabla \varphi\|_{L_2}^{-1} \int_\Omega \widetilde{G}(|u|) \, |\varphi| \, dx \right\}$$

$$\leq \sup_{\varphi \in H_1^0(\Omega) \setminus \{0\}} \left\{ \|\nabla \varphi\|_{L_2}^{-1} \|\varphi\|_{L_{\frac{p}{p-r}}} \right\} \cdot \|\widetilde{G}(|u|)\|_{L_{\frac{p}{r}}} \leq C(\tfrac{p}{p-r}) \cdot \|\widetilde{G}(|u|)\|_{L_{\frac{p}{r}}}.$$

Furthermore, $\psi(t) := \widetilde{G}\left(t^{\frac{1}{p}}\right)^{\frac{p}{r}}$ is concave, so that Jensen's inequality provides

$$\|\widetilde{G}(|u|)\|_{L_{\frac{p}{r}}} \leq \text{vol}(\Omega)^{\frac{r}{p}} \cdot \widetilde{G}\left(\text{vol}(\Omega)^{-\frac{1}{p}} \cdot \|u\|_{L_p}\right).$$

Consequently, (38) and (15) hold for

$$G(t) := C(\tfrac{p}{p-r}) \, \text{vol}(\Omega)^{\frac{r}{p}} \cdot \widetilde{G}\left(\text{vol}(\Omega)^{-\frac{1}{p}} \cdot t\right).$$

To illustrate our weak solution approach, we consider the L-shaped domain $\Omega := (0,1)^2 \setminus ([\frac{1}{2}, 1) \times (0, \frac{1}{2}])$ and the boundary value problem

$$-\Delta u - 56u = u^2 + \lambda \text{ on } \Omega, \quad u = 0 \text{ on } \partial\Omega. \tag{39}$$

The factor -56 is chosen to "spoil" (like any other sufficiently negative factor) the applicability of the "monotonicity" methods on the expected solution branch passing through ($\lambda = 0$, $u \equiv 0$).

For several values of λ, we computed approximate solutions using a Newton-iteration in combination with a Finite Element method. Since C_0-elements are sufficient, we chose (a uniform mesh of) 768 *quadratic* triangular elements (corresponding to 1488 global variables), and applied our existence and enclosure method. The following table contains the computed defect bound δ, the constant K satisfying (13) resp. (36), and the error bound α for $\|\nabla(u - \omega)\|_{L_2}$ provided by (17).

λ	$\|\omega\|_\infty \approx$	δ	K	α
2	0.2402	0.02270	3.417	0.08000
4	0.4831	0.04580	3.394	0.1659
6	0.7290	0.06932	3.370	0.2588
8	0.9780	0.09328	3.345	0.3607
10	1.230	0.1178	3.320	0.4744
12	1.486	0.1427	3.295	0.6047
14	1.746	0.1681	3.270	0.7615
16	2.009	0.1941	3.244	0.9724

Table 1: Results for problem (39)

The rather low precision of the computed error bounds α is partly due to the low-degree-elements, but mainly due to the singularity caused by the reentrant corner at $(\frac{1}{2}, \frac{1}{2})$, which has *not* been treated by mesh refinement techniques here.

5 Further applications of the abstract setting

Without mentioning details here, we refer to some other realizations of our abstract operator setting.

No fundamental difficulties arise, e.g., if F in (1) also depends on ∇u (which however requires additional growth restrictions), or if the Dirichlet boundary condition is replaced by a more general one.

If F in (1) depends, in addition, on a *parameter* λ, the situation may occur that the solution branches (u_λ) contain singular points such as *turning points* or *bifurcation points*. In (and close to) such points, the inverse of the linearization L does not exist or has a very large norm, so that the methods presented so far fail.

In (neighborhoods of) *turning points*, the singularity can be removed by a change of the coordinate system, e.g. in form of a *bordering equation*, which introduces a new parameter μ, generating a branch (u_μ, λ_μ) of solution pairs (u, λ) such that the linearization of the augmented problem is everywhere regular. Incorporating the bordering equation and the concept of solution pairs (u, λ) into our abstract operator setting, we obtain an existence and enclosure method which is applicable near turning points.

The situation is more difficult in (and near) *bifurcation points* which cannot be removed by change of coordinates. However, if the bifurcation is *symmetry-breaking*, we can enclose the symmetric branch and the crossing symmetry-breaking branch separately by appropriate realizations of our abstract setting.

Similar to the treatment of turning point problems, we can also deal with *non-selfadjoint eigenvalue problems*, where the variational methods mentioned earlier break down. Here, some normalizing equation for the eigenfunction serves as bordering equation, and "solutions" are now eigenpairs (u, λ). Regard that this concept makes the eigenvalue problem *nonlinear*, since the right-hand side contains a product of λ and u.

References

Behnke, H.; Goerisch, F. (1994): Inclusions for Eigenvalues of Selfadjoint Problems. In: Topics in validated Computations (ed. J. Herzberger). Elsevier (North-Holland), Amsterdam, pp. 277-322.

Breuer, B. (1998): Lösungseinschließungen bei einem nichtlinearen Randwertproblem mittels eines Fourierreihenansatzes. Diploma thesis, Univ. of Karlsruhe.

Choi, Y. S.; McKenna, P. J. (1993): A Mountain Pass Method for the Numerical Solution of Semilinear Elliptic Problems. Nonlinear Analysis, Theory, Methods and Applications, Vol. 20, No. 4: 417-437.

Collatz, L. (1952): Aufgaben monotoner Art. Arch. Math. 3: 366-376.

Klatte, R.; Kulisch, U.; Lawo, C.; Rauch, M.; Wiethoff, A. (1993): C-XSC - A C++ Class Library for Extended Scientific Computing. Springer, Heidelberg.

Nakao, M. T. (1993): Solving Nonlinear Elliptic Problems with Result Verification Using an H^{-1} Type Residual Iteration. Computing, Suppl. 9: 161-173.

Nakao, M. T.; Yamamoto, N. (1995): Numerical verifications for solutions to elliptic equations using residual iterations with higher order finite elements. J. Comput. Appl. Math. 60: 271-279.

Plum, M. (1991): Computer-assisted Existence Proofs for Two-Point Boundary Value Problems. Computing 46: 19-34.

Plum, M. (1992): Explicit H_2-estimates and pointwise bounds for solutions of second-order elliptic boundary value problems. J. Math. Anal. Appl. 165: 36-61.

Plum, M. (1994): Enclosures for Weak Solutions of Nonlinear Elliptic Boundary Value Problems. World Scientific Publishing Company, WSSIAA 3: 505-521.

Plum, M. (1997): Guaranteed Numerical Bounds for Eigenvalues. In: Spectral Theory and Computational Methods of Sturm-Liouville Problems (eds. D. Hinton, P. W. Schaefer). Marcel Dekker Inc., New York, pp. 313-332.

Schröder, J. (1980): Operator Inequalities. Academic Press, New York.

Schröder, J. (1991): Operator Inequalities and Applications. In: Inequalities, Fifty Years On from Hardy, Littlewood and Pólya (ed. W. N. Everitt). Marcel Dekker Inc., pp. 163-210.

Storck, U. (n.n.): An Adaptive Numerical Integration Algorithm with Automatic Result Verification for Definite Integrals. To appear in Computing.

Walter, W. (1970): Differential and Integral Inequalities. Springer, Berlin-Heidelberg.

Fast verification algorithms in Matlab

Siegfried M. Rump

1 Introduction

For the toolbox INTLAB, entirely written in Matlab, new concepts have been developed for very fast execution of interval operations to be used together with the operator concept in Matlab. The new implementation of interval arithmetic is strongly based on the use of BLAS routines. The operator concept of Matlab offers the possibility of easy and user-friendly access to interval operations, real and complex interval elementary functions, automatic differentiation, slopes, multiple-precision interval arithmetic and much more. Some of the new concepts are presented. The paper focusses on implementation and mainly on performance issues.

Hardware requirement for our approach is the IEEE 754 arithmetic standard (1985), which is implemented on many computers. If a special hardware supporting interval arithmetic or even elementary interval standard functions would be available, much of the below would be simpler and faster. However, our challenge was to design fast and easy to use algorithms running on standard computers, without additional hardware requirements. Furthermore, the algorithm should still be fast when written in Matlab.

Matlab (1997) is a widely used interactive programming environment for scientific computations. At a first glance it seems to be impossible to realize an operator concept in Matlab. This is because one of the main principles in Matlab is that *no* type declarations of variables are necessary but, by the interpretation principle, variables are automatically declared when used. Also, there is no distinction between scalars, vectors and matrices, whether they are real or complex; and a variable may frequently change its type.

The identification of new data types in Matlab works as follows. To define a new type, say TYPE, together with operators working on it, a subdirectory with name @TYPE is to be defined. This subdirectory has to be adjacent to the search path, i.e. the parent directory of @TYPE is in the search path of Matlab, whereas the subdirectory @TYPE itself is not in the search path.

Within the subdirectory @TYPE there has to be a routine named TYPE. This is the constructor for the new data type. The core of that routine, e.g. for TYPE being `intval`, could look as follows:

```
function A = intval(a)
    A.inf = a;
    A.sup = a;
    A = class(A,'intval')
```

The main statement is the last one, the "class-constructor", which tells the Matlab system that the output is a variable of "type" intval. From now on things are standard. Every operation involving a variable of the new data type calls a corresponding function in the subdirectory @TYPE, in our example `intval`, with fixed naming conventions. For example, names of operators are

```
+      plus
.*     times
*      mtimes
^      mpower
[ ... ]   horzcat
```

and many more. The operator concepts also includes a user-defined display routine. For example, the statement

```
A = intval(3.5);
5+A
```

calls the `intval` constructor in the first line: The variable A is now of type `intval`. The second statement calls the function `plus` for arguments 5 and A, and subsequently the `intval` display routine is called because the result of 5+A is of type `intval`.

Summarizing this is a really nice and easy way to define and to use new operators in Matlab. For further information see Matlab (1997).

2 Performance issues

The nice working in Matlab is, there may be a severe interpretation time penalty when using low-level operators. Matlab is a *Matrix Laboratory*, and extensive use of scalar operators causes much interpretation overhead. Consider the following four ways of writing a matrix multiplication in Matlab, timing for multiplication of two randomly generated 200×200 matrices included.

```
n = 200; A = rand(n); B = rand(n);

C = zeros(n);
tic
  for i=1:n
    for j=1:n
      for k=1:n
        C(i,j) = C(i,j) + A(i,k)*B(k,j);
      end
    end
  end
toc
```

```
            C = zeros(n);
         tic
            for i=1:n
               for j=1:n
                  C(i,j) = C(i,j) + A(i,:)*B(:,j);
               end
            end
         toc

            C = zeros(n);
         tic
            for i=1:n
               C(i,:) = C(i,:) + A(i,:)*B;
            end
         toc

            C = zeros(n);
         tic
            C = A*B;
         toc
```

The following table for a 300 MHz Pentium I Laptop shows the interpretation overhead in Mflop.

	3 loops	2 loops	1 loop	no loop
Mflop	0.05	1.9	36	44

Table 1. Interpretation overhead

The table clearly shows that minimization of interpretation overhead is mandatory if the system shall not be restricted to toy problems. In the following, consider interval matrix multiplication as our model problem.

Current implementations of interval matrix operations mostly use a top-down 3-loop approach, similar to the first one presented before. According to the above table this is much too slow in an interpretative system. However, even when using programming languages with highly optimized compilers such as Fortran or C, this top-down approach is very expensive in terms of computing time: The most inner loop is an interval operation, thus containing if-statements and case-distinctions. Such code can hardly be optimized by a compiler.

The effect of lack of optimization, of different sequencing of the loops and more subtle methods like unrolled loops shall be demonstrated in the following. For the moment, we restrict ourselve to pure floating point computations, no additional slow-down by interval computations present.

The first experiment is a scalar product $c = x^T y$ for $x, y \in \mathbb{R}^n$ for dimension $n = 1000$. The traditional loop is compared to an unrolled loop with five terms:

```
      for ( i=0, c=0; i<n; i++)
         c += x[i]*y[i] + x[i+1]*y[i+1] + x[i+2]*y[i+2]
            + x[i+3]*y[i+3] + x[i+4]*y[i+4];
```

The performance rates in Mflops for the standard loop and the unrolled loop, both without and with compiler optimization on a RS 6000 workstation are as follows.

Performance [Mflop]	standard loop	unrolled loop
w/o optimization	4.3	6.9
with optimization	12.7	19.9

We see quite some increase in performance by the unrolled loops and, as we expect, by the optimization of the compiler. Both is *not* possible for the standard implementation of interval matrix multiplication; in other words, performance is basically limited to the smallest number in the above table.

Another standard method for improving performance is the sequence of loops in matrix multiplication. Any of the six possibilities ijk, ikj, ..., kji computes the correct matrix product, however, with quite different performance. The following table shows performance for the six possibilities, both without and with compiler optimization.

Performance [Mflop]	jki	kji	ikj	kij	ijk	jik
w/o optimization	2.1	2.1	2.9	2.9	2.9	2.9
with optimization	5.8	5.4	27	25	62	62
BLAS 3	100					

The big differences in performance are mainly due to memory access and cache optimization. The last line, with another improvement of a factor 1.5 compared to the best possible achieved so far, gives the Basic Linear Algebra Subroutines (BLAS) performance. The BLAS library (Dongarra et al. 1990) is available for almost every computer today. The ingenious idea of BLAS was that only the function headers are specified, whereas the implementation for each individual computer is performed by the manufacturer.

The numbers presented show how high-level routines and, whereever possible, using BLAS improves performance significantly.

3 Interval arithmetic

The product of two point matrices is easy to implement using BLAS. We use a routine setround(m) with the property that after the call setround(-1) the rounding mode is permanently switched downwards, i.e. every following operation is performed with rounding downwards according to the IEEE 754 standard (1985). This remains true until the next call of setround. Accordingly, setround(m) shall switch the rounding upwards for m=1, and to nearest for m=0.

Then the product of two point matrices $A \in M_{n,k}(\mathbb{F}), B \in M_{k,m}(\mathbb{F})$, $\mathbb{F}$ denoting the set of double precision floating point numbers, can be realized as follows.

```
setround(-1)
C.inf = A*B;
setround(1)
C.sup = A*B;
setround(0)
```

It follows

```
C.inf <= A*B <= C.sup
```

for comparison in a componentwise sense. Note that the proof is a successive use of the fact that, in case rounding is switched downwards, the sum and the product of two floating point numbers yields a floating point result definitely being less than or equal to the correct (real) result, whereas the floating point result is definitely greater than or equal to the exact result for rounding switched upwards.

Note that this approach does not necessarily work for other composed operations. For example, it is not correct for triple matrix products. Similarly, the product of a point matrix $A \in M_{n,k}(\mathbb{F})$ and an interval matrix $\mathbf{B} \in \mathbb{I}M_{k,m}(\mathbb{F})$ cannot be computed by A*B.inf and A*B.sup. In Rump (1999a) a method was proposed for fast computation of $A*\mathbf{B}$ using BLAS. The idea is the intermediate use of midpoint-radius representation.

```
setround(1)
Bmid = B.inf + 0.5*(B.sup-B.inf);
Brad = Bmid - B.inf;
setround(-1)
C1 = A * Bmid;
setround(1)
C2 = A * Bmid;
Cmid = C1 + 0.5*(C2-C1);
Crad = ( Cmid - C1 ) + abs(A) * Brad;
setround(-1)
C.inf = Cmid - Crad;
setround(1)
C.sup = Cmid + Crad;
```

Algorithm 1. Point matrix times interval matrix

We mention that Algorithm 1 requires quite some additional memory. This could be reduced - at the expense of readability. If memory is a major issue, we would suggest to call a C or Fortran implementation, which may also avoid copying of matrices.

The first three lines calculate a matrix pair <Bmid,Brad> with the property

$$\forall B \in \mathbf{B} : \text{Bmid} - \text{Brad} \le B \le \text{Bmid} + \text{Brad}$$

for the comparision in the componentwise sense. This elegant way of transforming infimum-supremum representation into midpoint-radius representation in lines 1...3 is due to Oishi (1998). The next statements compute first an inclusion [C1,C2] of A*Bmid, and then take care of the radius Brad. The number of operations adds to $3n^3$ additions and $3n^3$ multiplications. This is 1.5 times more operations than necessary by the traditional implementation because the product of a floating point number and an interval can be performed with two multiplications (and a case distinction), and the interval addition requires two additions anyway. However, the following timings will demonstrate the vast improvement in performance of the new approach.

To our knowledge, there was only attempt to improve on the traditional implementation of interval matrix operations, namely the BIAS approach (Knüppel 1994). The following table gives the performance in "Miops" for multiplication of an $n \times n$ point times an $n \times n$ interval matrix for the traditional approach, for the BIAS approach and the above Algorithm 1. Timing is on a Convex SPP 200. Here the the matrix multiplication is counted as $2n^3$ interval operations, and a Miop are 1 Million interval operations.

Performance [Miops]	n=100	n=200	n=500	n=1000
traditional	6.4	6.4	3.5	3.5
BIAS	51	49	19	19
Algorithm 1	95	219	142	162

Table 2. Performance for point matrix times interval matrix

Obviously there is an immense improvement of performance by the new approach using BLAS routines. The decrease of performance of the BIAS library is due to cache misses. It could be improved by implementation of blocked algorithms, whereas the new Algorithm 1 uses BLAS, and therefore it uses blocked algorithms without effort on the part of the user. The varying performance of the new Algorithm 1 for different dimensions is also due to favourable and less favourable block sizes.

Another advantage of Algorithm 1 is that parallelization comes free of work, linking the parallel BLAS does the job. The BIAS approach can be parallelized as well, however, this has to be done by the user. The Convex SPP 200 allows us to use 4 processors, and performance data is as follows (for the traditional approach and the BIAS approach performance does not change unless special algorithms would be implemented).

Performance [Miops]	n=100	n=200	n=500	n=1000
Algorithm 1	142	551	397	526

Table 3. Parallel performance for point matrix times interval matrix using 4 processors

Comparing with the last row of Table 2 shows that the gain in performance is, for larger values of n, not too far from the magic factor 4. Note that this was achieved by merely linking the parallel BLAS library.

For the product of two interval matrices we also use an intermediate midpoint-radius representation. The performance numbers are even more impressing than before. However, there is a drawback to this, namely, that midpoint-radius product causes an overestimation of the true result whereas the infimum-supremum representation yields the sharp inclusion of the product of two interval matrices.

This would be a showstopper if overestimation could not be bounded. However, it can be shown that overestimation is globally bounded by a constant (Rump 1999a), and it is small if the input intervals are small. More precisely, define the relative precision prec(**A**) of an interval **A** by

$$\text{prec}(\mathbf{A}) := \min \left(\frac{\text{rad}(\mathbf{A})}{|\text{mid}(\mathbf{A})|}, 1 \right).$$

Let interval matrices **A** and **B** be given such that $\text{prec}(\mathbf{A}_{ij}) \le e$ and $\text{prec}(\mathbf{B}_{ij}) \le f$ for all i, j. Let **C** denote the result obtained by midpoint-radius arithmetic, and $\mathbf{A} * \mathbf{B}$ denote the narrowest interval matrix containing the power set product. Then the overestimation by midpoint-radius arithmetic satisfies

$$\frac{\text{rad}(\mathbf{C})_{ij}}{\text{rad}(\mathbf{A} * \mathbf{B})_{ij}} \le 1 + \frac{e \cdot f}{e + f} \le 1.5.$$

for all indices i, j. For example, input intervals with relative precision 1% suffer an overestimation of not more than 0.5% in radius.

If this overestimation is critical, the traditional interval matrix multiplication or some variant (Rump 1999a) may be used. Otherwise the following performance data apply. Again, matrix multiplication is counted as $2n^3$ interval operations.

Performance [Miops]	n=100	n=200	n=500	n=1000
traditional	4.7	4.6	2.8	2.8
BIAS	4.6	4.5	2.9	2.8
adapted Algorithm 1	91	94	76	99
adapted Algorithm 1 parallel	95	145	269	334

Table 4. Performance for interval matrix times interval matrix

Implementation of complex vector and matrix operations is not difficult. For various reasons we use circular arithmetic (midpoint-radius representation) in INTLAB. The implementation in Matlab is straightforward because real vector and matrix operations are already available. Thus the new approach also solves the problem of interpretation overhead.

Moreover, the above approach also applies to sparse matrices. As sparse matrices are already an intrinsic data type in Matlab, an implementation for sparse interval matrices comes without additional work.

A first simple application example is the check of nonsingularity of a given interval matrix $\mathbf{A} \in \mathbb{I}M_n(\mathbb{F})$. A well-known sufficient criterion for $\mathbf{A} \in \mathbb{I}M_n(\mathbb{F})$ being nonsingular is that for some matrix R and some interval vector **X**

$$(I - RA) \cdot \mathbf{X} \subseteq \text{int}(\mathbf{X})$$

is satisfied. A simple implementation of this criterion is

```
R = inv(A.mid);
C = eye(n) - R*A;
X = infsup(-1,1)*ones(n,1);
Y = C*X;
res = all( ( X.inf<Y.inf ) & ( Y.sup<X.sup ) );
```

If `res=1` after execution, every real matrix enclosed in the interval matrix
A is proved to be nonsingular. The above is only an example for ease of use.

4 Nonlinear problems

For application of verification methods to nonlinear problems especially the inclusion of derivatives of functions over a range is needed as well as interval
elementary functions. Gradients can be computed using an operator concept
and automatic differentiation (Rall 1981). The implementation is simplified by
the vector and matrix operations in Matlab.

When defining a function, a user friendly way would use the same source code
for evaluation of the function at some real or complex floating point number, for
the evaluation of the range of the function, for gradient information or for the
gradient of the function over a certain range. This causes a specific problem.
Consider the sample function

$$f(x) = \sin(\pi x).$$

A Matlab implementation is

```
function y = f(x)
  y = sin(pi*x);
```

There are no problems when inserting a (real or complex) floating point
number x, a gradient value or a slope value. Problems occur when inserting an
interval X. In this case the user may want to use an inclusion of the irrational
number π in the definition of the function. Otherwise a call `y = f(intval(1))`
may produce an interval not containing zero because the (intrinsic) floating point
approximation `pi` of π is used instead. A redefinition

```
function y = f(x)
  Pi = midrad(pi,1e-16);
  y = sin(Pi*x);
```

would calculate a correct inclusion `Pi` for π and deliver a correct inclusion
for zero when calling `f(intval(1))`. However, the simple call `f(1)` would yield
an unexpected interval answer. Obviously, the type of the result shall depend
on the type of the input argument x: For interval input the computation should
be performed in interval arithmetic with correct interval data for π, for floating
point input the Matlab internal approximation `pi` is sufficient and a floating
point approximate result should be delivered.

The solution to the dilemma are two functions to adjust the type of a constant. Consider

```
```

```
function y = f(x)
    Pi = typeadj( midrad(pi,1e-16) , typeof(x) );
    y = sin(Pi*x);
```

In this implementation `typeof(x)` returns type information about the input
parameter x. Especially, this is `intval` for interval input and `double` for floating
point input. The statement `typeadj(a,type)` adjusts the type of the input `a` to
the type `type`. Especially, in case `type` is `double`, the midpoint of a is returned.
This allows to write one source code for various applications, from pure floating
point computation to complex interval gradients and others.

The above implementation requires rigorous standard functions. This has
been indeed a major task in previous approaches. Following we present a simple method to implement rigorous standard functions over real and complex
intervals.

5 Standard functions

For single precision it is not difficult to test all values of a built-in standard
function for their accuracy and to add a suitable error margin. For double
precision this is not possible.

Usually the built-in standard functions are very accurate, and'it is seldom
that a value is not correctly rounded to least significant bit accuracy. In fact,
we rarely found cases where the computed result is off by more than one bit –
at least for arguments in a reasonable range. However, there is no proof for the
accuracy of the results, and in order to achieve truly rigorous results a guess of
accuracy is not sufficient.

The idea is to use a table approach together with some correction formulas.
Take, for example, the exponential. We suppose that multiple precision functions
$\underline{F}, \overline{F}$ are available such that for a given floating point number $x \in \mathbb{F}$ it is

$$\underline{F}(x) \le e^x \le \overline{F}(x)$$

with high accuracy. Define

$$R_{\exp} := \{\pm(0, 1, \dots, 2^{14} - 1) \cdot 2^{-14}\} \subseteq \mathbb{F}$$

as a reference set for the exponential. For given $x \in \mathbb{F}$ define $y = \exp_\square(x) \in \mathbb{F}$ to
be the floating approximation computed by the given (floating point) exponential
function. Then the error of such approximations over the reference set is defined
as follows.

$$\underline{\varepsilon} := \max_{x \in R_{\exp}} \{(y - \underline{F}(x))/|y| : \quad y = \exp_\square(x)\},$$

$$\overline{\varepsilon} := \max_{x \in R_{\exp}} \{(\overline{F}(x) - y)/|y| : \quad y = \exp_\square(x)\}.$$

A short computation yields

$$y - \underline{\varepsilon} \cdot |y| \le e^x \le y + \overline{\varepsilon} \cdot |y|.$$

Now lower and upper bounds for the left hand side and right hand side, respectively, are computable for every $x \in R_{\exp}$ by

```
ys = exp(x);
setround(-1)
y.inf = ys + (-eps)*abs(ys);
setround(1)
y.sup = ys + eps*abs(ys);
```

with the property

$$\texttt{y.inf} \le e^x \le \texttt{y.sup} \quad \forall x \in R_{\exp}$$

where $\texttt{eps} := \varepsilon_{\exp} = \max(\underline{\varepsilon}, \overline{\varepsilon})$. The advantage is that $\underline{F}(x), \overline{F}(x)$ for all $x \in R_{\exp}$ and $\varepsilon_{\exp}$ have to be computed *only once*. From then on the constant $\varepsilon_{\exp}$ is a system constant to be used in the further computations. For general $X \in \mathbb{F}$ inclusions of the exponential are computed as follows. In an initialization procedure we compute floating point numbers $E_\nu, \underline{E}_\nu, \overline{E}_\nu$ with

$$E_\nu + \underline{E}_\nu \le e^\nu \le E_\nu + \overline{E}_\nu \quad \text{for } \nu \in \mathbb{Z}, -744 \le \nu \le 709.$$

For $x \le -745$ or $x \ge 710$, e^x is outside the double precision floating point range. Otherwise for $X \in \mathbb{F}$, split $X := X_{\text{int}} + x$ with $X_{\text{int}} = \text{sign}(X) \cdot \lfloor |X| \rfloor, -744 \le X_{\text{int}} \le 709$, and $-1 < x < 1$. Furthermore, set

$$\begin{aligned} \tilde{x} &= 2^{-14} \cdot \lfloor 2^{14} x \rfloor \\ d &= x - \tilde{x}. \end{aligned}$$

Then $\tilde{x}$ has no more than 14 leading bits in its binary representation and $\tilde{x} \in R_{\exp}$. This implies

$$e^x = e^{\tilde{x}} \cdot e^d \quad \text{with} \quad \sum_{i=0}^{3} \frac{d^i}{i!} \le e^d \le \sum_{i=0}^{3} \frac{d^i}{i!} + e \cdot \frac{d^4}{4!}.$$

By the choice of the reference set it is $0 \le d < 2^{-14}$ and

$$e \cdot \frac{d^4}{4!} \le 0.68d \cdot \frac{d^3}{3!} < 1.6 \cdot 10^{-18}.$$

Putting things together yields rigorous and very sharp bounds for the value of the exponential over the entire floating point range.

Corresponding reference sets and formulas for splitting arguments have been developed for all elementary standard functions. A careful implementation of the formulas yields standard functions of very high accuracy, in fact always better than 3 ulp. For example, the relative error for the exponential, tested over some 50 million test cases, looks as follows. Here crosses depict the maximum relative error of the lower and upper bounds against each other over a certain domain, whereas the circles depict the average error.

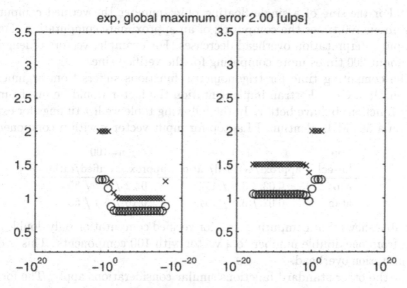

Graph 1. Accuracy exponential

The maximum error is not more than 2 ulps, whereas the average relative error is about 1 ulp. Even for the trigonometric functions, where problems with argument reduction may cause significant cancellation errors, high accuracy is achieved. For example, the error plot for the sine is as follows.

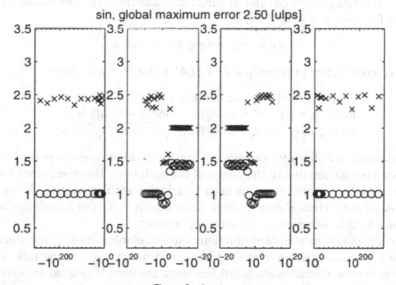

Graph 2. Accuracy sine

The test set for the sine comprises of some 100 million floating point numbers in the range from -10^{300} to 10^{300}.

Guaranteed accuracy does not come free of cost, especially when the entire implementation is performed in Matlab itself, suffering from interpretation overhead. For the sine of a single floating point number the verified computation takes about 700 times the computing of an approximate computation. For vector input interpretation overhead decreases. For example, vectors of length 100 take about 200 times more computing for the verified sine.

The computing time for trigonometric functions suffers from argument reduction. In a C- or Fortran implementation the factor would be much smaller. Other functions behave better. In the following table we list timings for exp and atan on a 300 MHz Pentium I Laptop for input vector x with n components.

| time | n=1 | n=100 |
[msec]	approx./verified/ratio	approx./verified/ratio
exp	0.03 / 3.9 / 152	0.22 / 7.7 / 35
atan	0.02 / 3.5 / 164	0.10 / 6.4 / 58

The table shows that computing time for verified computation only doubles when going from one double number to a vector with 100 components. This is due to interpretation overhead.

For the other standard functions similar considerations apply. The formulas must be developed carefully in order to maintain the anticipated accuracy of 3 ulp. This has been performed for exp, log, log10, the trigonometric functions and their inverse functions, and for the hyperbolic functions and their inverses. All are included in the new version INTLAB.

For standard functions over rectangular intervals, efficient and accurate algorithms have been developed by Braune (1987) and Krämer (1987). For complex interval functions in our circular arithmetic we use the results by Börsken (1978). Denote for given $a \in \mathbb{C}, 0 \leq r \in \mathbb{R}$

$$A = <a, r> := \{z \in \mathbb{C} : |z - a| \leq r\}.$$

Then Börsken defines the midpoint of $f(A)$ to be $f(a)$ and shows

$$\exp(<a, r>) \subseteq <\exp(a), |\exp(a)| \cdot (\exp(r) - 1)>$$
$$\log(<a, r>) \subseteq <\log(a), -\log(1 - r/|a|)>$$
$$\text{sqrt}(<a, r>) \subseteq <\text{sqrt}(a), |\sqrt{|a| - r} - \sqrt{|a|}|>.$$

These bounds are sharp for certain arguments, in other cases there may be quite some overestimation due to the choice of the midpoint. However, other formulas defining a different and more optimized midpoint with respect to, for example, the area of the inclusion may become quite involved. To our knowledge nothing is known in this direction, a research opportunity.

With those three standard function being available, all other mentioned elementary functions can be expressed by standard formulas. Doing this, another problem occurs. Complex standard functions are usually defined by some main value. This causes discontinuities. For example, $\sqrt{-4} = +2i$, but $\sqrt{-4 + \varepsilon i}$ for small $\varepsilon < 0$ yields a value near $-2i$. This causes problems when an interval approaches the negative real axis and $f(X)$, for a complex interval X, is defined to enclose the result of the usual power set operation

$$f(\mathbf{X}) := \{f(x) : x \in \mathbf{X}\}.$$

For the moment, we choose to use the above formulas. This implies that for a given function f the following weaker statement is true:

$$\mathbf{Y} = f(x) \Rightarrow \forall x \in \mathbf{X} \; \exists y \in \mathbf{Y} : f^{-1}(y) = x.$$

There is also space for future development and research. We note that for complex rectangles individual algorithms for each elementary standard function have been given in Braune (1987) and Krämer (1987), which cover also the above problem.

6 Long arithmetic

The above approach for the definition of standard functions requires some multiple precision arithmetic with error bounds. There are a number of such packages available, for example this of Aberth and Schaefer (1992), only to mention one. However, we choose to write a package in Matlab in order to maintain best possible portability.

The data type *long* has the structure

```
C.sign         in {-1,1}
C.mantissa     in 0 .. beta-1
C.exponent     representable integer ( -2^52+1 .. 2^52-1 )
C.error        nonnegative double, stored by C.error.mant and
               C.error.exp
```

representing the long number

$$\texttt{C.sign} \cdot \sum_{\nu=1}^{n} \texttt{C.mantissa}_\nu \cdot \beta^{\texttt{C.exponent}-\nu},$$

where C.mantissa is an array of length n corresponding to the precision in use. The field C.error is optional; if specified it represents the number

$$\texttt{C.error.mant} \cdot \beta^{\texttt{C.error.exp}},$$

where both C.error.mant and C.error.exp are nonnegative double numbers. It is interpreted as the radius of an interval with the above midpoint. Therefore the radius is stored in only two double numbers, whereas the midpoint is stored in an array of double numbers. The basis β is a system constant. It is a power of 2; usually 2^{25} is used.

The definition of multiple precision arithmetic is standard. For the current implementation we have two additional difficulties. First, it is no integer arithmetic but a long floating point arithmetic (to base β). This makes addition and subtraction more involved. Secondly, all long routines are written to support vector input in a vectorized computation. Treating vectors one component after the other causes a significant interpretation overhead. For example, given two long vectors X and Y of length 100 and precision of 500 decimal places, the calculation of the 100 products $X(i) * Y(i)$ by

```
                    for i=1:100, Z(i) = X(i)*Y(i); end
```

takes 25.9 sec, whereas the vectorized multiplication

```
                        Z = X.*Y;
```

takes only 0.6 sec on a 300 MHz Pentium I Laptop. Otherwise, the vectorized operations in Matlab can be used. For example, the multiplication of multiple precision numbers is a convolution and already built into Matlab.

7 An example of INTLAB code

Finally we give an example of INTLAB code to demonstrate its ease of use and readability. We print the full code to compute inclusions of multiple eigenvalues and corresponding invariant subspaces for a given (real or complex, not necessarily symmetric or Hermitian) matrix. We also give the full code how to call the algorithm. So the following is executable code in INTLAB under Matlab.

```
function [L,X] = VerifyEig(A,lambda,xs)
%VERIFYEIG Verification of eigencluster near (lamda,xs)
%
%    [L,X] = VerifyEig(A,lambda,xs)
%
%Input: an eigenvalue cluster near lambda, where xs(:,i), i=1:k
%   is an approximation to the corresponding invariant subspace.
%
%On output, L contains (at least) k eigenvalues of A, and X
%   includes a base for the corresonding invariant subspace.
%By principle, L is a complex interval.
%


% written   07/15/99 S.M. Rump
%

    [n k] = size(xs);

    [dummy, index] = sort(sum(abs(xs),2));    % choose normalization
                                              %  part
    u = index(1:n-k);
    v = index(n-k+1:n);
    midA = mid(A);

    % one floating point iteration
    R = midA - lambda*speye(n);
    R(:,v) = -xs;
    y = R\(midA*xs-lambda*xs);
    xs(u,:) = xs(u,:) - y(u,:);
    lambda = lambda - sum(diag(y(v,:)))/k;
```

```
R = midA - lambda*speye(n);
R(:,v) = -xs;
R = inv( R );
C = A - intval(lambda)*speye(n);
Z = - R * ( C * xs );
C(:,v) = -xs;
C = speye(n) - R * C;
Y = Z;
Eps = 0.1*abs(Y)*hull(-1,1) + midrad(0,realmin);
m = 0;
mmax = 15 * ( sum(sum(abs(Z(v,:))>.1)) + 1 );
ready = 0;
while ( ~ready ) & ( m<mmax ) & ( ~any(isnan(Y(:))) )
    m = m+1;
    X = Y + Eps;      % epsilon inflation
    XX = X;
    XX(v,:) = 0;
    Y = Z + C*X + R*(XX*X(v,:));
    ready = all(all(in0(Y,X)));
end

if ready
    M = abs(Y(v,:));        % eigenvalue correction
    [Evec,Eval] = eig(M);
    [rho,index] = max(abs(diag(Eval)));
    Perronx = abs(Evec(:,index));
    setround(1);
    rad = max( ( M*Perronx ) ./ Perronx );   % upper bound for
                                              % Perron root
    setround(0)
    L = tocmplx(midrad(lambda,rad));
    Y(v,:) = 0;
    X = xs + Y;
else
    disp('no inclusion achieved')
    X = NaN*ones(size(xs));
    L = NaN;
end
```

Algorithm 2. Rigorous inclusion of multiple eigenvalues

The algorithms follows Rump (2000) to be published in Linear Algebra and its Applications. It is based on the following.

For $\mathbb{K} \in \{\mathbb{R}, \mathbb{C}\}$ denote by $A \in M_n(\mathbb{K})$ an $n \times n$ matrix, by $\tilde{X} \in M_{n,k}(\mathbb{K})$ an approximation to an invariant subspace corresponding to a multiple or a cluster of eigenvalues near $\tilde{\lambda} \in \mathbb{K}$, such that $A\tilde{X} \approx \tilde{\lambda}\tilde{X}$.

The degree of arbitrariness is removed by freezing k rows of the approximation $\tilde{X}$. If the index set of the remaining rows is denoted by u, then we denote by $U \in M_{n,n-k}(\mathbb{R})$ the submatrix of the identity matrix with columns in u. Correspondingly, we set $v := \{1, \ldots, n\} \backslash u$ and define $V \in M_{n,k}(\mathbb{R})$ to comprise of the columns in v out of the identity matrix. That means $UU^T + VV^T = I$, and $V^T \tilde{X}$ is the normalizing part of $\tilde{X}$. Then the following is true (Rump 2000).

Theorem 1 *Let $A \in M_n(\mathbb{K}), \tilde{X} \in M_{n,k}(\mathbb{K}), \tilde{\lambda} \in \mathbb{K}, R \in M_n(\mathbb{K})$ and $\mathbf{X} \in \mathbb{I}M_{n,k}(\mathbb{K})$ be given, and let U, V partition the identity matrix as defined before. Define*

$$f(\mathbf{X}) := -R(A\tilde{X} - \tilde{\lambda}\tilde{X}) + \{I - R((A - \tilde{\lambda}I)UU^T - (\tilde{X} + UU^T \cdot \mathbf{X})V^T)\} \cdot \mathbf{X}.$$

Suppose

$$f(\mathbf{X}) \subseteq int(\mathbf{X}).$$

Then there exists $\hat{M} \in M_k(\mathbb{K})$ with $\hat{M} \in \tilde{\lambda}I_k + V^T\mathbf{X}$ such that the Jordan canonical form of $\hat{M}$ is identical to a $k \times k$ principal submatrix of the Jordan canonical form of A, and there exists $\hat{Y} \in M_{n,k}(\mathbb{K})$ with $\hat{Y} \in \tilde{X} + UU^T\mathbf{X}$ such that $\hat{Y}$ spans the corresponding invariant subspace of A.

Denote the k eigenvalues of $\hat{M}$ by $\mu_i, 1 \leq i \leq k$. Then the theorem implies that $\tilde{\lambda} + \mu_i$ are eigenvalues of A, and by Perron-Frobenius theory it is $|\mu_i| \leq \rho(\hat{M}) \leq \rho(|\hat{M}|)$ for $1 \leq i \leq k$. This proves that L is indeed an inclusion of (at least) k eigenvalues of A.

For successful termination of Algorithm 2, the matrix A must have a cluster of k eigenvalues near the input approximation lambda, which is well enough separated from the rest of the spectrum. The necessary degree of separation depends on the condition number of the cluster. In case of a multiple eigenvalue, the larger the maximum size of a corresponding Jordan block is, the larger the separation needs to be.

E.g., for a Jordan block of size m, the sensitivity of the eigenvalue to an ε-perturbation is $\varepsilon^{1/m}$, and practical experience shows that the algorithm terminates successfully if the separation is of the order $10\varepsilon^{1/m}$ (Rump 2000). Henceforth, the choice of the dimension k of the invariant subspace is important, as the separation and the sensitivity of the cluster depends on k.

Following we give two examples how to call the algorithm. The first one generates a random matrix, calculates approximations for the eigenvalues and eigenvectors and calls the algorithm for the first approximate eigenvalue/eigenvector pair. Note that [V,D] = eig(A) calculates a matrix V of eigenvectors and diagonal matrix D of eigenvalues, such that A*X is approximately equal to X*D. Furthermore, rand produces random numbers uniformly distributed in the interval [0,1] such that the entries of A are uniformly distributed within [-1,1].

```
n = 100; A = 2*rand(n)-1;
tic, [V,D] = eig(A); toc
tic, [L,X] = verifyeig(A,D(1,1),V(:,1)); toc
format long
L
```

This produces the following output:

```
elapsed_time =
    0.6100
elapsed_time =
    0.9900
intval L =
  -4.6875246581698_ +  3.4404126988075_i
```

The underscore in the output of the inclusion L of the eigenvalue indicates that the last digit of the real and the imaginary part is uncertain. More precisely, subtracting and adding one to the last displayed figure (before the underscore) yields a correct inclusion. Note that input/output is also rigorous by means of specific INTLAB routines for interval I/O (Rump 1999b).

The second example produces a triple eigenvalue 1 together with some randomly choosen 97 eigenvalues in $[-1, 1]$.

```
X = 2*rand(100)-1; A = X * diag([1 1 1 2*rand(1,97)-1]) * inv(X);
tic, [V,D] = eig(A); toc
index = find( abs(diag(D)-1)<1e-12 );
k = index(1);
tic, [L,X] = verifyeig(A,D(k,k),V(:,index)); toc
format long
L
```

The result is as follows.

```
elapsed_time =
    0.5000
elapsed_time =
    0.9900
intval L =
  1.000000000000__ -  0.000000000000__i
```

The inclusion fails if one of the 97 randomly chosen eigenvalues is, by chance, too close to 1 or, if X is too ill-conditioned such that eig delivers poor approximate eigenvectors and -values in V and D or, the size k of the cluster is incorrect such that the separation is too bad.

There are many more examples including ill-conditioned ones in the paper cited above (Rump 2000). Here our main objective is ease of use and readability. Note especially the index notation in Algorithm 2.

Conclusion

We presented some of the main ideas of the toolbox INTLAB for Matlab. INTLAB is available in its third release for PCs, a number of workstations and mainframes. More details can be found in Rump (1999b). The only machine dependency is the routine setround for switching the rounding mode. This assembly language routine is available for a number of machines. In Release 5.3

of Matlab under Windows even this is a built-in routine of Matlab. Then the entire toolbox is plain Matlab code.

All other code, some 362 functions and some 20 kLOC, is written in Matlab and therefore as portable as it can be. INTLAB is freely available for non-profit use from our homepage.

Acknowledgement

The author thanks the anonymous referee for his thorough reading and valuable comments.

References

Aberth, O., Schaefer, M. J. (1992): Precise computation using range arithmetic, via C++. ACM Trans. Math. Softw. 18(4): 481-491

Börsken, N. C. (1978): Komplexe Kreis-Standardfunktionen (Ph.D.), Freiburger Inter-vall-Ber. 78/2, Inst. f. Ange- wandte Mathematik, Universität Freiburg

Braune, K. D. (1987): Hochgenaue Standardfunktionen für reelle und komplexe Punkte und Intervalle in beliebigen Gleitpunktrastern (Ph.D.). Universität Karlsruhe

Dongarra, J. J., Du Croz, J. J., Duff, I. S., Hammarling, S. J. (1990): A set of level 3 Basic Linear Algebra Subprograms. ACM Trans. Math. Softw. 16: 1-17

ANSI/IEEE 754 (1985): Standard for Binary Floating-Point Arithmetic

Knüppel, O. (1994): PROFIL/BIAS - A fast interval library. Computing 53: 277-287

Krämer, W. (1987): Inverse Standardfunktionen für reelle und komplexe Intervallargu-mente mit a priori Fehlerab- schätzungen für beliebige Datenformate (Ph.D.). Uni-versität Karlsruhe

MATLAB User's Guide, Version 5 (1997): The Math Works Inc.

Oishi, S. (1998): private communication

Rall, L. B. (1981): Automatic Differentiation: Techniques and Applications. Lecture notes in Computer Science 120. Springer Verlag, Berlin-Heidelberg-New York

Rump, S. M. (1999a): Fast and parallel interval arithmetic. BIT 39(3):539-560

Rump, S. M. (1999b): INTLAB - INTerval LABoratory. In: Csendes, T. (ed.): De-velopements in Reliable Computing. Kluwer Academic Publishers, 77-104

Rump, S. M. (2000): Computational Error Bounds for Multiple or Nearly Multiple Eigenvalues. LAA, to appear

The Linear Complementarity Problem with Interval Data[1]

Uwe Schäfer

1 Introduction

Linear complementarity problems (LCP) model many important mathematical problems. Ferris and Pang (1997) gave an extensive documentation of complementarity problems in engineering and equilibrium modeling. Meanwhile, validation methods have been found for example by Alefeld et al. (1999) to give guaranteed bounds on the distance between the numerical solution and the exact solution of the LCP. But the question remains open, if and/or how one still gets those guaranteed bounds, when the input data itself are not known exactly. We do not only think of real numbers like $\sqrt{2}$ or $1/3$ that cannot be represented exactly on a computer, but we also think of mathematical problems for which even the exact real input data are not known.

2 Linear complementarity problems

Let $M \in \mathbf{R}^{n \times n}$ and $q \in \mathbf{R}^n$. Then the linear complementarity problem $LCP(M, q)$ is defined as follows: Determine (or conclude that there is no) $z \in \mathbf{R}^n$ with

$$q + Mz \geq 0, \quad z \geq 0, \quad (q + Mz)^T z = 0.$$

(Here, matrix inequalities are understood componentwise.) Cottle et al. (1992) have given more details concerning the LCP.

Let $[M]$ be an $n \times n$ interval matrix (we write $[M] \in \mathbf{IR}^{n \times n}$) and $[q] \in \mathbf{IR}^n$. Then we consider the family of linear complementarity problems

$$\left.\begin{aligned} q + Mz &\geq 0, \\ z &\geq 0, \\ (q + Mz)^T z &= 0, \end{aligned}\right\} \quad M \in [M], q \in [q].$$

Analogously to linear interval equations we define the solution set

$$\Sigma([M], [q]) := \left\{ z \in \mathbf{R}^n : q + Mz \geq 0, z \geq 0, (q + Mz)^T z = 0, \right.$$
$$\left. M \in [M], q \in [q] \right\}.$$

I.e. for each fixed $M \in [M]$ and for each fixed $q \in [q]$ each solution z of $LCP(M, q)$ is an element of $\Sigma([M], [q])$. We recall some definitions: Let $[a] = [\underline{a}, \overline{a}]$ be an interval. We set

$$\langle [a] \rangle := \min\{|a| : a \in [a]\}, \quad \|[a]\| := \max\{|a| : a \in [a]\}.$$

---

[1] This paper sums up the main results of the author's thesis. The proofs will be published elsewhere.

For $[A] \in \mathbf{IR}^{n \times n}$ the comparison matrix $\langle [A] \rangle := (c_{ij})$ is defined by

$$c_{ij} := \begin{cases} -\|[a_{ij}]\|, & \text{if } i \neq j, \\ \langle [a_{ij}] \rangle, & \text{if } i = j. \end{cases}$$

If $\langle [A] \rangle$ is nonsingular with $\langle [A] \rangle^{-1} \geq 0$, then $[A]$ is an H-matrix.

Theorem 2.1 (Schäfer 1999) *Let $[M] \in \mathbf{IR}^{n \times n}$ be an H-matrix with $0 < \underline{m}_{ii}$, $i = 1, ..., n$. Then the solution set $\Sigma([M], [q])$ is nonempty and compact for every $[q] \in \mathbf{IR}^n$.*

Example (Schäfer 1999): For

$$[M] = \begin{pmatrix} [\frac{1}{8}, 1] & [-\frac{1}{4}, -\frac{1}{5}] \\ [-\frac{1}{4}, -\frac{1}{10}] & 1 \end{pmatrix}, \quad [q] = \begin{pmatrix} [-3, -1] \\ [1, 2] \end{pmatrix}$$

$\Sigma([M], [q])$ can be illustrated as in Figure 1. Notice that the line $(1, 0) - (20, 0)$

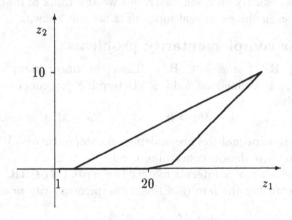

Figure 1: The shape of a solution set $\Sigma([M], [q])$

belongs to $\Sigma([M], [q])$. So, $\Sigma([M], [q])$ is not necessarily convex. The proof for the shape of the solution set in Figure 1 has used the ideas of Beeck (1972). An Oettli-Prager-type description of the solution set $\Sigma([M], [q])$ has been already given by Gwinner (1988). However, it is not easy to see how to get the shape of the solution set $\Sigma([M], [q])$ by Gwinner's results.

3 Enclosure methods

3.1 First idea

If the matrix $[M]$ is an H-matrix with $0 < \underline{m}_{ii}$, $i = 1, ..., n$, then $\Sigma([M], [q])$ is compact and the first idea is to find $[S] \in \mathbf{IR}^n$ with $\Sigma([M], [q]) \subseteq [S]$.

Theorem 3.1 (Schäfer 1999) *Let* $[q] \in \mathbf{IR}^n$ *and* $[M] \in \mathbf{IR}^{n \times n}$. *If* $[M]$ *is an H-matrix with* $0 < \underline{m}_{ii}$, $i = 1, ..., n$, *then the function*

$$f([x]) := IGA\Big(I + [M], \ (I - [M])abs([x]) - [q]\Big),$$

$$abs([x]) = (abs(x_i)) = ([\langle [x_i] \rangle, |[x_i]|]),$$

has exactly one fixed interval vector $[x^*]$, *i.e.* $f([x^*]) = [x^*]$, *and it holds*

$$\emptyset \neq \Sigma([M], [q]) \subseteq abs([x^*]) + [x^*] =: [S].$$

For each $[x^{(0)}] \in \mathbf{IR}^n$ *the iteration* $[x^{(k+1)}] := f([x^{(k)}])$, $k \geq 0$ *is feasible and it converges to* $[x^*]$.

(I.G.A. is an abbreviation and stands for interval Gaussian algorithm. This algorithm is applied to an interval matrix $[A] \in \mathbf{IR}^{n \times n}$ and an interval vector $[b] \in \mathbf{IR}^n$. The result is $IGA([A], [b])$. Mayer (1991) has stated the main results concerning its feasibility.)

3.2 Second idea

The second idea is based on the following equivalence:

$$H(z; M, q) := \min(z, q + Mz) = 0 \Leftrightarrow q + Mz \geq 0, \ z \geq 0, \ (q + Mz)^T z = 0.$$

So, $z \in \Sigma([M], [q])$ means that there are $M \in [M]$ and $q \in [q]$ with $H(z; M, q) = 0$. If we define $H(z; [M], [q])$ componentwise by

$$(H(z; [M], [q]))_i = \begin{cases} z_i & \text{if} \quad z_i < q_i + \overline{(Mz)}_i, \\ [q_i] + ([M]z)_i & \text{if} \quad z_i > \overline{q_i} + \overline{(Mz)}_i, \\ [\underline{q}_i + \underline{([M]z)}_i, z_i] & \text{if} \quad z_i \in [q_i] + ([M]z)_i, \end{cases}$$

and if we assume that we know an interval matrix $G(x, [z], [q], [M]) \in \mathbf{IR}^{n \times n}$ with

$$H(x; M, q) - H(y; M, q) = G(x, y, q, M)(x - y) \in G(x, [z], [q], [M])(x - y)$$

for all $y \in [z]$, $M \in [M]$ and $q \in [q]$, then we define the operator

$$N(x, [z], [q], [M]) := x - IGA(G(x, [z], [q], [M]), H(x; [M], [q]))$$

with arbitrary $x \in [z]$.

Theorem 3.2 *Let* $[z]$, $[q] \in \mathbf{IR}^n$, $[M] \in \mathbf{IR}^{n \times n}$ *and* $x \in [z]$. *If there exist* $z \in [z]$, $q \in [q]$ *and* $M \in [M]$ *with* $H(z; M, q) = 0$ *and if the I.G.A. is feasible for* $G(x, [z], [q], [M])$, *then it holds*

$$z \in N(x, [z], [q], [M]).$$

By induction one can show the following corollary.

Corollary 3.3 *Let $[z^{(0)}]$, $[q] \in \mathbf{IR}^n$ and $[M] \in \mathbf{IR}^{n \times n}$. If there exist $z \in [z^{(0)}]$, $q \in [q]$ and $M \in [M]$ with $H(z; M, q) = 0$, then it holds*

$$z \in [z^{(k+1)}] := N(x^{(k)}, [z^{(k)}], [q], [M]) \cap [z^{(k)}],$$

for every $k \geq 0$, if the I.G.A. is feasible for $G(x^{(k)}, [z^{(k)}], [q], [M])$. $x^{(k)} \in [z^{(k)}]$ is arbitrary.

Alefeld et al. (1999) have proved this theorem and determined the interval matrix $G(x, [z], [q], [M])$ for the special case $[M] = M \in \mathbf{R}^{n \times n}$ and $[q] = q \in \mathbf{R}^n$. The new thing Schäfer (1999) has done is the determination of the interval matrix $G(x, [z], [q], [M])$ in the general case. (I.e. $[M] \in \mathbf{IR}^{n \times n}$ and $[q] \in \mathbf{IR}^n$.) In addition, for a class of interval matrices the feasibility of the iteration step given in Corollary 3.3 is guaranteed by the following theorem.

Theorem 3.4 (Schäfer 1999) *If $[M] \in \mathbf{IR}^{n \times n}$ is an H-matrix with $0 < \underline{m}_{ii}$, $i = 1, ..., n$ (and w. l. o. g. $\overline{m}_{ii} \leq 1$, $i = 1, ..., n$), then $G(x, [z], [q], [M])$ is an H-matrix and the I.G.A. is feasible for $G(x, [z], [q], [M])$ for every $[z], [q] \in \mathbf{IR}^n$ and for every $x \in [z]$.*

4 The LCP with interval data arising from a free boundary problem

We shall combine the main ideas of Cryer and Lin (1985) with the main ideas of Hansen (1969) in order to show that linear complementarity problems with interval data occur. We shall consider free boundary problems. Collatz (1981) has given a simple example for a free boundary problem. The so-called rope problem is illustrated in Figure 2. The rope path coincides to the graph of a

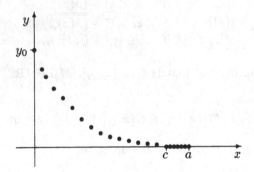

Figure 2: The rope problem

function $y(x)$ with

$$
\begin{aligned}
y''(x) &= \sqrt{1 + (y'(x))^2}, & \text{if } x \in [0, c], \\
y(x) &= 0, & \text{if } x \in [c, \infty),
\end{aligned}
$$

(where we have set a constant equal to 1 for the sake of simplicity, here). c is not known. It has to be determined as part of the solution. The general case is the following: Let $f : [0, \infty) \times \mathbf{R} \times \mathbf{R} \to \mathbf{R}$ be a function and $y_0 > 0$. Then the free boundary problem is defined as follows:

$$
\left.
\begin{array}{ll}
\text{Find } c \in \mathbf{R} \quad \text{and} \quad y(\cdot) : [0, \infty) \to \mathbf{R} \text{ with} & \\
y''(x) = f(x, y(x), y'(x)) & \text{if } x \in [0, c], \\
y(x) > 0 & \text{if } x \in [0, c), \\
y(x) = 0 & \text{if } x \in [c, \infty), \\
y'(c) = 0, & \\
y(0) = y_0. &
\end{array}
\right\}
\tag{1}
$$

We shall not discuss the existence of a solution of (1) and refer to Thompson and Walter (1992) concerning that. We assume that (1) has a unique solution $\left(\tilde{c}, \tilde{y}(\cdot)\right)$ and we assume that we know an $a \in \mathbf{R}$ with $\tilde{c} \leq a$. Then we choose an $n \in \mathbf{N}$ and subdivide the interval $[0, a]$ into n subintervals via $x_0 := 0$, $h := a/(n+1)$, $x_{i+1} := x_i + h$, $i = 0, ..., n$. The aim of the following theorem is to show that

$$
\tilde{y} := \begin{pmatrix} \tilde{y}(x_1) \\ \vdots \\ \tilde{y}(x_n) \end{pmatrix} \in \Sigma([M], [q])
\tag{2}
$$

for some $[M] \in \mathbf{IR}^{n \times n}$ and some $[q] \in \mathbf{IR}^n$. In other words: We are searching for $[M] \in \mathbf{IR}^{n \times n}$ and $[q] \in \mathbf{IR}^n$, such that $M \in [M]$ and $q \in [q]$ exist with

$$
q + M\tilde{y} \geq 0, \quad \tilde{y} \geq 0, \quad (q + M\tilde{y})^T \tilde{y} = 0.
$$

Theorem 4.1 (Schäfer 1999) *The free boundary problem (1) is considered. It is assumed that (1) has a unique solution $\left(\tilde{c}, \tilde{y}(\cdot)\right)$ and it is assumed that an $a \in \mathbf{R}$ with $\tilde{c} \leq a$ is known. $n + 2$ points are determined by $x_0 := 0$, $h := a/(n+1)$, $x_{i+1} := x_i + h$, $i = 0, ..., n$.*
Let f fulfill the following conditions:

- *$f(x, s, t) : [0, \infty) \times \mathbf{R} \times \mathbf{R} \to \mathbf{R}$ is continuously differentiable.*

- *There exists $[F] = [\underline{F}, \overline{F}]$ with*
  $\{\mu \in \mathbf{R} : \mu = f(x, \tilde{y}(x), \tilde{y}'(x)), \ x \in [0, a]\} \subseteq [F]$ and $\underline{F} \geq 0$.

- *There exists $D \in \mathbf{R}$ with*
  $|f_x(x, \tilde{y}(x), \tilde{y}'(x)) + f_s(x, \tilde{y}(x), \tilde{y}'(x))\tilde{y}'(x) + f_t(x, \tilde{y}(x), \tilde{y}'(x))\tilde{y}''(x)| \leq D, \ x \in [0, \tilde{c}]$.

Then (2) holds with

$$
[q] = \frac{1}{2} \begin{pmatrix} [\frac{1}{2}, 1]h^2[F] + \frac{1}{2}h^3[-D, D] - y_0 \\ [\frac{1}{2}, 1]h^2[F] + \frac{1}{2}h^3[-D, D] \\ \vdots \\ [\frac{1}{2}, 1]h^2[F] + \frac{1}{2}h^3[-D, D] \end{pmatrix} \in \mathbf{IR}^n
$$

and

$$[M] = M = \begin{pmatrix} 1 & -\frac{1}{2} & 0 & \cdots & 0 \\ -\frac{1}{2} & 1 & -\frac{1}{2} & \cdots & \vdots \\ 0 & \ddots & \ddots & \ddots & 0 \\ \vdots & \ddots & -\frac{1}{2} & 1 & -\frac{1}{2} \\ 0 & \cdots & 0 & -\frac{1}{2} & 1 \end{pmatrix} \in \mathbf{R}^{n \times n}.$$

The proof uses Taylor's formula with remainder where one has to take into account that it is not necessary that $\tilde{y}(x) : [0, \infty) \to \mathbf{R}$ is twice differentiable at $x = \tilde{c}$. In addition, we want to emphasize that it is not possible to fulfill (2) with $[M] = M \in \mathbf{R}^{n \times n}$ and $[q] = q \in \mathbf{R}^n$ (even if $f \equiv 1$).

In order to apply Theorem 4.1 we need to calculate a, $D \in \mathbf{R}$ and $[F] \in \mathbf{IR}$ there. We are able to do that very easily, if $f : [0, \infty) \times \mathbf{R} \times \mathbf{R} \to \mathbf{R}$ fulfills the following conditions (Schäfer 1999):

- $f(x, s, t) : [0, \infty) \times [0, y_0] \times \mathbf{R} \to \mathbf{R}$ is continuously differentiable and there are interval extensions for f, f_x, f_s, f_t.

- There exists $\kappa > 0$ with $f(x, \tilde{y}(x), \tilde{y}'(x)) > \kappa$, $x \in (0, \tilde{c})$.

- There exists $K \geq 0$ with $f(x, s, 0) \leq K$, $(x, s) \in [0, \tilde{c}] \times [0, y_0]$.

- There exists $L \geq 0$ with $|f(x, s, t_1) - f(x, s, t_2)| \leq L|t_1 - t_2|$, $(x, s, t_1), (x, s, t_2) \in [0, \tilde{c}] \times [0, y_0] \times \mathbf{R}$.

5 Application and numerical results

Our aim is to find an interval vector $[S]$ with $[S] \ni \tilde{y}$, where $\tilde{y}$ is defined in (2). Theorem 3.1 and Corollary 3.3 leave us two ways to find such an interval vector. We omit the details here, but we want to stress how we apply Corollary 3.3.

If f fulfills the conditions stated after Theorem 4.1, then it is easy to see that $\tilde{y}(\cdot)$ is decreasing. So, we have

$$\tilde{y} \in \begin{pmatrix} [0, y_0] \\ \vdots \\ [0, y_0] \end{pmatrix} =: [z^{(0)}] \in \mathbf{IR}^n.$$

It is well-known that the matrix M considered in Theorem 4.1 is an H-matrix. So, due to Corollary 3.3 and Theorem 3.4 we have for every $k \geq 0$

$$\tilde{y} \in [z^{(k)}] =: [S].$$

After we have calculated $[S]$ (for a very large $k > 0$), we include the free boundary $\tilde{c}$ in the following way:

Let T be the smallest index $(< n + 1)$ with $[S_T] = 0$. Then we have $\tilde{y}(x_T) = 0$ and $\tilde{c} \leq x_T$. So, we define $a := x_T$ anew and repeat the whole procedure (until

a cannot be improved anymore). Afterwards let t be an index with $\underline{S}_t > 0$ and $\underline{S}_{t+1} \le 0$. Then we know that $\tilde{y}(x_t) > 0$ and $x_t < \tilde{c}$. I.e. $\tilde{c} \in [x_t, a]$.

1. Example (The rope problem): The free boundary problem (1) with

$$f(x, s, t) = \sqrt{1 + t^2}, \quad y_0 = 0.1$$

has the solution $\tilde{c} \approx 0.443568254385$ and

$$\tilde{y}(x) = \begin{cases} \cosh(x - \tilde{c}) - 1, & x \in [0, \tilde{c}), \\ 0, & x \in [\tilde{c}, \infty). \end{cases}$$

For $n = 300$ we got $\tilde{c} \in [0.4412705576, 0.4472135955]$.

2. Example: We have considered (1) with

$$f(x, s, t) = s + \arctan(xt) + \pi, \quad y_0 = 0.1.$$

For $n = 300$ we got $\tilde{c} \in [0.2501330156, 0.3568248233]$. Here, we found $[S_{220}] = 0$. So, we restarted with $a := x_{220}$ and finally we got

$$\tilde{c} \in [0.2505055169, 0.2564699340]$$

after a had been improved twice.

Remark: It must be confessed that the computing time is still large. Improving the algorithm with respect to this will be part of future research.

Acknowledgement

The author would like to thank the referees for their helpful comments.

References

Alefeld, G., Chen, X., Potra, F. (1999): Numerical validation of solutions of linear complementarity problems. Numer. Math. 83: 1-23

Beeck, H. (1972): Über Struktur und Abschätzungen der Lösungsmenge von linearen Gleichungssystemen mit Intervallkoeffizienten. Computing 10: 231-244

Collatz, L. (1981): Differentialgleichungen. Teubner Studienbücher, Stuttgart

Cottle, R. W., Pang, J. S., Stone, R. E. (1992): The linear complementarity problem. Academic Press

Cryer, C. W., Lin,Y. (1985): An alternating direction implicit algorithm for the solution of linear complementarity problems arising from free boundary problems. Appl. Math. Optim. 13: 1-17

Ferris, M. C., Pang, J. S. (1997): Engineering and economic applications of complementarity problems. Siam Rev. Vol 39, No. 4: 669-713.

Gwinner, J. (1988): Acceptable solutions of linear complementarity problems. Computing 40: 361-366

Hansen, E. (1969): On solving two-point boundary-value problems using interval arithmetic. In: Hansen, E. (ed.): Topics in Interval Analysis. Clarendon Press, Oxford, pp. 74-90

Mayer, G. (1991): Old and new aspects for the interval Gaussian algorithm. In: Kaucher, E., Markov, S. M., Mayer, G. (eds.): Computer Arithmetic, Scientific Computation and Mathematical Modelling. J. C. Baltzer AG, Basel, pp. 329-349

Oettli, W., Prager W. (1964): Compatibility of approximate solutions of linear equations with given error bounds for coefficients and right-hand sides. Numer. Math. 6: 405-409

Schäfer, U. (1999): Das lineare Komplementaritätsproblem mit Intervalleinträgen. Thesis. Universität Karlsruhe

Thompson, R. C., Walter, W. (1992): An existence theorem for a parabolic free boundary problem. Differential and Integral Equations 5: 43-54

Some Numerical Methods for Nonlinear Least Squares Problems

Stepan Shakhno

1 Introduction

Nonlinear least-square problems appear in estimating parameters and checking the hypotheses of mathematical statistics; in estimating parameter of physical process from measurement, and in managing of different objects, processes, etc. For solving such problems the Gauss-Newton method and Levenberg-Marquardt method (Dennis and Schnabel 1983, Ortega and Rheinboldt 1970, Schwetlick 1991) are usually applied. Iterative-difference analogue to Gauss-Newton method is investigated by Shakhno and Gnatyshyn (1999). Some updatings Gauss-Newton method are proposed by Bartish and Shakhno (1993). In this work we propose methods for solving nonlinear least-squares problems. These methods are constructed by using a combination of known iterative methods with the aim of obtaining greater efficiency in regards to the number of iterations and the number of calculations. The theorems about convergence conditions of the methods as well as speed of iteration convergence are formulated and proved. A comparison is made between these methods and the Gauss-Newton method. The results of extensive numerical experiments are demonstrated on the basis of tested problems, which are widely known though rather complex. Conclusions have been made on the basis of these experimental results.

2 Parametric iterative methods for nonlinear least squares problems

The wide-spread and effective methods for solving nonlinear least-squares problems:

$$\text{Find} \quad \min_{x \in R^n} f(x) = \frac{1}{2}F(x)^T F(x) = \frac{1}{2}\sum_{i=1}^{m} F_i^2(x), \ m \geq n, \tag{1}$$

where $F : R^n \to R^m$ is a nonlinear function, are Gauss-Newton method and its modifications (Dennis and Schnabel 1983). For solving the problem (1) we propose the following method:

$$x_{k+1} = x_k - \left[F'(\bar{x}_k)^T F'(\bar{x}_k)\right]^{-1} F'(\bar{x}_k)^T F(x_k), \tag{2}$$

where $\bar{x}_k = (1 - \mu)x_k + \mu\varphi(x_k)$, $k = 0, 1, 2, ...$, $0 \leq \mu \leq 1$, $\varphi : R^n \to R^n$ is an auxiliary operator such that

$$x_* = \varphi(x_*). \tag{3}$$

Obviously, if $\mu = 0$, then (2) reduces to the well-known Gauss-Newton method. We will show that we can choose a parameter μ and an appropriate operators

$\varphi(x)$ to obtain a method with the best speed of convergence among the methods of the class (2). We shall consider some variants in the choice of operator $\varphi(x)$.

2.1 Zero residual

If $F(x_*) = 0$, then we can obtain an operator $\varphi(x)$ from arbitrary n equations $F_i(x) = 0$, $i = 1, \ldots, m$. For instance, we can put $\varphi(x) = x_i - F_i(x)$, $i = 1, \ldots, n$. Conditions and speed of convergence of iterations (2) to the exact solution x_* of problem (1) is established by the following theorem.

Theorem 1   *Let $F : R^n \to R^m$, $m \geq n$ and function $f(x) = \frac{1}{2}F(x)^T F(x)$ be twice continuously differentiable in an open convex set $D \subset R^n$. Assume that*

$$\|F'(x)\| \leq \gamma, \ \|F''(x)\| \leq L, \ F''(x) \in Lip_N(D), \ \|\varphi'(x)\| \leq \alpha$$

for all $x \in D$ and there exists a point $x_* \in D$ such that $F(x_*) = 0$. Let $\lambda > 0$ be a minimal eigenvalue for $F'(x_*)^T F'(x_*)$. Then there exists $r_0 > 0$ such that, for all $x_0 \in \Omega(x_*, r_0)$ the sequence $\{x_k\}$ generated by method (2) is well defined, converges to x_* and satisfies the inequality:

$$r_k = \|x_k - x_*\| \leq (h_0 r_0)^{2^k - 1} r_0, \tag{4}$$

where

$$h_0 = \frac{\gamma}{2\lambda}\left[\frac{N}{3}r_0\mu^2(1+\alpha)^2 + L(|2\mu - 1| + 2\mu\alpha)\right] < \frac{1}{r_0},$$

$$r_0 = \|x_0 - x_*\|, \ \Omega(x_*, r_0) = \{x : \|x - x_0\| < (1 + 2\alpha)r_0\}, \ k = 0, 1, \ldots.$$

Proof: It follows from (2) that

$$x_1 - x_* = x_0 - x_* - \left[F'(\bar{x}_0)^T F'(\bar{x}_0)\right]^{-1} F'(\bar{x}_0)^T F(x_0) = -\left[F'(\bar{x}_0)^T F'(\bar{x}_0)\right]^{-1}$$

$$\times F'(\bar{x}_0)^T\left[\int_0^1 (F''(\bar{x}_0 + \tau(x_0 - \bar{x}_0)) - F''(\bar{x}_0 + \tau(x_* - \bar{x}_0)))(1 - \tau)d\tau(x_0 - \bar{x}_0)^2\right.$$

$$\left. + \int_0^1 (F''(\bar{x}_0 + \tau(x_* - \bar{x}_0)))(1 - \tau)d\tau(x_0 - x_*)(x_0 - 2\bar{x}_0 + x_*)\right]. \tag{5}$$

Furthermore, we have

$$\|x_0 - \bar{x}_0\| = \|x_0 - (1 - \mu)x_0 - \mu\varphi(x_0)\| = \|\mu(\varphi(x_0) - x_0)\|$$

$$= \|\mu(\varphi(x_0) - \varphi(x_*) + x_* - x_0)\| \leq \mu(1 + \alpha)\|x_0 - x_*\|; \tag{6}$$

$$\|\bar{x}_0 - x_*\| \leq (1 - \mu + \mu\alpha)\|x_0 - x_*\|; \tag{7}$$

$$\|x_0 - 2\bar{x}_0 - x_*\| \leq (|2\mu - 1| + 2\mu\alpha)\|x_0 - x_*\|. \tag{8}$$

From (5)-(8), we obtain

$$\|x_1 - x_*\| \le \frac{\gamma}{2\lambda}\left(\frac{N}{3}\|x_0 - x_*\|^3\mu^2(1+\alpha)^2 + L(|2\mu - 1| + 2\mu\alpha)\|x_0 - x_*\|^2\right).$$

Further, by induction, we get the estimation (4). The theorem is proved.

We shall investigate the influence of parameter μ on the speed of convergence of the iterative process (2). We can increase the speed of convergence by two ways: by increasing the rate of convergence or by minimizing the denominator of convergence $h_0 r_0$. The denominator $h_0 r_0$ includes the function $\chi(\mu) = |2\mu - 1| + 2\mu\alpha$, whose minimum is attained when $\mu = 1/2$ in the case $\alpha < 1$, and when $\mu = 0$ in the case $\alpha \ge 1$. Then, the iterative process (2) will have a maximal speed of convergence and the most favorable conditions for choosing the initial approximation x_0 (a wider region of convergence). As we can see from (4), under the small value r_0, the process (2) with the value $\mu = 1/2$ and the operator $\varphi(x)$, which satisfies the condition $\|\varphi'(x)\| \le \alpha < 1$ has a higher speed of convergence than the Gauss-Newton method.

2.2 Nonzero residual

In the case, when $F(x_*) \ne 0$, the condition (3) is satisfied by the operator $\varphi(x) = x - \beta F'(x)^T F(x)$, $\beta > 0$, which corresponds to the gradient method of minimization of the function $f(x) = \frac{1}{2}F(x)^T F(x)$.

Theorem 2 *Let $F : R^n \to R^m$, $m \ge n$ and the function $f(x) = \frac{1}{2}F(x)^T F(x)$ be twice continuously differentiable in an open convex set $D \subset R^n$. Assume that*

$$\|F'(x)\| < \gamma, \quad \|(F'(x) - F'(x_*))^T F(x_*)\| \le \sigma\|x - x_*\|,$$

$$\|F''(x)\| \le L, \quad F''(x) \in Lip_N(D) \tag{9}$$

for all $x \in D$, where $\lambda > \sigma(1 + \delta) \ge 0$, $\delta = \mu\beta(\gamma^2 + \sigma)$.

Then for any $c \in (1, \lambda/((1 + \delta)\sigma))$ there exists $\epsilon > 0$ such that for all $x_0 \in \Omega(x_*, \epsilon)$ the sequence (2) is well defined, converges to x_* and satisfies the inequalities

$$\|x_{k+1} - x_*\| \le \frac{c\sigma(1+\delta)}{\lambda}\|x_k - x_*\| + \frac{c\gamma}{2\lambda}\left[\frac{N}{3}\|x_k - x_*\|\delta^2 + L(1+2\delta)\right]\|x_k - x_*\|^2, \tag{10}$$

$$\|x_{k+1} - x_*\| \le \frac{c\sigma(1+\delta) + \lambda}{2\lambda}\|x_k - x_*\| < \|x_k - x_*\|. \tag{11}$$

Proof: The proof we shall carry out is analogous to Dennis et al.(1983). Let c be a constant chosen from an interval $(1, \lambda/((1+\delta)\sigma))$. Then there exists $\epsilon_1 > 0$ such that

$$\left\|[F'(\bar{x}_0)^T F'(\bar{x}_0)]^{-1}\right\| \le \frac{c}{\lambda} \tag{12}$$

for all $x_0 \in \Omega(x_*, \epsilon_1)$.

Let

$$\epsilon = \min\left\{\epsilon_1, \frac{\lambda - c\sigma(1+\delta)}{c\gamma\left[\frac{N}{3}\|x_0 - x_*\|\delta^2 + L(1+2\delta)\right]}\right\}. \tag{13}$$

Then x_1 is well defined and

$$x_1 - x_* = -\left[F'(\bar{x}_0)^T F'(\bar{x}_0)\right]^{-1}\left\{F'(\bar{x}_0)^T\right.$$

$$\times \left[\int_0^1 (F''(\bar{x}_0 + \tau(x_0 - \bar{x}_0)) - F''(\bar{x}_0 + \tau(x_* - \bar{x}_0)))(1-\tau)d\tau(x_0 - \bar{x}_0)^2\right.$$

$$\left. + \int_0^1 (F''(\bar{x}_0 + \tau(x_* - \bar{x}_0))(1-\tau)d\tau(x_0 - x_*)(x_0 - 2\bar{x}_0 + x_*)\right]$$

$$\left. + (F'(\bar{x}_0) - F'(x_*))^T F(x_*)\right\}. \tag{14}$$

Since

$$F'(x)^T F(x) = F'(x)^T F(x) - F'(x_*)^T F(x_*) = F'(x)^T F'(\tilde{x})^T (x - x_*)$$

$$+ (F'(x)^T - F'(x_*)^T)F(x_*), \quad (\tilde{x} = x + \theta(x - x_*), \quad 0 < \theta < 1),$$

we obtain $\|F'(x)^T F(x)\| \le (\gamma^2 + \sigma)\|x - x_*\|$ for all $x \in \Omega(x_*, \epsilon)$.

Now we shall obtain the estimates

$$\|x_0 - \bar{x}_0\| \le \mu\beta(\gamma^2 + \sigma)\|x_0 - x_*\| = \delta\|x_0 - x_*\|; \tag{15}$$

$$\|\bar{x}_0 - x_*\| \le (1 + \mu\beta(\gamma^2 + \sigma))\|x_0 - x_*\| = (1+\delta)\|x_0 - x_*\|; \tag{16}$$

$$\|x_0 - 2\bar{x}_0 - x_*\| \le (1 + 2\mu\beta(\gamma^2 + \sigma))\|x_0 - x_*\| = (1+2\delta)\|x_0 - x_*\|. \tag{17}$$

Taking into account estimations (12), (15) - (17), we have from (14)

$$\|x_1 - x_*\| \le \frac{c\sigma(1+\delta)}{\lambda}\|x_0 - x_*\| + \frac{c\gamma}{2\lambda}\left[\frac{N}{3}\|x_0 - x_*\|^3\delta^2 + L\|x_0 - x_*\|^2(1+2\delta)\right], \tag{18}$$

which proves (10) for $k = 0$. From (12) and (18) it follows

$$\|x_1 - x_*\| \le \|x_0 - x_*\|\left\{\frac{c\sigma(1+\delta)}{\lambda} + \frac{c\gamma}{2\lambda}\left[\frac{N}{3}\|x_0 - x_*\|\delta^2\right.\right.$$

$$\left.\left. + L(1+2\delta)\right]\|x_0 - x_*\|\right\} \le \|x_0 - x_*\|\left[\frac{c\sigma(1+\delta)}{\lambda} + \frac{\lambda - c\sigma(1+\delta)}{2\lambda}\right]$$

$$= \frac{c\sigma(1+\delta) + \lambda}{2\lambda}\|x_0 - x_*\| < \|x_0 - x_*\|.$$

This proves (11) for $k = 0$. Further proof is carried out by induction.

Consequence   *Let the conditions of the theorem 2 be satisfied. If $F(x_*) = 0$ then there exists $\epsilon > 0$ such that for all $x_0 \in \Omega(x_*, \epsilon)$ the sequence $\{x_k\}$, generated by (2), is well defined and converges quadratically to x_*.*

It follows from Theorem 2 that the speed of convergence of method (2) is decreasing with the increasing of the relative nonlinearity and with the value of the residual in the solution of this problem.

In the case, when $F'(\bar{x}_k)$ doesn't have a full column range $\left([F'(\bar{x}_k)^T F'(\bar{x}_k)]^{-1}\right.$ is not defined) we consider a method which is a generalization of the Levenberg-Marquardt method

$$x_{k+1} = x_k - \left[F'(\bar{x}_k)^T F'(\bar{x}_k) + \tau_k I\right]^{-1} F'(\bar{x}_k)^T F(x_k), \tag{19}$$

where $\bar{x}_k = (1 - \mu)x_k + \mu\varphi(x_k),\ 0 \le \mu \le 1,\ k = 0, 1, 2, \dots.$

The properties of local convergence of a method (19) are similar to properties of method (2) and are given in the following theorem.

Theorem 3   *Let the conditions of Theorem 2 be satisfed and sequence of non-negative real numbers $\{\tau_k\}$ be bounded above by number b. If $\sigma(1 + \delta) < \lambda$, for any $c \in \left(1, \frac{\lambda + b}{(1+\delta)\sigma + b}\right)$ there exists $\epsilon > 0$ such that for all $x_0 \in \Omega(x_*, \epsilon)$ the sequence generated by method (19), is well defined and satisfies the inequalities*

$$\|x_{k+1} - x_*\| \le \frac{c(\sigma(1 + \delta) + b)}{\lambda + b}\|x_k - x_*\|$$

$$+ \frac{c\gamma}{2(\lambda + b)}\left[\frac{N}{3}\|x_k - x_*\|\delta^2 + L(1 + 2\delta)\right]\|x_k - x_*\|^2,$$

$$\|x_{k+1} - x_*\| \le \frac{c(\sigma(1 + \delta) + b) + (\lambda + b)}{2(\lambda + b)}\|x_k - x_*\| < \|x_k - x_*\|.$$

If $F(x_*) = 0$ and $\tau_k = O\left(\|F'(\bar{x}_k)^T F(x_k)\|\right)$, then $\{x_k\}$ converges quadratically to x_*.

The proof of the theorem is similar to the proof of the Theorem 2.

3 Numerical Examples

The proposed method has been tested on many complex problems . Comparisons have been made with the methods of Newton, Gauss-Newton, Brown and other (Yermakov and Kalitkin 1981).

Example 1:

$$\begin{cases} f_1(x, y) = 2x - \sin 0.5(x - y) = 0, \\ f_2(x, y) = 2y - \cos 5(x + y) = 0, \\ f_3(x, y) = 2x + y^2 + 0.0778599 = 0; \end{cases} \qquad \varphi(x) = \begin{pmatrix} 0.5 \sin 0.5(x - y) \\ 0.5 \cos 5(x + y) \end{pmatrix};$$

$$x_* = (-0.1605, 0.4931), \quad \epsilon = 10^{-6}.$$

Table 1 shows the results of methods (2) with operator $\varphi(x)$, presented above.

Table 1

(x_0, y_0)	μ	Iterations	Time, msec
(10,10)	0.00	8	38
	0.25	6	16
	0.50	5	16
	0.75	7	21
	1.00	10	27
(0,0.5)	0.00	4	10
	0.25	4	10
	0.50	3	5
	0.75	3	10
	1.00	4	27
(-10,10)	0.00	8	21
	0.25	6	16
	0.50	5	16
	0.75	8	21
	1.00	7	21

The examples 2 and 3 below are solved by method (2), where $\varphi(x)$ is given by formula $\varphi(x) = x - \beta F'(x)^T F(x), \quad \beta \in (0, 1]$.

Example 2:

$$\begin{cases} 10\frac{x}{y} - 5(x + y) - 2(1 - y) = 0, \\ 10(x - y)^2 + x - \frac{x}{y} = 0, \\ 4xy - y^2 - 3 = 0; \end{cases} \qquad x_* = (1, 1), \quad \epsilon = 10^{-6}.$$

Table 2

(x_0, y_0)	μ	Iterations	Time, msec
(-2,10)	0.00	10	32
	0.25	9	32
	0.50	7	21
	0.75	7	21
	1.00	8	27
(10,-10)	0.00	37	109
	0.25	36	104
	0.50	9	27
	0.75	41	120
	1.00	17	54

Table 2 shows the results of Example 2.

Example 3 [Dennis and Schnabel 1983)]: Let $F : R^1 \to R^3$, $F_i(x) = e^{t_i x} - y_i$, $i = 1, 2, 3$, $f(x) = \frac{1}{2} F(x)^T F(x)$, where $t_1 = 1$, $y_1 = 2$, $t_2 = 2$, $y_2 = 4$, $t_3 = 3$ and let y_3 and x_0 have the values shown in Table 3.

Table 3

$y_3, x_0,$ $x_*, f(x_*)$	μ	Time, msec Method (2)		Iterations required to achive $\|f(x_k)\| \le 10^{-10}$ by:	
				Secant method	Newton's method
$y_3 = 8,$	0.00	0.16	5		
$x_0 = 1,$	0.25	0.16	5		
$x_* = 0.69315,$	0.50	0.10	3	6	7
$f(x_*) = 0$	0.75	0.21	6		
	1.00	0.16	5		
$y_3 = 3,$	0.00	0.38	5		
$x_0 = 1,$	0.25	0.27	5		
$x_* = 0.44005,$	0.50	0.21	3	8	9
$f(x_*) = 1.6390$	0.75	0.32	6		
	1.00	0.32	5		

Examples 4 - 7 below are taken from the work of Yermakov and Kalitkin (1981). These problems are solved by methods (2) and (19), where $\varphi(x)$ is given by formula $\varphi(x) = x - \beta F'(x)^T F(x)$, $\beta \in (0, 1]$. The method (19) was used with $\tau_k = \alpha \|F(x_k)\|^2$, where α is given in Table 4. In the top part of the Table 4 the results from the work of Yermakov and Kalitkin (1981) are submitted. In the bottom part of this table there are our results with accuracy 10^{-8}.

Example 4: The system of N equations (for the values $N = 4, 8, 16, 32$) is given for the function

$$f_i(x) = x_i + \sum_{j=1}^{N} x_j - N - 1, \quad 1 \le i \le N - 1, \quad f_N(x) = \prod_{j=1}^{N} x_j - 1.$$

The vector with $x_i^0 = 0.5$ was chosen as the initial approximation. The solution of this systems is not unique, besides the obvious first solution $x_i^* = 1$ there is also at least one more (its components do not have a simple form). In this example Brown's method converges to the first root of the system, and the other methods to the second solution.

Example 5: A system of two equations is given for the functions

$$f_1(x) = x_1^2 - x_2 - 1, \quad f_2(x) = (x_1 - 2)^2 + (x_2 - 0.5)^2 - 1$$

and the initial approximation $x^0 = (0.1, 2.0)$ was chosen. This system has two solutions: $x^* \approx (1.5463, 1.3912)$ and $x^* \approx (1.0673, 0.1392)$. In this example all the methods converge to the second solution.

Example 6: A system of two equations is given for the functions

$$f_1(x) = x_1 - 13 + x_2[x_2(5 - x_2) - 2], \quad f_2(x) = x_1 - 29 + x_2[x_2(x_2 + 1) - 14]$$

and the initial approximation $x_0 = (15, -2)$ was chosen. This system has the solution $x_* = (5, 4)$. The determinant of its Jacobi matrix $\det f'(x) = 6x_2^2 - 8x_2 - 12$ vanishes on the singular lines $x_2 \approx 2.53$ and $x_2 \approx -0.90$. It is obvious that on moving from the chosen initial approximation to the solution the trajectory intersects both these lines.

Example 7: It is required to find the minimum of a scalar function of two variables, which has a relief of the ravine type with very steep slopes:

$$\Phi(x) = x_1^2 + 100(x_1^3 - x_1 - x_2)^2, \quad x^* = (0,0), \quad x^0 = (1,1).$$

Table 4

Method	Example						
	4				5	6	7
	$N=4$	$N=8$	$N=16$	$N=32$			
Newton	18	∞	∞	∞	24	57	8
Brown	6	7	8	–	10	10	–
Regularized analog of Newton's method with optimal step [Yermakov...]							
$\alpha = 10^{-4}$	12	2	4	3	8	∞	50
$\alpha = 10^{-2}$	3	4	5	3	7	∞	67
Partly regul.anal.Newt.m.o.s.							
$\alpha = 10^{-3}$	4	4	3	5	8	12	43
Method (2)							
$\mu = 0$	7	10	3	4	8	29	∞
$\mu = 0,25$	8	11	3	4	6	15	41
$\mu = 0,50$	6	9	3	4	5	10	27
$\mu = 0,75$	4	5	5	12	10	∞	35
$\mu = 1,00$	8	7	3	9	∞	5	36
Method (19), $\alpha = 10^{-3}$							
$\mu = 0$	3	4	4	4	7	26	38
$\mu = 0,25$	4	4	4	4	4	13	42
$\mu = 0,50$	5	5	4	4	5	14	15
$\mu = 0,75$	6	6	5	4	5	∞	38
$\mu = 1,00$	5	8	11	4	5	8	∞

Conclusion

The theoretical results that have been obtained and calculations from experiments permit us to make the following conclusions:

1. The proposed methods of constructing combined iterative processes enable us to receive highly effective and hopeful algorithms.

2. Among the compared methods, one of the most hopeful methods which has the highest convergence speed is the proposed method (2) with the parameter

$\mu = \frac{1}{2}$ and with step regulation. The method (19) with $\mu = \frac{1}{2}$ has the similar character.

3. For the evalution of $\varphi(x)$ the exact one-dimension optimization is not necessary. It is sufficient to find the first meaning of β_k, for which the decreasing of the residual $\|F'(\bar{x}_k)F(x_k)\|$ is obtained.

4. The proposed algorithm efficiently solves the systems of equations, which correspond to the minimization of functions, which has a relief of the ravine type with very steep slopes.

5. The use of Cholesky's modified factorization has properties of regularization in singular cases.

References

Bartish, M.Ya., Shakhno, S.M. (1993): Some methods of solving nonlinear least square problems. Visnyk Lviv university. Ser. mech.-math. 39: 3-7 (in Ukrainian)

Bartish, M.Ya., Shachno S.M. (1996): On the Iterative Steffensen Type Methods. Zeitschrift für Angewandte Mathematik und Mechanik 76, S1, pp.351-352

Dennis J.M.,Jr., Schnabel R.B. (1983): Numerical Methods for Unconstrained Optimization and Nonlinear Equations. Prentice-Hall, Inc., Englewood Cliffs, New Jersey

Ortega, J.M., Rheinboldt, W.C. (1970): Iterative Solution of Nonlinear Equations in Several Variables. Academic Press, New York

Schwetlick, H. (1991): Nichtlineare Parameterschätzung: Modelle, Schätzkriterien und numerische Algorithmen. GAMM-Mitteilungen. Heft 2, 13-51

Shachno, S.M. (1996): Numerical methods for solving nonlinear least squares problems. In: 9th Conference of the European Consortium for Mathematics in Industry. Technical University of Denmark Lyngby /Copenhagen, Denmark, June 25-29,1996, Book of abstracts, pp. 543-545

Shachno, S.M., Gnatyshyn, O.P. (1999): Iterative-Difference Methods for Solving Nonlinear Least-Squares Problem. In: Progress in Industrial Mathematics at ECMI 98, Teubner, Stuttgart, pp.287-294

Yermakov, V.V., Kalitkin, N.N. (1981): The optimal step and the regularization of Newton's method. Zhurnal vychislit. math. i math. physiky. v.21, N2, pp. 491 - 497 (in Russian)

A New Insight of the Shortley-Weller Approximation for Dirichlet Problems

Tetsuro Yamamoto

1 Introduction

The method of finite differences is one of fundamental techniques for solving boundary value problems of ordinary and partial differential equations, where ordinary and partial derivatives are replaced by divided differences.

The mathematical theory of the finite difference method (FDM) can be found in Courant et al. (1928), Forsythe and Wasaw (1960), Strikwerda (1986), Hackbusch (1992), etc. However, it appears that the study has not yet been completed. In fact, new results including superconvergence properties of the Shortley-Weller (S-W) approximaiton for Dirichlet problems in a bounded domain of $\mathbb{R}^2$ have recently been obtained by Yamamoto (1998), Chen et al. (1999), Matsunaga (1999), Matsunaga and Yamamoto (2000), Yamamoto et al. (2000), etc.

In this paper, first in §2, we consider the S-W approximation applied to the two-point boundary value problem

$$-\frac{d^2u}{dx^2} = f(x), \quad a < x < b \tag{1.1}$$

$$u(a) = u(b) = 0, \tag{1.2}$$

and asserts that the S-W approximation $\{U_i\}_{i=1}^n$ with not necessarily equidistant nodes $a = x_0 < x_1 < \cdots < x_n < x_{n+1} = b$ applied to (1.1)–(1.2) is closely related to the expression

$$u(x) = \int_a^b G(x,y)f(y)dy$$

with the Green function

$$G(x,y) = \begin{cases} \frac{1}{b-a}(x-a)(b-y) & (x \leq y) \\ \frac{1}{b-a}(b-x)(y-a) & (x > y). \end{cases} \tag{1.3}$$

That is, we have

$$U_i = \sum_{j=1}^n G(x_i, x_j)f_j \frac{x_{j+1} - x_{j-1}}{2}, \quad i = 1, 2, \ldots, n.$$

Next, in §3, we state a superconvergence property for the S-W approximation to the problem

$$-\Delta u = f(x,y) \quad \text{in } \Omega \tag{1.4}$$
$$u = g(x,y) \quad \text{on } \Gamma, \tag{1.5}$$

where Ω is a bounded domain of $\mathbb{R}^2$ and f, g are given function, provided that the solution $u \in C^4(\overline{\Omega})$ or $u \in C^{3,1}(\overline{\Omega})$. Furthermore, in §4, we treat the case where $u \notin C^{3,1}(\overline{\Omega})$, but $u \in C(\overline{\Omega}) \cap C^4(\Omega)$. Some results are stated. (The proof is given in Yamamoto et al. (2000).) They imply that different situations will occur if $u \notin C^{3,1}(\overline{\Omega})$: In the case $\Omega = (0,1) \times (0,1)$ and $u \in C(\overline{\Omega}) \cap C^\infty(\Omega)$, but $u \notin C^1(\overline{\Omega})$, there are examples in which superconvergence does not occur anywhere and does occur near a part of Γ or a corner $(1,1)$.

Finally, some remarks are given in §5.

2   The S-W approximation to the two-point boundary value problems

Consider the two-point boundary value problem

$$-u'' = f(x), \quad a < x < b, \tag{2.1}$$
$$u(a) = u(b) = 0. \tag{2.2}$$

Let $a = x_0 < x_1 < \cdots < x_n < x_{n+1} = b$ and

$$h_i = x_i - x_{i-1}, \quad i = 1, 2, \ldots, n+1.$$

Then the S-W approximation for (2.1) and (2.2), which is a generalization of the usual centered three-point formula, is given by

$$-\frac{2}{h_i(h_i + h_{i+1})}U_{i-1} + \frac{2}{h_i h_{i+1}}U_i - \frac{2}{h_{i+1}(h_i + h_{i+1})}U_{i+1} = f_i \quad i = 1, 2, \ldots, n \tag{2.3}$$

where $f_i = f(x_i)$ and U_i denotes the approximation for $u_i = u(x_i)$. The system (2.3) may be written in the form

$$AU = f, \tag{2.4}$$

where

$$A = \begin{pmatrix}
\frac{2}{h_1 h_2} & -\frac{2}{h_2(h_1+h_2)} & & & \\
-\frac{2}{h_2(h_2+h_3)} & \frac{2}{h_2 h_3} & -\frac{2}{h_3(h_2+h_3)} & & \\
& \ddots & \ddots & \ddots & \\
& & \ddots & \ddots & \ddots \\
& & & \ddots & \ddots \\
& & & -\frac{2}{h_n(h_n+h_{n+1})} & \frac{2}{h_n h_{n+1}}
\end{pmatrix}, \tag{2.5}$$

$U = (U_1, \ldots, U_n)^t$ and $f = (f_1, \ldots, f_n)^t$.

If $h_i = h \ \forall i$, then (2.5) reduces to the $n \times n$ matrix

$$\frac{1}{h^2}\begin{pmatrix} 2 & -1 & & & \\ -1 & 2 & -1 & & \\ & \ddots & \ddots & \ddots & \\ & & \ddots & \ddots & -1 \\ & & & -1 & 2 \end{pmatrix}.$$

We now write (2.5) as $A = DA_0$, where

$$D = \begin{pmatrix} \frac{2}{h_1+h_2} & & & \\ & \ddots & & \\ & & \ddots & \\ & & & \frac{2}{h_n+h_{n+1}} \end{pmatrix}$$

and

$$A_0 = \begin{pmatrix} \frac{1}{h_1}+\frac{1}{h_2} & -\frac{1}{h_2} & & & \\ -\frac{1}{h_2} & \frac{1}{h_2}+\frac{1}{h_3} & -\frac{1}{h_3} & & \\ & \ddots & \ddots & \ddots & \\ & & \ddots & \ddots & \\ & & & -\frac{1}{h_n} & \frac{1}{h_n}+\frac{1}{h_{n+1}} \end{pmatrix}.$$

Then it can be shown that

$$A_0^{-1} = \frac{1}{b-a}(\alpha_{ij}) \tag{2.6}$$

with

$$\alpha_{ij} = \begin{cases} \displaystyle\sum_{\lambda=1}^{i} h_\lambda \sum_{\mu=j+1}^{n+1} h_\mu & (i \leq j) \\ \displaystyle\sum_{\lambda=1}^{j} h_\lambda \sum_{\mu=i+1}^{n+1} h_\mu & (i > j). \end{cases}$$

Hence

$$\frac{1}{b-a}\alpha_{ij} = \begin{cases} \frac{1}{b-a}(x_i - a)(b - x_j) & (i \leq j) \\ \frac{1}{b-a}(x_j - a)(b - x_i) & (i > j) \end{cases}$$

$$= G(x_i, x_j),$$

where $G(x, y)$ denotes the Green function for $-\frac{d^2}{dx^2}$ subject to (1.2), which is defined by (1.3). The proof of (2.6) is done by verifying $A \cdot (\alpha_{ij}) = (b - a)I$ directly. Hence we have from (2.4)

$$U = A^{-1}f = A_0^{-1}D^{-1}f = A_0^{-1} \begin{pmatrix} \frac{h_1+h_2}{2} & & \\ & \ddots & \\ & & \ddots \\ & & & \frac{h_n+h_{n+1}}{2} \end{pmatrix} f$$

so that

$$U_i = \sum_{j=1}^{n} G(x_i, x_j) \frac{h_j + h_{j+1}}{2} f_j \tag{2.7}$$

$$= \frac{1}{2} \left[\sum_{j=1}^{n} G(x_i, x_j) f_j h_j + \sum_{j=1}^{n} G(x_i, x_j) f_j h_{j+1} \right]$$

If $h_i = h \; \forall i$, then (2.7) reduces to

$$U_i = \sum_{j=1}^{n} G(x_i, x_j) f_j h. \tag{2.8}$$

Therefore, if $f \in C^2[a, b]$, then $u \in C^4[a, b]$ and we have $u_i - U_i = O(h^2)$. This implies

$$U_i = \int_a^b G(x_i, x) f(x) dx + O(h^2). \tag{2.9}$$

Hence, (2.7)–(2.9) give a new insight of the fact that the S-W formula

$$-\frac{2}{h_W(h_E + h_W)} u(x - h_W) + \frac{2}{h_E h_W} u(x) - \frac{2}{h_E(h_E + h_W)} u(x + h_E)$$

for $-u''(x)$ which employs three points $x - h_W, x, x + h_E$ is a natural generalization of the usual centered difference formula

$$-\frac{1}{h^2} u(x - h) + \frac{2}{h^2} u(x) - \frac{1}{h^2} u(x + h)$$

for the case $h_E = h_W = h$.

3 Superconvergence of the S-W Approximation for $u \in C^{3,1}(\overline{\Omega})$

Consider the S-W approximation for the Dirichlet problem

$$-\Delta u = f(x, y) \quad \text{in } \Omega, \tag{3.1}$$

$$u = g(x, y) \quad \text{on } \Gamma = \partial \Omega, \tag{3.2}$$

where Ω is a bounded domain of $\mathbb{R}^2$ and functions f and g are given. It is assumed that a solution u for (3.1)–(3.2) exists, which also implies the uniqueness of the solution.

We now construct a net over $\overline{\Omega} = \Omega \cup \Gamma$ by the grid points (x_i, y_j) in Ω with the mesh size h, $x_{i+1} - x_i = y_{j+1} - y_j = h$, where Ω is not necessarily square, rectangle or their union. The grid point $(x_i, y_j) \in \Omega$ and the set of grid points in Ω are denoted by P_{ij} and Ω_h, respectively. $\mathscr{P}_\Gamma$ stands for the set of points P_{ij} such that at least one of four points $(x_i \pm h, y_j)$, $(x_i, y_j \pm h)$ does not belong to Ω and $\mathscr{P}_0$ is defined by $\mathscr{P}_0 = \Omega_h \setminus \mathscr{P}_\Gamma$. Furthermore, the set of points of intersection of grid lines with Γ is denoted by Γ_h. Let $\hat{\Gamma}$ be a part or the whole of Γ and κ be a constant with $\kappa > 1$, which is arbitrarily chosen independently of h. We then put

$$\mathscr{S}_h(\kappa, \hat{\Gamma}) = \{P_{ij} \in \Omega_h \mid \mathrm{dist}(P_{ij}, \hat{\Gamma}) \leq \kappa h\}.$$

If $\hat{\Gamma} = \Gamma$, then we write $\mathscr{S}_h(\kappa)$ in place of $\mathscr{S}_h(\kappa, \Gamma)$. Four points P_E, P_W, P_S, P_N in $\overline{\Omega}_h = \Omega_h \cup \Gamma_h$ which are adjacent to $P \in \Omega_h$ and on horizontal and vertical grid lines through P are called the neighbors of P. Their distance to P are denoted by h_E, h_W, h_S and h_N, respectively.

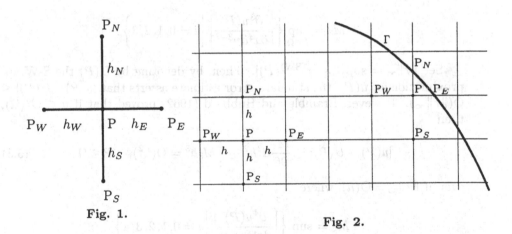

Fig. 1.

Fig. 2.

Then the S-W formula for approximating $-\Delta u(P)$ by employing five points $(x, y), (x - h_W, y), (x + h_E, y), (x, y - h_S), (x, y + h_N)$ is then given by

$$- \Delta_h^{(SW)} U(P) = (\frac{2}{h_E h_W} + \frac{2}{h_S h_N}) U(P) - \frac{2}{h_E(h_E + h_W)} U(P_E)$$
$$- \frac{2}{h_W(h_E + h_W)} U(P_W) - \frac{2}{h_S(h_S + h_N)} U(P_S) - \frac{2}{h_N(h_S + h_N)} U(P_N).$$

This reduces to the usual centered five point formula $-\Delta_h U(P)$ if $h_E = h_W = h_S = h_N$. The local truncation error $\tau^{(SW)}(P)$ of this formula at $P \in \Omega_h$ then yields

$$\tau^{(\mathrm{SW})}(P) = \begin{cases} O(h) & \text{if } (h_E, h_W, h_S, h_N) \neq (h, h, h, h) \\ O(h^2) & \text{if } h_E = h_W = h_S = h_N = h, \end{cases}$$

if $u \in C^4(\overline{\Omega})$. In spite of this fact, it was shown by Bramble and Hubbard (1962) that $|u_{ij} - U_{ij}| = O(h^2)$ holds for every grid point $P_{ij} \in \Omega_h$. This means that the finite difference method can be applied to any shaped domain Ω to obtain numerical solution of $O(h^2)$-accuracy at every grid point. If $u \in C^{3,1}(\overline{\Omega})$ (the set of functions $\in C^3(\overline{\Omega})$ whose third order partial derivatives are Lipschitz continuous in $\overline{\Omega}$), then it is easy to see the local truncation error $\tau^{(\mathrm{SW})}(P) = -[\Delta_h^{(\mathrm{SW})} u(P) - \Delta u(P)]$ of the S-W formula at P is estimated by

$$|\tau^{(\mathrm{SW})}(P)| \leq \begin{cases} \frac{2}{3}Lh^2 = O(h^2) & \text{if } h_E = h_W = h_S = h_N = h \\ \frac{2}{3}M_3 h = O(h) & \text{otherwise}, \end{cases}$$

where $\max(h_E, h_W, h_S, h_N) \leq h$ and L is a Lipschitz constant common to all third order derivatives, i.e., $|\frac{\partial^3 u}{\partial x^i \partial y^{3-i}}(P) - \frac{\partial^3 u}{\partial x^i \partial y^{3-i}}(Q)| \leq L \operatorname{dist}(P, Q)$, $P, Q \in \overline{\Omega}$, and

$$M_3 = \sup_{P \in \Omega} \left\{ \left| \left| \frac{\partial^3 u(P)}{\partial x^i \partial y^{3-i}} \right| \right| i = 0, 1, 2, 3 \right\}.$$

Set $\|\tau\|_\infty = \sup_{P \in \Omega_h} |\tau^{(\mathrm{SW})}(P)|$. Then, by denoting by $U(P)$ the S-W approximation for $u(P)$, the standard error estimate asserts that $|u(P) - U(P)| \leq O(\|\tau\|_\infty)$. However, Bramble and Hubbard (1962) proved that if $u \in C^4(\overline{\Omega})$, then

$$|u(P) - U(P)| \leq \frac{M_4}{96} d^2 h^2 + \frac{2}{3} M_3 h^3 = O(h^2), \quad P \in \Omega_h \tag{3.3}$$

even if $\|\tau\|_\infty = O(h)$, where

$$M_4 = \sup_{P \in \Omega} \left\{ \left| \left| \frac{\partial^4 u(P)}{\partial x^i \partial y^{4-i}} \right| \right| i = 0, 1, 2, 3, 4 \right\}$$

and d stands for the diameter of any sphere with center O which covers Ω.

We recall here that the notation $C^{l,\alpha}(\overline{\Omega})$ stands for the set of functions $\in C^l(\overline{\Omega})$ whose l-th order partial derivatives are Hölder continuous in $\overline{\Omega}$, that is,

$$\sup_{\substack{P, Q \in \Omega \\ P \neq Q}} \frac{|u(P) - u(Q)|}{[\operatorname{dist}(P, Q)]^\alpha} < \infty, \quad 0 < \alpha < 1.$$

This definition is extended to the case $\alpha = 1$ (Lipschitz continuous case).

Then the following result holds which improves Bramble-Hubbard's result (3.3):

Theorem 3.1 (Yamamoto 1998, Matsunaga and Yamamoto 2000).
If $u \in C^{l+2,\alpha}(\overline{\Omega})$, $l = 0$ or 1 and $0 < \alpha \leq 1$. Then

$$|u(P) - U(P)| \leq \begin{cases} O(h^{l+\alpha+1}) & (P \in \mathscr{S}_h(\kappa)) \\ O(h^{l+\alpha}) & \text{(otherwise)}. \end{cases}$$

The theorem implies that if $u \in C^{3,1}(\overline{\Omega})$, then we have

$$|u(P) - U(P)| \leq O(h^3) \quad \text{at } P \in \mathscr{S}_h(\kappa)$$

and

$$|u(P) - U(P)| \leq O(h^2) \quad \text{at the other grid points,}$$

even if $\tau^{(\mathrm{SW})}(P) = O(h)$ at $P \in \mathscr{P}_\Gamma$.

Definition 3.1. We say that the S-W approximation has a superconvergence property near $\hat{\Gamma} \subseteq \Gamma$ if, for some constant $\gamma > 0$,

$$|u(P) - U(P)| \leq \begin{cases} O(h^{\gamma+1}) & P \in \mathscr{S}_h(\kappa, \hat{\Gamma}) \\ O(h^\gamma) & \text{otherwise.} \end{cases}$$

According to this definition, we can now say the S-W approximation has a superconvergence property near Γ if $u \in C^{3,1}(\overline{\Omega})$.

On the other hand, Collatz's method employs the linear interpolation

$$U(P) = \frac{1}{h_E + h_W}[h_W U(P_E) + h_E P(U_W)] + \frac{1}{h_S + h_N}[h_N U(P_S) + h_S P(U_N)]$$

at $P \in \mathscr{P}_\Gamma$ and the usual centered five point formula at $P \in \mathscr{P}_0$. Then it is known that the error $u(P) - U(P)$ is again $O(h^2)$-accuracy at every $P \in \Omega_h$. As was recently reported by Matsunaga (1999) that the Collatz method does not have any superconvergence property. Finite element method does not have in general the property, too. Hence this property is a nice feature of the S-W approximation, which might be useful to the computation of the Neumann data from the solution of Dirichlet problems.

4 Behavior of the S-W approximation for $u \notin C^{3,1}(\overline{\Omega})$

In the previous section, we stated that the S-W approximation applied to (3.1)–(3.2) has a superconvergence property near Γ if $u \in C^{3,1}(\overline{\Omega})$. Then a question arises naturally: Does any superconvergence of the S-W approximation occur for the problem having a solution whose derivatives have singularities on the whole Γ or a part of Γ?

The following examples show that there are different types of convergence in the case $u \notin C^{3,1}(\overline{\Omega})$.

Example 4.1 (Q. Fang). Let f and g be chosen so that the function $u = \sqrt{x(1-x)} + \sqrt{y(1-y)}$ is the solution of (3.1)–(3.2). Observe that $u \in C(\overline{\Omega}) \cap C^\infty(\Omega)$, but $u \notin H^1(\Omega)$. Then, the numerical experiment done by Fang shows that

$$|u(P) - U(P)| = O(h^{\frac{1}{2}}) \quad \forall p \in \Omega_h \tag{4.1}$$

and superconvergence does not occur at any point in Ω_h. It should also be remarked that $u^{(4)}(Q) = O(h^{\frac{1}{2}-4})$ if Q is near Γ and the local truncation error $\tau(Q)$ approaches to infinity as Q approaches to Γ.

Example 4.2 (X. Chen). Let f and g be chosen so that the function $u = \sqrt{x} + y \in C(\overline{\Omega}) \cap C^\infty(\Omega)$ is the solution of (3.1)–(3.2). Then, the results of computation done by Fang show that with $\hat{\Gamma} = \{(1,y) \mid 0 \le y \le 1\}$

$$|u(P) - U(P)| \le \begin{cases} O(h^{\frac{3}{2}}) & P \in \mathscr{S}_h(\kappa, \hat{\Gamma}) \\ O(h^{\frac{1}{2}}) & \text{otherwise,} \end{cases} \tag{4.2}$$

where $\kappa > 1$. That is, a superconvergence occurs near the side $x = 1$ of Γ.

Example 4.3. Let f and g be chosen so that the function $u = \sqrt{x} + \sqrt{y} \in C(\overline{\Omega}) \cap C^\infty(\Omega)$ is the solution of (3.1)–(3.2). Then the results of computation done by Fang show that with $\hat{\Gamma} = \{(x,1) \mid 1 - 2h \le x \le 1\} \cup \{(1,y) \mid 1 - 2h \le y \le 1\}$

$$|u(P) - U(P)| \le \begin{cases} O(h^{\frac{3}{2}}) & P \in \mathscr{S}_h(\kappa, \hat{\Gamma}) \text{ with } \kappa = 2 \\ O(h^{\frac{1}{2}}) & \text{otherwise.} \end{cases} \tag{4.3}$$

That is, the superconvergence occurs only near the corner $(1,1)$.

Mathematical proofs for the estimates (4.1)–(4.3) are given in Yamamoto et al. (2000).

5 Observations

In (3.1)–(3.2), if Ω is a general domain, then there is in general a grid point P such that $(h_E, h_W, h_S, h_N) \ne (h, h, h, h)$. At such a point P, the truncation error $\tau^{(\mathrm{SW})}(P)$ is $O(h)$. Therefore, it appears that there has been a misleading among some researchers that the finite difference method works well only for square, rectangle, their union, etc. However, by Theorem 3.1, the S-W approximation works well for any bounded domain Ω as long as $u \in C^{3,1}(\overline{\Omega})$.

Furthermore, Examples 4.1–4.3 suggest us that the S-W approximation can converge to the exact solution as $h \to 0$ even if $u \notin C^{3,1}(\overline{\Omega})$ and $\tau^{(\mathrm{SW})}(P) \to \infty$.

It is also interesting to compare the S-W approximation with the finite element method (FEM). A generalization of Theorem 3.1 to the three dimensional space is straightforward and an easy task, while it appears that a construction of a triangular mesh of FEM in $\mathbb{R}^3$ is a not necessarily easy task. An advantage of FEM is that mesh refinement is easier than FDM. However, the author asserts that adaptive mesh refinement is possible in FDM (S-W approximation).

If one wants to use a smaller mesh size in a subdomain of Ω, then we can do it easily. For example, we can apply this technique to Example 4.1 to obtain $O(h^2|\log h|)$ accuracy. The detail of this technique as well as its proof will be reported elsewhere.

References

Bramble, J. H., Hubbard, B. E. (1962): On the formulation of finite difference analogues of the Dirichlet problem for Poisson's equation. Numer. Math. 4: 313–327

Chen, X., Matsunaga, N., Yamamoto, T. (1999): Smoothing Newton methods for nonsmooth Dirichlet problems. In: Fukushima, M., Qi, L. (eds.): Reformulation—Nonsmooth, piecewise smooth, semismooth and smoothing methods. Kluwer, Dordrecht, pp. 65–79

Courant, R., Friedrichs, K. O., Lewy, H. (1928): Üeber die partiellen differenzengleichungen der mathematischen physik. Math. Annal. 100: 32–74 (English translation: (1967): On the partial difference equations of mathematical physics. IBM J. 11: 215–234)

Forsythe, G. E., Wasaw, W. R. (1960): Finite difference methods for partial differential equations. John Wiley & Sons, Inc., New York

Hackbusch, W. (1992): Elliptic differential equations. Springer Verlag, Berlin

Matsunaga, N. (1999): Comparison of three finite difference approximations for Dirichlet problems. Information 2: 55–64

Matsunaga N., Yamamoto, T. (2000): Superconvergence of the Shortley-Weller approximation for Dirichlet problems. Journal Comp. Appl. Math. 116: 263–273

Strikwerda, J. C. (1989): Finite difference schemes and partial differential equations. Wadsworth, Inc., Belmont

Yamamoto, T. (1998): On the accuracy of finite difference solution for Dirichlet problems. In: RIMS Kokyuroku 1040, RIMS, Kyoto University, pp. 135–142

Yamamoto, T., Fang, Q., Chen, X. (2000): Superconvergence and nonsuperconvergence of the Shortley-Weller approximation for Dirichlet problems. (submitted)

Yamamoto, T., Ikebe, Y. (1979): Inversion of band matrix. Linear Algebra Appl. 24: 105–111

How Orthogonality is Lost in Krylov Methods

Jens–Peter M. Zemke

1 Introduction

In this paper we examine the behaviour of finite precision Krylov methods for the algebraic eigenproblem. Our approach presupposes only some basic knowledge in linear algebra and may serve as a basis for the examination of a wide variety of Krylov methods. The analysis carried out does not yield exact bounds on the accuracy of Ritz values or the speed of convergence of finite precision Krylov methods, but helps understanding the principles underlying the behaviour of finite precision Krylov methods.

We are only concerned with Krylov methods that compute approximations to a few eigenpairs of the eigenproblem $AV = V\Lambda$ for some $A \in \mathbb{K}^{n \times n}$ where $\mathbb{K} \in \{\mathbb{R}, \mathbb{C}\}$. The examination of Krylov methods for the solution of linear systems will be delayed to a subsequent paper. The eigenvalues of A will be denoted by λ_i, the corresponding left and right eigenvectors will be denoted by v_i and w_i, respectively:

$$v_i^H A = \lambda_i v_i^H, \qquad A w_i = w_i \lambda_i.$$

All eigenvectors are assumed to be normalized to unit length.

In infinite precision a Krylov method reduces the eigenproblem to some simpler problem, i.e. a problem with exploitable structure. This problem involves a matrix $C_k \in \mathbb{K}^{k \times k}$ containing information of the action of A on some Krylov subspace

$$\mathcal{K}(A, q) = \mathcal{K}_k(A, q) = \operatorname{span}\{q, Aq, A^2 q, \ldots, A^{k-1} q\}.$$

The capital C in C_k shall remind on computed and/or condensed.

Depending on whether a short-term recurrence or a long-term recurrence is used in the investigated Krylov method, the matrix C_k has either tridiagonal form and will be denoted by T_k, or Hessenberg form and will be denoted by H_k.

The eigenvalues of C_k are used as approximations to the eigenvalues of A. They are referred to as Ritz values, even if they were computed in finite precision.

The eigenvalues of C_k will be denoted by θ_j^k, the corresponding right and left eigenvectors will be denoted by s_j^k and t_j^k, respectively:

$$C_k s_j^k = s_j^k \theta_j^k, \qquad \left(t_j^k\right)^H C_k = \theta_j^k \left(t_j^k\right)^H.$$

The columns of a given matrix Q will be denoted by q_i, the components of a given vector s_j will be denoted by s_{ij}. The identity matrix of dimension k will be denoted by I_k with columns e_j and components δ_{ij}.

The rest of the paper is divided as follows. In Section 2 we repeat basic facts about Krylov methods for the standard eigenproblem. In Section 3 we derive a unified framework for Krylov methods for the standard eigenproblem. The following three sections are devoted to the numerical behaviour of the three most important Krylov methods for the standard eigenproblem, namely the method of Arnoldi, the symmetric Lanczos method and the nonsymmetric Lanczos method. In Section 7 we draw conclusions.

All numerical experiments and plots were performed using Matlab 5.3 with machine precision 2^{-53} on an IBM RS/6000 workstation.

2 Krylov eigensolvers

The three well known Krylov methods used to compute approximations to the solution of the standard eigenproblem are Arnoldi's method and the symmetric as well as the nonsymmetric Lanczos method. For details concerning derivation, implementation and convergence properties of infinite precision Krylov methods we refer the reader to the book of Golub and van Loan (1996).

The Arnoldi method (Arnoldi 1951) uses a long-term recurrence to compute an orthonormal basis for the Krylov subspace spanned by some starting vector $q_1 \in \mathbb{K}^n$. Denoting by $Q_k = [q_1, \ldots, q_k]$ the matrix with the computed columns q_l, the governing equation of Arnoldi's method can be expressed in matrix form by

$$AQ_k - Q_k H_k = h_{k+1,k} q_{k+1} e_k^T + F_k^A. \tag{1}$$

The matrix F_k^A covers errors due to finite precision computation. Error analysis (Paige 1969) shows that $\|F_k^A\|$ can be bounded by a small multiple of the machine precision times some function of k. Numerical experiments justify the assumption that the norm of the error matrix is comparable to $\|A\| \epsilon$ where ϵ denotes machine precision.

In infinite precision the orthogonality condition

$$Q_k^H Q_k = I_k \tag{2}$$

holds. The picture changes drastically turning to finite precision. In finite precision the computed q_l are no longer orthogonal, in fact the computed q_l may even become linearly dependent.

The algorithm breaks down when a zero divisor $h_{k+1,k}$ occurs. But in this case $AQ_k - Q_k H_k = F_k^A$ and in infinite precision we conclude that all eigenvalues of H_k are eigenvalues of A.

The symmetric Lanczos method (Lanczos 1950/1952) for Hermitian A can be derived in infinite precision from the method of Arnoldi using the fact that $A = A^H$. The symmetric Lanczos method uses a short-term recurrence. The resulting symmetric tridiagonal matrix will be denoted by

$$T_k = \text{tridiag}\,(\beta, \alpha, \beta)$$

with diagonals $\alpha = (\alpha_1, \ldots, \alpha_k)^T$ and $\beta = (\beta_1, \ldots, \beta_{k-1})^T$.

Like in Arnoldi's method, the governing equation can be expressed in matrix form

$$AQ_k - Q_k T_k = \beta_k q_{k+1} e_k^T + F_k^L, \tag{3}$$

using some error matrix F_k^L to balance the equations in finite precision. The error analysis for symmetric Lanczos was carried out by Paige in his Ph.D. thesis (Paige 1971). Numerical experiments confirm that $\|F_k^L\|$ it will be of size $\|A\|\,\epsilon$.

In infinite precision the orthogonality condition

$$Q_k^H Q_k = I_k \qquad (4)$$

holds similar to Arnoldi's method. In finite precision the computed q_l frequently become linearly dependent.

Again we conclude that in case of a zero divisor $AQ_k - Q_k T_k = F_k^L$ holds and thus in infinite precision all eigenvalues of T_k are eigenvalues of A.

The third method is the nonsymmetric Lanczos method (Lanczos 1950/1952). It is applicable for general matrices, uses a short-term recurrence, but has the drawback that the method may break down without result.

The nonsymmetric Lanczos method computes Q_k and $P_k = [p_1, \ldots, p_k]$ spanning intimately related Krylov subspaces $\mathcal{K}(A, q_1)$ and $\mathcal{K}(A^H, p_1)$.

The resulting general tridiagonal matrix will be denoted by

$$T_k = \mathrm{tridiag}\,(\gamma, \alpha, \beta)$$

with diagonals α and β like in the symmetric case and $\gamma = (\gamma_1, \ldots, \gamma_{k-1})$.

The underlying recurrence can be expressed in matrix form by the two relations

$$AQ_k - Q_k T_k = \beta_k q_{k+1} e_k^T + F_k^Q, \qquad (5a)$$
$$A^H P_k - P_k T_k^H = \gamma_k p_{k+1} e_k^T + F_k^P \qquad (5b)$$

with two matrices F_k^P, F_k^Q accounting for roundoff. As the matrices Q_k, P_k are no longer assumed to have orthonormal columns, the error matrices will no longer be of size $\|A\|\,\epsilon$. Nevertheless $\|f_l^P\| \approx \|p_l\|\,\epsilon$, $\|f_l^Q\| \approx \|q_l\|\,\epsilon$ holds in most cases.

The method breaks down in case we have either a zero divisor β_k and/or γ_k and can conclude that all eigenvalues of T_k are eigenvalues of A, or in case we have $\beta_k \neq 0 \neq \gamma_k$ but $q_{k+1}^H p_{k+1} = 0$.

The last case is known as serious breakdown because the algorithm terminates with no gain of information about the location of the eigenvalues.

In infinite precision, provided no serious breakdown occurs, the method computes biorthogonal bases

$$Q_k^H P_k = I_k \qquad (6)$$

for the two Krylov subspaces. In finite precision this (bi)orthogonality condition no longer remains valid.

All methods have various generalizations like block variants, variants with look-ahead and variants using explicit and implicit restart techniques. It is possible to extend the analysis to these cases.

The finite precision behaviour may depend on the method of orthogonalization. This includes reorthogonalization and semiorthogonalization techniques used in practical algorithms.

3 Unified description

All three methods use one or two matrix equations of the form

$$AQ_k - Q_k C_k = c_{k+1,k} q_{k+1} e_k^T + F_k, \tag{7}$$

which is pictorially

For simplicity we assume A and C_k diagonalizable, such that equation (7) can be diagonalized via $(v_i^H \cdot (7) \cdot s_j^k)$ yielding

$$(\lambda_i - \theta_j) v_i^H Q_k s_j^k = v_i^H q_{k+1} c_{k+1,k} s_{kj}^k + v_i^H F_k s_j^k. \tag{8}$$

We note that equation (7) is a Sylvester equation for Q_k, which is nothing than a linear system of higher dimension

$$\left(I_k \otimes A - C_k^T \otimes I_n\right) \text{vec}\,(Q_k) = \text{vec}\,\left(c_{k+1,k} q_{k+1} e_k^T + F_k\right).$$

Rewriting (7) in Kronecker form shows that we can use the standard perturbation theory results for linear systems. For an introduction to the Kronecker product and the vec($\cdot$) operator we refer the reader to the book of Horn and Johnson (1994).

If A or C_k are not diagonalizable it is possible to perturb both matrices by a small quantity that does not affect the size of the error term such that the perturbed matrices are diagonalizable.

We denote the approximation to the right eigenvector, also called Ritz vector, by $y_j^k = Q_k s_j^k$, and obtain from (8)

$$v_i^H q_{k+1} = \frac{\left(\lambda_i - \theta_j^k\right) v_i^H y_j^k - v_i^H F_k s_j^k}{c_{k+1,k} s_{kj}^k}. \tag{9}$$

It reveals the impacts of the local errors. We will call (9) the *local error formula*.

This formula consists of four ingredients. *First* we have some measure on the *convergence*, i.e.

$$(\lambda_i - \theta_j^k) \, v_i^H y_j^k. \tag{10}$$

Assuming that the Ritz value will only converge if the corresponding Ritz vector y_j^k approaches the right eigenvector w_i, the condition numbers $\kappa(\lambda_i) = 1/v_i^H w_i$ of the eigenvalues come into play. From equation (8) we see that the best possible achievement of the eigenvalues is given by $|\lambda_i - \theta_j^k| \approx \kappa(\lambda_i) |v_i^H F_k s_j^k|$.

If the error term is of size $\|A\| \, \epsilon$, Ritz values can only come as close as $\kappa(\lambda_i) \|A\| \, \epsilon$ to λ_i corresponding to standard perturbation results. We will call this level the *attainable level*.

We denote the *local error terms*, which are the *second ingredient*, by

$$f_{i,j,k} \equiv v_i^H F_k s_j^k. \tag{11}$$

Assuming that the errors are non-structured, i.e. that they are not in direction of the left eigenvectors of A and the right eigenvectors of C_k, we see that these quantities most probably will be of size $\max_l (\|q_l\|) \|A\| \, \epsilon$. Note that f_l has negligible influence on f_k, $l < k$.

The second ingredient is subtracted from the first one and the result is amplified by the reciprocal of the *third ingredient*

$$r_{j,k} \equiv c_{k+1,k} s_{kj}^k. \tag{12}$$

As $\|Ay_j^k - y_j^k \theta_j\| \leq \|q_{k+1}\| \cdot |r_{j,k}| + |f_{i,j,k}|$, we will call $r_{j,k}$ the *estimated residuals*. They all together have influence on the *fourth ingredient*, the *actual eigenpart*

$$v_i^H q_{k+1} \tag{13}$$

in direction of the left eigenvector v_i of the vector q_{k+1} extending the basis of the Krylov subspace.

To understand the principles underlying finite precision Krylov methods, we need to examine in detail the behaviour of the four ingredients of the local error formula.

We observe numerically that convergence continues to hold in slightly disturbed Krylov methods for a number of steps.

We can distinguish *three phases* in the run of a finite precision Krylov eigensolver. The *first phase* is indicated by a moderate deviation from the exact process. In this phase the method converges to at least one eigenpair/eigentriplet.

The *second phase* begins when one eigenvalue has converged to the attainable level. In this phase the component of q_{k+1} in direction of v_i is amplified by the inverse of the estimated residual $r_{i,j}$.

The *third phase* can only occur if the amplification is sufficient to raise the component of q_{k+1} in direction of the right eigenvector to a level where a new Ritz value has to emerge. This phase can not occur if the (bi)orthogonality is forced by the orthogonalization method used.

The loss of orthogonality in direction v_i can not continue to grow forever. Of course, a new Ritz value appears whensoever $|q_{k+1}^T w_i|$ is close or equal to one.

Using these three phases and some (major) simplifications we have a *simple model* for the behaviour of finite precision Krylov methods.

This model consists of lines and reflections w.r.t. the lines. The lines are mainly given by the norms of the columns of Q_k, the attainable level and some value in between.

We now discuss some of the impacts of our finite precision Krylov method model on the analysis of the three methods.

4 The method of Arnoldi

In the method of Arnoldi all vectors q_k have to be stored because of the long-term recurrence used. The vector q_{k+1} is explicitly orthogonalized against the previous ones.

Arnoldi was first used with classical Gram-Schmidt orthogonalization, GS, and later with modified Gram-Schmidt, MGS. Another variant is Arnoldi using Householder transformations.

Using any of these variants, the errors can be amplified so that the vectors q_k lose orthogonality once some Ritz value has converged to the attainable level. This was the reason for inventing Arnoldi with multiple orthogonalization.

Kahan proved in a straightforward manner that orthogonalizing two times is enough to maintain orthogonality, see Parlett (1998). It is only necessary to orthogonalize for a second time if severe cancellation occurs in some step.

The variant using double orthogonalization will be called Arnoldi-DO.

Loss of orthogonality can (and will) only occur in direction of the left eigenvector corresponding to the converging Ritz value.

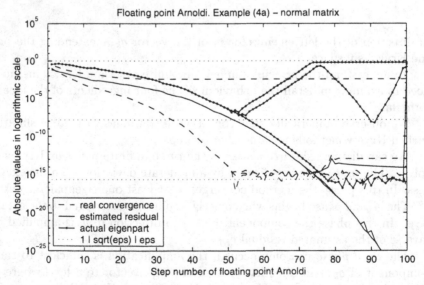

Fig. 1. Arnoldi-GS, Arnoldi-MGS and Arnoldi-DO for a normal matrix with equidistant eigenvalues

As numerical example (4a) we use a normal matrix with equidistant eigenvalues in $[0, 1]$. By *real convergence* we denote the minimal distance of a Ritz value to some eigenvalue of interest. See figure (1) for the convergence to the eigenvalue of maximal modulus for Arnoldi-GS, Arnoldi-MGS and Arnoldi-DO.

The figure contains the plots of all three variants. In steps 1 to 50 the three plots can not be distinguished, i.e. there seems to be only one line for real convergence, one line for estimated residual and one line for actual eigenpart.

The *dashed lines* plot the distance of the two closest Ritz values to the eigenvalue of maximal modulus, i.e. $\lambda_{max} = 1$. The three dashed lines corresponding to the converging Ritz values of the three methods decrease geometrically until they reach the attainable level and start to differ. The line where distance increases in steps 75 to 79 corresponds to the numerically unstable Gram-Schmidt variant.

The *solid lines* correspond to the convergence of the estimated residuals. Only the curve of Arnoldi-GS changes direction when reaching the attainable level. The small upward jumps between step 86 and 91 correspond to Arnoldi-DO.

The *dash-dotted lines* correspond to the actual eigenpart. In Arnoldi-DO the eigenpart converges to the machine precision. In Arnoldi-GS and Arnoldi-MGS the eigenpart increases when convergence reaches the attainable level, the upper curve corresponds to unstable Arnoldi-GS. The more stable Arnoldi-MGS manages to decrease the level again, but nevertheless orthogonality is lost.

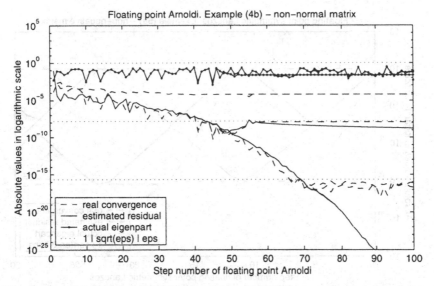

Fig. 2. Arnoldi-GS, Arnoldi-MGS and Arnoldi-DO for a non-normal matrix with equidistant eigenvalues

Figure (2) shows the behaviour of the same three methods for example (4b), a non-normal matrix with the same eigenvalues. The picture differs mainly in the size of the (left) eigenpart. It remains at a high level and thus the estimated residual is closer to the real convergence.

In this example only Arnoldi-GS has trouble with the non-normal matrix. Convergence stops at the square root of the machine precision.

5 The symmetric Lanczos method

The symmetric Lanczos method is one of the best examined Krylov methods. Beginning with the roundoff error analysis due to Paige (1971) many researchers added more and more parts to the picture.

Paige proved that the Ritz values computed in finite precision symmetric Lanczos lie close to what is predicted by the estimated residuals.

Greenbaum (1989) proved that every finite precision recurrence may be continued to give a residual of size zero introducing some new error terms $\tilde{f}_{k+i}$ in every step. The size of these error terms is comparable to the size of the already computed ones so that $\|F_k\| \approx \|\tilde{F}_{k+n}\|$ where $\tilde{F}_{k+n}$ denotes the matrix consisting of F_k as the first k columns followed by the constructed error terms. She used the result of Paige to conclude that finite precision symmetric Lanczos behaves like infinite precision symmetric Lanczos applied to some larger matrix with eigenvalues located in small intervals around the original eigenvalues.

As numerical example (5) we use a symmetric non-negative matrix with random entries in $[0,1]$. The matrix has one eigenvalue of maximal modulus (Perron root) which is well separated from other eigenvalues and thus becomes approximated very fast by Kaniel-Paige-Saad theory (Kaniel 1966, Paige 1971, Saad 1980).

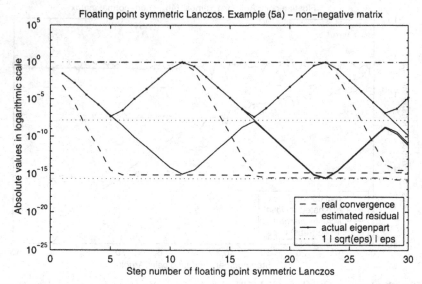

Fig. 3. Symmetric Lanczos for a non-negative matrix – convergence behaviour for eigenvalue of largest modulus

The behaviour of the estimated residuals, the eigenpart and the real convergence is shown for the eigenvalue of largest modulus in figure (3) and for the eigenvalue of second largest modulus in figure (4).

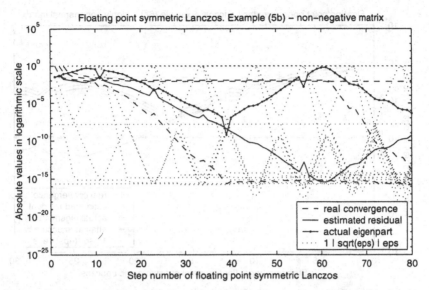

Fig. 4. Symmetric Lanczos for a non-negative matrix – convergence behaviour for eigenvalue of second largest modulus

All curves depending on $\|A\|$ have been scaled such that the attainable level is equal to the machine precision.

Figure (3) shows the ideal model behaviour. The distance to a Ritz value converges until it hits the ground, i.e. the attainable level, which is not obvious from (9). Then it does not move any longer. The actual eigenpart and the estimated residual converge similarly.

When convergence reaches the ground, the eigenpart starts to increase like the estimated residual decreases. The estimated residual converges to the attainable level.

The eigenpart continues to grow until it hits the ceiling, i.e. until it is of order one. At that time a new Ritz value occurs as can be seen in the figure at step 11 and the situation looks familiar.

Figure (4) shows how this affects the second largest eigenvalue. Convergence is disturbed every time the estimated residual of a Ritz value corresponding to the largest eigenvalue hits the ground.

6 The nonsymmetric Lanczos method

Most part of the error analysis of nonsymmetric Lanczos grounds on the results obtained for symmetric Lanczos. We used nonsymmetric Lanczos without any look-ahead.

Nonsymmetric Lanczos only needs the biorthogonality $q_i^H p_j = \delta_{ij}$ of the computed vectors, where δ_{ij} denotes Kronecker delta. To fix the recurrence we used nonsymmetric Lanczos with $|\gamma_i| = |\beta_i| \ \forall i$. This ensures that if the vectors used to build up the bases of the Krylov subspaces have the same length then all computed vectors have the same length $\|q_i\| = \|p_i\| \ \forall i$.

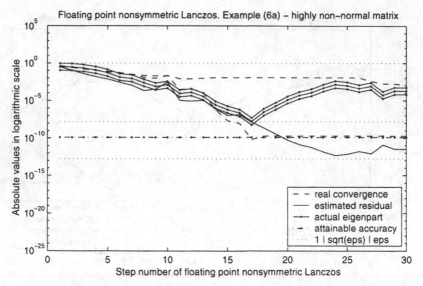

Fig. 5. Nonsymmetric Lanczos for a highly non-normal matrix

Figure (5) shows a run of nonsymmetric Lanczos with a matrix whose entries are zero below the fourth subdiagonal and randomly chosen between $[0, 1]$ elsewhere. This ensures that the matrix is highly non-normal. The behaviour looks like in the symmetric case, except that we now have a 'left' and 'right' eigenpart, i.e. the two quantities $v_i^H q_{k+1}$ and $w_i^H p_{k+1}$. They both and their geometric mean are plotted with dotted lines.

Convergence stops when the distance reaches the attainable level, that is $\left|\lambda_i - \theta_j^k\right| \approx \kappa\left(\lambda_i\right) \|A\| \epsilon$.

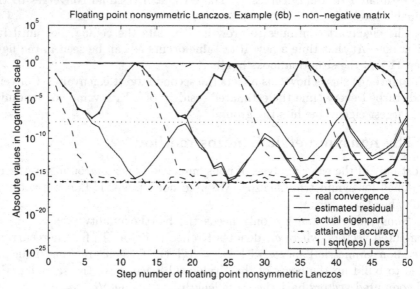

Fig. 6. Nonsymmetric Lanczos for a non-negative matrix

The estimated residual which gives information on the convergence of the Ritz vector continues to converge in this example to the level $\sqrt{\kappa\left(\lambda_i\right)}\,\|A\|\,\epsilon$.

Figure (6) shows the behaviour for a nonsymmetric non-negative matrix with entries randomly chosen between $[0,1]$. Again we observe fast convergence and many Ritz values representing one eigenvalue. But in contrast to the symmetric case only the first two Ritz values converge to the attainable level which is comparable to the attainable level in the symmetric case.

7 Conclusion

In general the use of any bound on the error matrices from above and below is not enough to predict the behaviour after the first Ritz value converged, i.e. Krylov methods are not forward stable.

If multiple precision is used, it is not sufficient to increase the working precision when the loss of orthogonality begins to raise to a certain level. *All* columns of the error matrix influence the attainable level in the actual step.

To compute the exact condensed matrix C_k, the three methods have to be implemented using exact number computations. This is possible as the methods only use rational operations.

Interval arithmetic can only be used as long as the estimated residuals are above some certain threshold. Then the diameters of the intervals used for normalization will grow, such that the intervals contain zero.

Acknowledgements

The author wishes to thank all people at Dagstuhl for the nice time and an anonymous referee for helpful comments.

References

Arnoldi, W. E. (1951): The Principle of Minimized Iterations in the Solution of the Matrix Eigenvalue Problem. Quarterly of Applied Mathematics. 9:17–29

Golub, G. H., van Loan, C. F. (1996): Matrix Computations (3rd edition). The John Hopkins University Press, Baltimore

Greenbaum, A. (1989): Behaviour of Slightly Perturbed Lanczos and Conjugate-Gradient Recurrences. Linear Algebra and its Applications. 113:7–63

Horn, R. A., Johnson, C. R. (1994): Topics in Matrix Analysis. Cambridge University Press, Cambridge

Kaniel, S. (1966): Estimates for Some Computational Techniques in Linear Algebra. Math. Comp. 20:369–378

Lanczos, C. (1950): An Iteration Method for the Solution of the Eigenvalue Problem of Linear Differential and Integral Operators. Journal of Research of the National Bureau of Standards. 45(4):255–282

Lanczos, C. (1952): Solution of Systems of Linear Equations by Minimized Iterations. Journal of Research of the National Bureau of Standards. 49(1):33–53

Paige, C. C. (1969): Error Analysis of the Generalized Hessenberg Processes for the Eigenproblem. Tech. Note ICSI 179, London University Institute of Computer Science

Paige, C. C. (1971): The Computation of Eigenvalues and Eigenvectors of Very Large Sparse Matrices. PhD thesis, London University Institute of Computer Science

Parlett, B. N. (1998): The Symmetric Eigenvalue Problem. Classics in Applied Mathematics. SIAM, Philadelphia. Unabridged, corrected republication of 1980 edition

Saad, Y. (1980): On the Rates of Convergence of the Lanczos and the Block-Lanczos Methods. SIAM J. Numer. Anal. 17(5):687–706

SpringerMathematics

Adi Ben-Israel,

Robert P. Gilbert

Computer-Supported Calculus

2001. Approx. 800 pages.
Hardcover DM 128,–, öS 896,–
(recommended retail price)
ISBN 3-211-82924-5
Texts and Monographs in Symbolic Computation

This is a new type of calculus book: Students who master this text will be well versed in calculus and, in addition, possess a useful working knowledge of how to use modern symbolic mathematics software systems for solving problems in calculus. This will equip them with the mathematical competence they need for science and engineering and the competitive workplace.

MACSYMA is used as the software in which the example programs and calculations are given. However, by the experience gained in this book, the student will also be able to use any of the other major mathematical software systems, like for example AXIOM; MATHEMATICA, MAPLE; DERIVE, or REDUCE, for "doing calculus on computers".

♞ SpringerWienNewYork

A-1201 Wien, Sachsenplatz 4–6, P.O. Box 89, Fax +43.1.330 24 26, e-mail: books@springer.at, Internet: www.springer.at
D-69126 Heidelberg, Haberstraße 7, Fax +49.6221.345-229, e-mail: orders@springer.de
USA, Secaucus, NJ 07096-2485, P.O. Box 2485, Fax +1.201.348-4505, e-mail: orders@springer-ny.com
Eastern Book Service, Japan, Tokyo 113, 3–13, Hongo 3-chome, Bunkyo-ku, Fax +81.3.38 18 08 64, e-mail: orders@svt-ebs.co.jp

SpringerMathematics

Dongming Wang

Elimination Methods

2001. XIII. 244 pages. 12 figures.
Softcover DM 108,–, öS 756,–
(recommended retail price)
ISBN 3-211-83241-6
Texts and Monographs in Symbolic Computation

This book provides a systematic and uniform presentation of elimination methods and the underlying theories, along the central line of decomposing arbitrary systems of polynomials into triangular systems of various kinds. Highlighting methods based on triangular sets, the book also covers the theory and techniques of resultants and Gröbner bases.

The methods and their efficiency are illustrated by fully worked out examples and their applications to selected problems such as from polynomial ideal theory, automated theorem proving in geometry and the qualitative study of differential equations. The reader will find the formally described algorithms ready for immediate implementation and applicable to many other problems.

Suitable as a graduate text, this book offers an indispensable reference for everyone interested in mathematical computation, computer algebra (software), and systems of algebraic equations.

SpringerWienNewYork

A-1201 Wien, Sachsenplatz 4–6, P.O. Box 89, Fax +43.1.330 24 26, e-mail: books@springer.at, Internet: www.springer.at
D-69126 Heidelberg, Haberstraße 7, Fax +49.6221.345-229, e-mail: orders@springer.de
USA, Secaucus, NJ 07096-2485, P.O. Box 2485, Fax +1.201.348-4505, e-mail: orders@springer-ny.com
Eastern Book Service, Japan, Tokyo 113, 3–13, Hongo 3-chome, Bunkyo-ku, Fax +81.3.38 18 08 64, e-mail: orders@svt-ebs.co.jp